AQUATIC BIODIVERSITY CONSERVATION

AQUATIC BIODIVERSITY CONSERVATION

Edited by

Dr. Prashant Telvekar

Delve PUBLISHING

www.delvepublishing.com

Aquatic Biodiversity Conservation

Dr. Prashant Telvekar

Delve Publishing
4164 Lakeshore Road
Burlington, ON L7L 1A4
Canada
www.delvepublishing.com
Email: orders@arclereducation.com

ISBN: 978-1-77956-406-1 (Hardcover)

ABOUT THE EDITOR

Dr. Prashant A. Telvekar (1979) is presently serving as Assistant profe
Department of Fisheries Extension, Economics and statistics, College of F
Science, Nagpur. He has obtained Bachelor of Fisheries Science (B.F.Sc.) in 20
Dr. Balasaheb Sawant Konkan Krishi Vidyapeeth, Dapoli. He has done his Ma
Fisheries Science (MF.F.Sc.) in 2002) from the Central Institute of Fisheries Edu
Mumbai (CIFE) and Ph. D. in Fish and Fisheries Science with specializa
Fisheries Resources Management from CIFE, Mumbai in the year 2007. He
his career as Assistant Professor of Fisheries Biology at College of Fishery S
Nagpur in the year 2006. Since 2007, he has been working as Assistant profe
Fisheries Resources, Economics, Statistics and Extension Education at the Col
Fishery Science, Nagpur under the Maharashtra Animal and Fishery Science Univ
Nagpur. His field of specialization is Fisheries Resources Management and Fis
Extension. He has published 21 Research Papers in the National and Interna
journals of repute. He has participated in many National, State seminars

TABLE OF CONTENTS

LIST OF FIGURES

LIST OF TABLES

PREFACE

Aquatic ecosystems play a crucial role in maintaining the earth's biological diversity. They serve as essential reservoirs of life, supporting a significant portion of the planet's biological productivity. The relationship between aquatic resources and their biodiversity is complex and deeply interconnected, with each contributing to and depending on a variety of ecological functions. These functions are vital for the sustainability of biotic communities, which encompass all living organisms within these ecosystems.

Unfortunately, biodiversity within both freshwater and marine environments is experiencing a continuous decline. This decline can be attributed to several factors, including the overexploitation of species, the introduction of exotic plants and animals, pollution from urban areas, industries, and agricultural zones, as well as the loss and alteration of ecological niches. These threats have severe implications for the health and stability of aquatic ecosystems.

To address these challenges, conservation and management efforts are imperative. This involves the establishment of bio-reserves and the implementation of bioregional management strategies. Additionally, global monitoring systems are necessary to effectively protect and preserve aquatic biodiversity. These efforts are essential for maintaining ecological balance and ensuring the long-term sustainability of aquatic environments.

This textbook provides comprehensive information on biodiversity found in aquatic habitats, covering both marine and freshwater ecosystems. It offers an in-depth examination of aquatic biodiversity through 10 chapters. Chapter 1 defines the scope of aquatic biodiversity, highlights the importance of aquatic ecosystems, and categorizes various habitats such as freshwater, marine, and brackish environments. It provides an overview of global aquatic biodiversity, setting the stage for deeper exploration.

Chapter 2 explores the essential services provided by aquatic biodiversity, including nutrient cycling, water purification, habitat provision, food security, and climate regulation through carbon sequestration. This chapter underscores the integral role of aquatic species in maintaining healthy ecosystems.

In Chapter 3, the book addresses significant threats such as pollution (chemical, plastic, and thermal), habitat destruction from activities like deforestation and damming, overfishing, and the impacts of invasive species. This chapter provides a critical analysis of the human-induced challenges facing aquatic ecosystems.

Chapter 4 examines the effects of climate change on aquatic environments, including ocean acidification, rising water temperatures, and sea level rise. It also discusses adaptation and resilience strategies for aquatic species in response to these changes.

The focus shifts in Chapter 5, which covers various methods for assessing aquatic biodiversity, including bioindicators, ecological surveys, remote sensing, and GIS. The chapter also highlights the role of citizen science in data collection.

Chapter 6 discusses ecological restoration principles and techniques for restoring rivers, wetlands, and coral reefs. It includes success stories and lessons learned from past restoration efforts.

In Chapter 7, the book examines sustainable practices, including aquaculture, and the benefits of ecotourism for conservation.

Chapter 8 reviews in-situ and ex-situ conservation methods, the role of Marine Protected Areas (MPAs) and freshwater reserves, sustainable fisheries management, and habitat rehabilitation strategies.

Chapter 9 outlines key international conventions such as the Convention on Biological Diversity (CBD), the Convention on International Trade in Endangered Species of Wild Fauna and Flora (CITES), and the Ramsar Convention. It also discusses national policies and the role of NGOs in aquatic conservation.

Finally, Chapter 10 emphasizes the importance of local and indigenous knowledge, participatory conservation methods, and the balance between conservation and human needs. It highlights the benefits and challenges of community involvement in conservation efforts.

Through this detailed exploration, readers will gain a deeper understanding of the importance of aquatic biodiversity and the efforts required to protect it. This text is aimed at a diverse readership, including students, researchers, and professionals in environmental science, marine biology, and ecology. It is particularly valuable for those involved in conservation, environmental management, and sustainability. Policymakers and practitioners in governmental and non-governmental organizations focused on aquatic ecosystems will find the book useful for developing and implementing effective conservation strategies. Additionally, educators and academic institutions seeking comprehensive resources for teaching and research on aquatic biodiversity and ecosystem management will benefit from its in-depth coverage and practical insights. The book also appeals to enthusiasts and advocates of aquatic conservation and sustainability.

Chapter 1

Introduction to Aquatic Biodiversity

Contents

INTRODUCTION

The quantity of distinct native species, or species richness, is known as biodiversity. On Earth, there are currently over 170 million recognized species, and every year, scientists discover new ones. According to experts, there could be millions or even tens of millions of additional undiscovered species. The United States is home to about 200,000 known species of plants and animals, or more than 10% of the world's total known biodiversity richness. In addition to species richness, biodiversity also refers to genetic diversity and the variety of habitats and ecosystems that are home to both plants and animals. Together with the nonliving components (soils, sunlight, and water) that support living things, ecosystems are home to both plants and animals. The vast and amazing range of flora and fauna that inhabit aquatic environments, such as crayfish, catfish, mayflies, tadpoles, and trout, is known as aquatic biodiversity. Numerous animal and plant species are found in water; some, like fish, live there their entire lives, while others, like salamanders and toads, may only use surface waters as young or during the spring mating season. While some aquatic animals, like water striders, spend their entire lives skimming the water's surface, others live their whole lives in the deep ocean.

Our rivers, streams, ponds, lakes, bays, streams, swamps, and marshes, along with the creatures that live there, comprise aquatic ecosystems (habitats and organisms). Nearly 10% of all known animal species on Earth live in freshwater ecosystems, despite making up only slightly more than 2% of the planet's surface (wetlands excluded). Although marine ecosystems cover a larger area, their species richness is similar. Almost half of all salamander species in North America, almost half of all crocodilian species, most temperate freshwater turtle species worldwide, and some of the world's most diverse freshwater bivalve populations can be found in the Southeast United States. The region is also home to a variety of other aquatic species. It has taken millions of years for aquatic species to evolve and adapt to their aquatic environments. Fish and plants that live in aquatic environments need food, water, shelter, and space to survive.

The more varied the range of habitats, whether they are aquatic or terrestrial, the higher the level of biological diversity. Coastal estuaries and mangrove swamps, for instance, represent "edge" environments that bridge saltwater and freshwater, capture nutrients, and fosters a wealth of aquatic plant and animal species. In general, the more intricate or extensive the ecosystem, the greater its biodiversity. Biodiversity in a confined space, like a single water droplet, is significantly less than that found in the ocean. Tendencies suggest that species diversity increases from the poles towards the equator and with rising rainfall and decreasing altitude. Typically, islands exhibit lower species diversity than the surrounding mainland. The United States serves as a global hub for freshwater biodiversity. Rich in fish, mussels, crayfish, snails, salamanders, frogs, and toads, the United States boasts one of the world's most diverse populations of aquatic animals, many of which reside in our own backyards. The United States leads the world in the number of freshwater mussels, crayfish, snails, and various aquatic insect species (mayflies, caddisflies, dragonflies, and damselflies). The country ranks seventh in fish diversity, most of which are found in its Southeastern rivers and streams. Furthermore, the United States holds the top spot for threatened fish species. Sharing this commonality with our tropical rainforest counterparts, we are confronted with a concerning vulnerability to development, habitat loss, and species declines.

1.1. DEFINITION AND SCOPE OF AQUATIC BIODIVERSITY

Within the broad category of freshwater ecosystems, which includes wetlands, rivers, streams, ponds, lakes, and reservoirs, is the term "aquatic biodiversity." Algal colonies, coral reefs, salt marshes, estuaries, mangroves, and oceans comprise the marine ecosystems, which make up the other portion of aquatic biodiversity. Another significant component of aquatic biodiversity are various forms of phytoplanktons, zooplanktons, aquatic plants, insects, fish, birds, and mammals. Although they are rich and valuable, marine ecosystems are also delicate. A large portion of the marine ecosystem—roughly 90%—and 10% of all marine animal species are found in extended sea and ocean water. Freshwater resources are sparsely distributed and poorly utilized, and their quantities pale in comparison to those of other water systems on the planet.

Since ancient times, human life has been intimately linked to bodies of water for a variety of purposes. Aquatic biodiversity and water reservoirs are closely related, and both ecological units depend on one another. The process of water circulation from the atmosphere to the lithosphere is largely dependent on the evaporation of ocean surface water. The ocean's enormous capacity to transfer heat from the earth helps to mitigate harsh natural disasters by interacting reciprocally with the atmosphere and maintaining the ideal temperature for the occurrence and growth of organisms. Oceans play a major role in the weather patterns and climatic shifts that occur worldwide. A diverse ecosystem is created by the emergence and expansion of unusual ocean life, which is made up of a wide range of organisms. The relationship between climate change and the ocean is receiving a lot of attention these days. The ocean ecosystem serves as a "carbon sink," absorbing large amounts of carbon in addition to storing a significant amount of water. An annual net primary production of about 50 billion tons of carbon can be processed by marine phytoplanktons; this quantity is roughly equivalent to the primary production of terrestrial plants. Primary production turnovers happen quickly, and there is a lot of activity in the matter transportation process.

The sea floor areas next to the shallow coastal water and the photic zones, which extend down to a depth of approximately 200 meters, are home to the primary producers of the oceans. There are whole other kinds of life in the deep-sea zone. It is a well-established fact that the oceans and seas serve as a vital support system for a wide range of biologically diverse species, all of which are crucial for maintaining ecological diversity. Almost 200,000 of the 222,000–230,000 marine species that are listed are members of the Animalia phylum. Because marine habitats are so sensitive to changes in their physical surroundings, they are brittle.

More biodiversity than in any other terrestrial ecosystem is found in freshwater resources. The growing proportion of plant and animal communities within the water regime is causing the freshwater habitat to become more vulnerable. The world's freshwater ecosystems are home to over 10,000 different species of fish, accounting for 40% of all fish communities and 14% of all vertebrate species. The freshwater habitat is the only preferred biological location for all vertebrates, as demonstrated by the vast numbers of amphibians, aquatic reptiles, and mammals that coexist with the total quantity of freshwater fish. Comprehensive data regarding the overall species diversity of freshwater resources is lacking, particularly with regard to microbes and invertebrates as well as in tropical regions of the world that are home to a variety of global species. Within the last 10 years, 5,778 species have been identified within the phylum of amphibians. There is a global "hotspot" for regional river fish biodiversity, and the Mekong drainage in Cambodia is one of them. A fish wealth of roughly 1700 species has been estimated by recent research. There seems to be less interest in freshwater biological zones than in terrestrial ones. There isn't any information on the diversity of invertebrate animals in tropical freshwaters. Nonetheless, the crustacean, mollusk, and aquatic arthropod groups do have significant endemism and species richness at the regional habitat. Microorganisms play a crucial role in the functioning of the earth's biogeochemical cycles, but information on their biodiversity in freshwater bodies is likewise poorly documented. There is still much to learn about the majority of the prokaryotic taxonomic groups.

The maximum of aquatic biodiversity is found in tropical regions, and it increases with latitude. The Amazon River offers some of the best illustrations of the biodiversity of aquatic ecosystems. The Great Barrier Reef, which is located on the Australian continent, is the world's largest coral reef ecosystem.

It is home to more than 700 different species of coral and a wide variety of fish and mollusk species. The systems with the highest diversity of marine life are coral reefs. According to a study conducted in the Gulf of Aqaba, Red Sea region, aquatic animals' egg-releasing phases vary throughout the year in this particular marine ecosystem. This provides a glaringly different detail from the Great Barrier Reef, where coral spawning and aquatic animal spawning happen simultaneously. In addition to marine systems, a variety of freshwater ecosystems are also ideal biological niches for rare species. This is because freshwater habitats are remote locations in the natural world, and difficult to tolerate in terms of weather and distance. These kinds of ecological phenomena encourage the emergence of new species in various freshwater habitats.

1.1.1. Scope of Aquatic Biodiversity

The range of organisms that live in freshwater and marine aquatic habitats is referred to as aquatic biodiversity. These ecosystems, which make up more than 70% of the Earth's surface, depend on this diversity for their health and survival. A vast variety of organisms, from larger species like fish, marine mammals, and aquatic plants to microscopic phytoplankton and zooplankton, make up the aquatic biodiversity. These organisms participate in the ecological processes that keep life on Earth alive by forming intricate food webs. The biodiversity of freshwater ecosystems encompasses species found in lakes, rivers, streams, and wetlands. Fish, amphibians, reptiles, insects, mollusks, crustaceans, and a wide variety of plant species are all included in this incredibly diverse group. Because freshwater ecosystems offer a variety of habitats, from swift-moving streams to still ponds, they are frequently hotspots for biodiversity. These habitats are home to species that have adapted to a broad variety of circumstances, ranging from the nutrient-rich but frequently oxygen-poor waters of wetlands to the oxygen-rich waters of mountain streams. Because so many species are involved in filtering pollutants, cycling nutrients, and promoting the growth of aquatic plants, freshwater biodiversity is essential to preserving the quality of the water.

The diversity of species found in seas, oceans, and coastal regions makes up even greater marine biodiversity. From deep-sea vents and open ocean environments to coral reefs and mangrove forests, marine ecosystems are extraordinarily diverse. For instance, coral reefs are home to thousands of different species of fish, invertebrates, and algae, making them one of the planet's most diverse ecosystems. These reefs are essential for tourism, fishing, and coastal protection. Mangrove forests are found in tropical and subtropical areas. They are rich in species and act as nurseries for a variety of marine organisms, including fish and crustaceans that are important to the global economy. The richness of life forms in the pelagic zone, or open ocean region distant from the coast, is one of the most remarkable aspects of marine biodiversity. From the minuscule plankton that forms the foundation of the marine food web to the largest animals on the planet, like the blue whale, this zone is home to a wide variety of species. Particular communities of organisms that can survive in harsh environments like high pressure, low temperatures, and total darkness can be found in deep-sea ecosystems, such as hydrothermal vents and cold seeps. These deep-sea species frequently display amazing adaptations, like chemosynthesis and bioluminescence, that enable them to endure and procreate in such hostile conditions.

Globally, aquatic biodiversity is not dispersed equally. Because of their high species richness and endemism, some areas, like the Great Lakes of Africa, the Amazon Basin in South America, and the Coral Triangle in Southeast Asia, are known as biodiversity hotspots. These regions are frequently the focus of conservation efforts due to their enormous ecological significance. However, human activity poses a serious threat to aquatic biodiversity. Overfishing, pollution, habitat loss, and climate change are a few of the main issues that aquatic ecosystems are currently facing. Water quality can be deteriorated and aquatic life can be harmed by pollution from domestic, agricultural, and industrial sources. Ecosystem services and biodiversity can be lost as a result of habitat destruction, which includes damming rivers, draining wetlands, and destroying coral reefs. Numerous fish populations have declined as a result of overfishing, upsetting food chains and changing the composition of marine ecosystems. Because of the potential effects of rising temperatures, ocean acidification, and sea level rise on species distribution, behavior, and the health of ecosystems like coral reefs and mangroves, climate change presents a particularly serious threat to aquatic biodiversity. A multifaceted strategy is needed to conserve aquatic biodiversity, including pollution reduction, climate change mitigation, sustainable fisheries management, and the preservation of important habitats. Because they give species a haven and support the maintenance of ecosystem functions, marine protected areas (MPAs) are a crucial instrument for preserving marine biodiversity. In a similar vein, freshwater protected areas can preserve important species' habitats in wetlands, lakes, and rivers.

Aquatic ecosystem resilience can also be increased by restoration initiatives like removing invasive species, restoring coral reefs, and replanting mangroves. Education and public awareness campaigns are essential for advancing the preservation of aquatic biodiversity. Involving stakeholders, local communities, and legislators in conservation efforts can support the sustainable use of aquatic resources. Since many aquatic species cross national borders and the wellbeing of aquatic ecosystems frequently depends on regional and global efforts, international cooperation is also essential. Understanding the condition of aquatic biodiversity and the effects of human activity requires research and monitoring. Technological innovations like genetic analysis, satellite remote sensing, and autonomous underwater vehicles are opening up new avenues for aquatic life research and conservation. Initiatives for citizen science, in which volunteers assist with data collection and monitoring, are also becoming more and more significant in conservation efforts (Pan et al., 2013; Dieter and Cameron, 2014; Roberto, 2016).

1.1.2. Threats to Aquatic Biodiversity

The diversity of life forms found in aquatic environments, or aquatic biodiversity, is seriously threatened by many factors brought about by human activity. Chemicals, plastics, and other waste products contaminate water bodies, posing a serious threat to aquatic life and degrading their habitat. Fish and other marine organism populations have drastically decreased as a result of overfishing and unsustainable fishing methods, upsetting food chains and ecosystem balance. An additional important factor is climate change, which affects the survival and breeding patterns of species by bringing about rising sea temperatures, acidification of the ocean, and changes to water cycles. The degradation of habitat resulting from projects like dam building, coastal development, and deforestation intensifies the decline of aquatic ecosystems. Human-introduced invasive species frequently outcompete native species, which reduces biodiversity in the area.

Additionally, the ability of aquatic ecosystems to withstand natural disasters and human impacts has been weakened by the loss of wetlands and mangroves, which are essential breeding grounds and protective buffers. Addressing these risks and preserving the wide variety of life in aquatic habitats require effective conservation measures (Figure 1.1).

Nations came together to fulfill the water-related targets of the 2000 Millennium Declaration, and the resolution demanded that serious attention be paid to global water issues and development efforts. The proposal's Agenda 21, which is widely acknowledged as every person's fundamental right, was adopted with serious concern in order to guarantee that everyone on the planet has access to clean, safe drinking water and sanitary restrooms. The United Nations "Water for Life" resolution was put into effect concurrently with threats to aquatic biodiversity and biological resources from human activity. Although the reasons behind the threats to the world's aquatic biodiversity are readily apparent, the extent of the harm varies from place to place. Threat identification may not have much of an impact on conservation, so it is best to take practical steps to reduce the threats as a first step toward restoration.

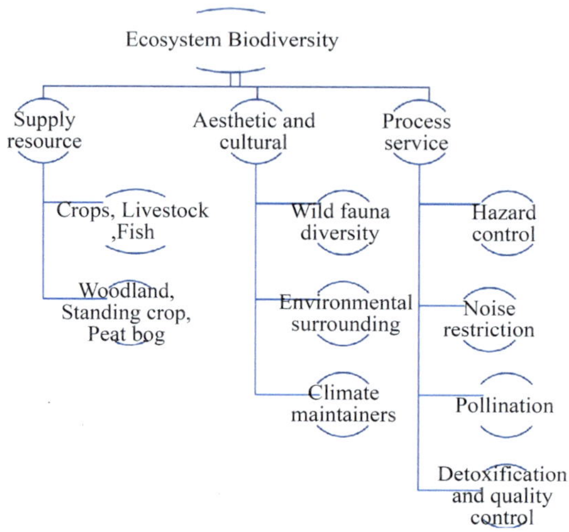

Figure 1.1. *Showing various uses of ecosystem biodiversity.*

Source: https://html.scirp.org/file/1-1380986x2.png.

The Red Sea, which is home to over 1000 species of fish and more than 50 genera of reef-forming (hermatypic) corals, has long been recognized as a place with high marine biodiversity and endemism. Certain predator fish can still be preserved if Saudi Arabia's Red Sea reefs are brought back to their pre-crisis state, according to recent project studies that have revealed the possibility of the extinction of numerous large fish in the coral reef ecosystem. A vast expanse of coral reefs supports a diverse range of endemic species that are flourishing in the Red Sea's marine ecosystem. Additionally dependent on the Red Sea are communities of sea grasses, meadows, and mangrove systems. Marine water desalination, fishing, shipping, the oil and gas industry, urban and industrial discharge, plastic waste, human activity, and climate change are all putting pressure on this area of the marine ecosystem.

Ecosystems that provide Trinidad and Tobago with sufficient freshwater resources were constantly threatened, and there was obvious harm being done to the freshwater habitats. Deforestation, the use of pesticides in agriculture and water for irrigation, the growth of the urban and industrial sectors, the regulation of water for dam and hydroelectricity production, the extraction of minerals and petroleum, solid waste dumps, bottom dredging and water flow channelization, fishing, and the aquarium market are some of the major impacts that have caused extensive damage to water resources and the loss of aquatic biodiversity. Each and every one of these freshwater systems was listed as a regional conservation priority. In Chile, the preservation of freshwater resources was aimed at safeguarding the habitats of rare Salmonid species. The preservation of native wildlife and the equilibrium of ecosystem processes were both benefited by this management of biodiversity. A similar situation also occurred in New Zealand, where the development of habitat protection has been greatly impacted by the conservation of salmonids and other valuable fish biodiversity.

Whether intentional or not, human activity-induced colonization of new species in the aquatic system is becoming a major environmental problem on a worldwide scale. In the United States, the issue of introduced species in freshwater resources is becoming more prevalent. Six exotic fish were brought into the freshwater in 1920; by 1945, a few more had been added, and other species had been discovered up until 1980. Because they begin to breed and expand their populations outside of their designated area, recently introduced species cause numerous issues for the established species within the ecosystem. Once a species has been introduced, it is very difficult to reverse the process of introduction. Chemical contaminations in the aquatic ecosystem can be removed at the point of source, or the habitat can be restored after pollution peaks. In addition, introduced or alien species may have detrimental effects on native organisms in the water resources through animal predation, competition, niche modification, interspecific breeding, and the development of parasitic infections and diseases. Certain species, such as aquaculture, are introduced to new environments in order to benefit economically or to meet developmental needs. Due to their invasive growth, this introduction may initially be economically profitable but may eventually result in negative economic and ecological effects. As an illustration, consider oysters, which have been introduced to coastal waters worldwide as a food source. The role of oysters in the estuarine ecosystem involves many intricate changes. In addition, when oysters merge with non-native oysters, the resulting accreted shells lead to a form of aquatic system mismanagement. Because of their rapid growth and other negative effects that disrupt the environment both ecologically and economically, introduced invasive species pose a serious threat to the ecosystem's biodiversity in the marine environment.

- **Overexploitation of Species:** Overusing a species to satisfy human greed can lead to a loss of genetic diversity, which in turn reduces the relative abundance of that species in individual cases or in groups where species interact. Because of the disruption in the age distribution and sex composition, this process may result in a smaller population. Because some alleles are lost from the genome as a result of fishing exploitation, the genetic makeup of fish populations is disturbed and genetic diversity is decreased.

- **Habitat Modification:** A species' extinction may also result from structural changes made to its habitat, such as the building of dams, the clearing of forests, the rerouting of water courses for agricultural purposes, and the conversion of marsh land and small bodies of water for other purposes. Building dams on rivers disrupts fish migration, causes populations to shift from their natural spawning grounds, and splits the population into smaller subpopulations. Another example is the result of deforestation, which causes the catchment area to gradually deteriorate as a result of soil erosion that results in siltation and sedimentation. In addition to having a detrimental effect on aquatic organisms' breeding zones, this process causes small fish to have gill blockage.
- **Organic Pollutants Drainage/Suspended Solids/Sewage Discharge:** This includes poisonous and toxic substances such as metals, agrochemicals, phenolic discharge, and acidic discharge, which increase the death rate and inhibit aquatic animals' ability to reproduce. Fish are more vulnerable to infections from aquatic pathogens due to suspended solids' disruption of respiratory physiology, suffocation, and mucus production. When sewage is released into an aquatic system, it creates a eutrophic environment, which depletes oxygen and increases the mortality rate of aquatic life and submerged aquatic plants.
- **Thermal Pollution:** Results in temperature variations in the surrounding water and a decrease in the water column's dissolved oxygen concentration, which kills small, delicate aquatic life. Aquatic biodiversity is impacted by the acceleration of the death rate in aquatic organisms, which can lead to a decrease in population or loss of species. Scientists who have observed coral bleaching in the ocean fear that it may continue past 2017 due to a notable increase in ocean water temperature caused by the El Niño phenomenon. Worldwide coral reefs are becoming more and more vulnerable, and the primary cause of this rising temperature is widely acknowledged to be climate change and global warming (Pan et al., 2013; Dieter and Cameron, 2014; Roberto, 2016).

1.1.3. Framework for Considering Aquatic Biodiversity in Forests

The structure we use in this case (Figure 1.2) draws inspiration and expansion from concepts created by others. Because of their complexity and variety, aquatic ecosystems in forests are best understood as a collection of distinct ecological states that occur within a domain both spatially and temporally (cup, Figure 1.2), which includes the variety of ecosystem conditions or states. A domain's ecological states vary and are constantly changing due to stochastic processes, local conditions, and minor disturbances (moving cup, Figure 1.2). Every domain, and corresponding state, is made up of different elements (balls in a cup), such as specific biotic and abiotic circumstances, activities, and processes at a location during a time period. Within and between forested watersheds, these conditions change over time and space, symbolically represented by varying ball sizes and colors.

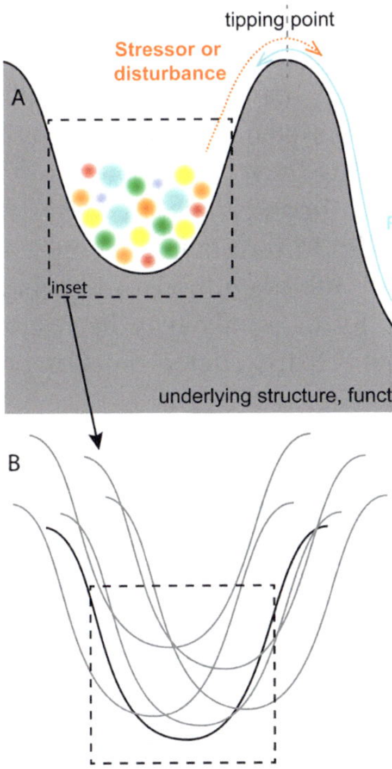

Figure 1.2. *Multi-state concept for changes in forested aquatic ecosystems over time.*

Source: https://www.fs.usda.gov/pnw/lwm/aem/docs/penaluna/2016_penaluna_etal_aquatic_ecosystem_services_biodivcons.pdf.

The ability of an ecosystem to withstand natural disruptions and human stressors while adjusting to change and maintaining its place within a domain's dynamics is known as resilience. Ecosystems can be redefined as a result of significant changes brought about by human stressors and natural disturbances. This involves changes to the ecosystem's processes, components, and functions that transfer it from one domain to another (Figure 1.2 passing a "tipping point"). Through this transition, some ecosystem components might survive, while others might not, resulting in a new set of states in an uncharted territory. For instance, biodiversity may disappear if taxa are unable to adapt to changing conditions over time. This could result in altered system components, a new state within the domain (ecosystem), or even the emergence of a whole new domain with distinct structural components for biodiversity and ensuing variations in ecological processes and functions. Concern over aquatic biodiversity's overall lack of resilience to modern stresses is growing, given its seeming increased susceptibility to losses. Because aquatic biota is essential to ecological processes and functions, the loss of aquatic biodiversity may become a "weak link" in the resilience of ecosystems. Furthermore, ecosystems' ability to provide crucial ecosystem services to people may alter as they pass tipping points to new domains. We further define these problems for aquatic forest ecosystems below. We start by looking at aquatic biodiversity in forests in relation to ecological services and roles. Secondly, we classify the main modern threats to aquatic biodiversity in forests, which individually or jointly may be responsible for tipping points in ecosystems (Pan et al., 2013; Dieter and Cameron, 2014; Roberto, 2016).

1.2. IMPORTANCE OF AQUATIC ECOSYSTEMS

All living things and inanimate objects working together to form a stable system is called an ecosystem. The study of ecology focuses on the relationships between organisms and their surroundings, or ecosystems. Since living and non-living things depend on and have an impact on one another, it is crucial to comprehend this connection. Energy exchange keeps ecosystems running on a daily basis. Between the physical and biological components of an ecosystem, energy is recycled. Plants in an ecosystem use solar energy to produce food, which animals then graze on before being eaten by predators. When waste materials are broken down into substances that plants can use as food, microorganisms like bacteria and fungi within an ecosystem exchange energy as well. Thus, for the ecosystem to survive, every component depends on every other element.

Rivers, streams, lakes, estuaries, wetlands, and oceans are examples of aquatic ecosystems. Fish, plants, and microorganisms are just a few of the living creatures that can be found in these aquatic environments. Pollution can quickly upset these ecosystems, which are extremely delicate. The same watershed is shared by all organisms in an ecosystem. When water flows from a common body of water, like a lake or pond, it is referred to as a watershed. All of us reside in a drainage basin or watershed. The size of a watershed can range from a farm with a pond to the Mississippi River drainage basin. There could be towns, farms, houses, shops, or mountains within your watershed. Every other living creature in the ecosystem shares your watershed.

An excellent illustration of how the living and nonliving components of an ecosystem are interdependent is a watershed. Any changes made to a watershed will have an impact on all of the local wildlife. Paving over land and putting up buildings are two ways that people can change a watershed. This could result in dangerous materials washing straight into the water and altering how water moves across the land. The creatures that rely on the water to survive will be impacted by this. Fish that consume aquatic organisms are one example. These organisms may perish in contaminated water, depriving the fish of food. We must thus comprehend these connections and safeguard our water supplies. Unlike terrestrial ecosystems found on land, aquatic ecosystems are found in and around bodies of water. Communities of organisms, or aquatic life, that depend on one another and their surroundings are found in aquatic ecosystems. Freshwater and marine ecosystems are the two primary categories of aquatic ecosystems. Pools, ponds, and lakes are examples of lentic ecosystems; streams and rivers are examples of lotic ecosystems; and wetlands are areas where the soil is submerged or saturated for a portion of the year.

1.2.1. Types

Freshwater ecosystems and marine ecosystems are the two main categories of aquatic ecosystems. Different aquatic ecosystems further subdivide both freshwater and marine ecosystems.

1.2.1.1. Marine Ecosystems

The largest aquatic ecosystems on Earth are marine ecosystems, which are found in waters with a high salinity content. Freshwater ecosystems are different from these systems in that they contain less salt. More than 70% of the Earth's surface is made up of marine waters, which also supply 97% of the planet's water and 90% of its habitable area. The average salinity of seawater is 35 parts per thousand of water. The salinity levels in various marine ecosystems differ. Depending on the shoreline features and depth of the water, marine ecosystems can be classified into several zones. The vast open area of the ocean where creatures like whales, sharks, and tuna reside is known as the oceanic zone. Many invertebrates live on substrates below water, which make up the benthic zone. The space between high and low tides is known as the intertidal zone. Mudflats, seagrass meadows, mangroves, rocky intertidal systems, salt marshes, coral reefs, and lagoons are examples of additional near-shore (neritic) zones. Hydrothermal vents may arise in deep water at the base of the food web, which is formed by chemosynthetic sulfur bacteria.

The physical environment and the biological community of organisms with which they are associated define marine ecosystems. Brown algae, dinoflagellates, corals, cephalopods, echinoderms, and sharks are some of the classes of organisms that can be found in marine environments. A sizable fraction of the world's population depends on marine ecosystems for food, employment, and ecosystem services. The stability of marine ecosystems is seriously threatened by human activities within them as well as pollution. The overfishing of certain species, unsustainable resource exploitation, marine pollution, climate change, and coastal area development are some of the environmental issues affecting marine ecosystems. Furthermore, a significant portion of the carbon dioxide responsible for global warming and the heat it absorbs is absorbed by the ocean; as a result, processes like ocean acidification are altering the chemistry of the ocean, endangering marine ecosystems.

- **Marine Coastal Ecosystem:** An ecosystem found where land meets the ocean is called a marine coastal ecosystem. Marine coastal ecosystems comprise a wide variety of distinct marine habitat types, each with unique traits and species compositions. They are distinguished by high productivity and biodiversity levels.
- **Marine Surface Ecosystem:** The term "neuston" refers to organisms that are free-living at the ocean's surface. These include floating barnacles, marine snails, nudibranchs, cnidarians, and keystone organisms like the golden seaweed Sargassum, which forms the Sargasso Sea. Numerous fish species that are significant to both ecology and the economy reside in or depend on neuston. The distribution of species at the ocean's surface is not uniform; the surface serves as home to distinct neustonic communities and ecoregions that are restricted to a few latitudes and ocean basins. However, the surface is also directly affected by pollution and climate change. Worlds are connected by life on the ocean's surface. Many terrestrial and marine species depend on the surface ecosystem and the organisms found there, from shallow waters to the deep sea, the open ocean to rivers and lakes.

1.2.1.2. Freshwater Ecosystems

An area of Earth's aquatic ecosystems is made up of freshwater ecosystems. These consist of wetlands, bogs, rivers, streams, lakes, and ponds. They are comparable to marine ecosystems, which contain more salt. There are several ways to categorize freshwater habitats, such as through considerations of temperature, light penetration, nutrients, and vegetation. The three main categories of freshwater ecosystems are lentic ecosystems, which have slower-moving water and include pools, ponds, and lakes; lotic ecosystems, which have faster-moving water and include streams and rivers; and wetlands, which are regions where the soil is submerged or saturated for a portion of the year. 41% of all fish species are found in freshwater environments. The characteristics of freshwater ecosystems have been affected by significant changes that have occurred over time. Risks to human health, such as cholera outbreaks brought on by sewage contamination, motivated early efforts to study and track freshwater ecosystems. Prioritizing chemical indicators in the beginning, bacteria were next, and then algae, fungus, and protozoa. Measuring the stream conditions that are related to various groups of organisms (fish, macroinvertebrates, and macrophytes) is a novel approach to monitoring.

Overexploitation, water pollution, altered flows, habitat destruction or degradation, and exotic species invasion are some of the threats facing freshwater biodiversity. There are three basic types of freshwater ecosystems: lotic (faster-moving water, such as streams and rivers), lentic (slow-moving water, such as pools, ponds, and lakes), and wetlands (areas where the soil is saturated or inundated for at least part of the time). Climate change is placing additional pressure on these ecosystems because there has already been a significant decline in ice coverage and an increase in water temperatures of about 1°C. Freshwater ecosystems are the subject of limnology, and its subfield, freshwater biology.

1.2.1.2.1. Lentic Ecosystems

A lacustrine or lake ecosystem consists of both abiotic (non-living) physical and chemical interactions as well as biotic (living) plants, animals, and microorganisms. Much of this article pertains to lentic ecosystems generally. Lentic ecosystems, which include ponds, lakes, and wetlands, are characterized by their stationary or relatively still freshwater. A prime example of these ecosystems are lakes. The term "lentic" comes from the Latin word "lentus," which means sluggish. Lotic ecosystems, which include moving terrestrial waters like rivers and streams, are comparable to lentic ecosystems. These two ecosystems together serve as illustrations of freshwater ecosystems.

Although the general distinction between pools/ponds and lakes is not clear, according to Brown, ponds and pools have their entire bottom surfaces exposed to light, while lakes do not. Lentic systems range in size from a small, temporary rainwater pool a few inches deep to Lake Baikal, which has a maximum depth of 1642 m. Certain lakes also develop seasonal stratification. There are two regions in ponds and pools: the benthic zone, which includes the shoreline and bottom areas, and the pelagic open water zone. Lake systems feature a deeper zone known as the profundal because of their dark, deep bottom areas. These three locations may have quite distinct abiotic circumstances, which is why they are home to species that have evolved there especially.

1.2.1.2.2. Lotic Ecosystems

Waterways that flow through an ecosystem are called rivers, and they contain both abiotic (nonliving) physical and chemical interactions between the many components of the ecosystem as well as biotic (living) interactions between plants, animals, and microorganisms. Lower headwater streams flow into mid-size streams, which eventually flow into larger river networks. River ecosystems are a component of larger watershed networks or catchments. The riverbed gradient or the current's velocity define the main zones in river ecosystems. The dissolved oxygen concentrations in faster-moving turbulent water are usually higher than in slower-moving pool water, supporting a higher level of biodiversity. Rivers are classified into lowland and upland categories based on these differences.

Wider streams and those without a canopy get most of their food from algae, but the majority of streams in riparian forests get their food from the trees. Another important source of nutrients are anadromous fish. Rivers face environmental threats from introduced species, chemical pollution, dams, and water loss. A dam has detrimental effects that stretch down the watershed. The two main detrimental effects are the decrease in spring flooding, which harms wetlands, and the retention of sediment, which causes deltaic wetlands to disappear. Lotic ecosystems include, for the most part, river ecosystems. From the Latin lotus, which means cleansed, the word *locic* means flowing water. Small springs with a few millimeters of width to large rivers with kilometers of width are examples of littoral waters. This article covers a lot of ground regarding lotic ecosystems in general, which includes related lotic systems like streams and springs. Lettic ecosystems are distinguished from lotic ecosystems by the presence of relatively still terrestrial waters, like those found in wetlands, lakes, and ponds. These two ecosystems combined make up the broader field of study known as freshwater or aquatic ecology.

1.2.1.2.3. Wetlands

An ecosystem known as a wetland is unique among semi-aquatic ecosystems, characterized by groundcovers that are permanently submerged in water for extended periods, or only temporarily during certain seasons. Anoxic (low-oxygen) processes occur as a result of flooding, particularly in the soil. Because the roots of their vegetation have adapted to oxygen-poor, waterlogged soils, wetlands provide a transitional zone between waterbodies and dry lands. This sets them apart from other terrestrial or aquatic ecosystems. As homes to a diverse array of aquatic and semi-aquatic plants and animals, they are regarded as some of the most biologically diverse ecosystems on Earth. The plants' ability to eliminate excess nutrients, like phosphates and nitrates, has also been shown to improve water quality.

Each continent is home to wetlands. Fresh, brackish, or saltwater can be found in wetlands. Based on the predominant plant species and the water source, the main types of wetlands are categorized. For instance, emergent herbaceous vegetation like sedges, cattails, and reeds predominate in marshes, which are wetlands.

Trees and shrubs that are woody in nature predominate in swamps; however, in Europe, reed swamps are dominated by reeds rather than trees. Mangals are marshes containing mangroves, which are halophytic woody plants that have adapted to live in salty environments. Estuaries, which receive their water from a combination of river and tidal sources, floodplains, which receive their water from overflowing rivers or lakes, bogs, and vernal ponds, are a few examples of wetlands categorized according to their water sources. Certain wetlands are hard to categorize since they have a variety of plant species and receive water from several sources. The largest wetlands in the world are found in the West Siberian Plain, South America's Pantanal, the Ganges-Brahmaputra delta's Sundarbans, and the Amazon River basin (Pan et al., 2013; Dieter and Cameron, 2014; Roberto, 2016).

1.2.2. Functions

Many crucial environmental tasks are carried out by aquatic ecosystems. They replenish groundwater, clean water, reduce flooding, recycle nutrients, and create habitats for various wildlife species. The aquatic biota, which includes bacteria, protists, fish, higher plants, invertebrates, microorganisms, and aquatic fungi, is responsible for the ecosystem's ability to purify itself. Water filtration and the breakdown of organic matter are just two of the many self-purification processes in which these organisms actively participate. Because they serve as homes for the species that live there, aquatic ecosystems must be able to sustain themselves on a regular basis. Aquatic ecosystems serve many environmental purposes, but they are also vital to human recreation and the tourism sector, particularly in coastal areas. Lakes are utilized for ecological research, among other educational and religious purposes. Christians, for example, worship the Jordan River.

1.2.3. Biotic Characteristics (Living Components)

The organisms that are present mostly dictate the biotic properties. Snails or geese may graze the vegetation, leaving behind sizable mud flats, or wetland plants may create dense canopies that cover sizable areas of silt. Because oxygen levels in aquatic environments are generally low, the organisms that live there must adapt. To deliver oxygen to their roots, for instance, aerenchyma is a necessary component of many wetland plants. Predation, mutualism, and competition's relative importance are examples of other biotic traits that are more subtle and challenging to quantify. Predation by coastal herbivores, such as sharks, geese, and mammals, appears to be a dominant biotic factor in an increasing number of cases.

1.2.3.1. Autotrophic Organisms

Producers that convert inorganic material into organic compounds are known as autotrophic organisms. Algae are arguably the most significant autotrophic organisms in aquatic environments because they use solar energy to convert carbon dioxide into biomass. The biomass contribution from rooted and floating vascular plants increases with increasing water depth. Together, these two sources provide the exceptional production of wetlands and estuaries, where the autotrophic biomass is transformed into aquatic species such as fish, birds, and amphibians. Benthic marine environments contain bacteria that can synthesize chemicals. These organisms can consume hydrogen sulfide found in water originating from volcanic vents. Animals that consume these bacteria can be found in large groups near volcanic vents. For instance, there are 30 cm-long clams (*Calyptogena magnifica*) and enormous tube worms (*Riftia pachyptila*) measuring 1.5 m in length.

1.2.3.2. Heterotrophic Organisms

In addition to eating autotrophic organisms, heterotrophic organisms use the organic compounds in their bodies as raw materials for biomass production and as sources of energy. While stenohaline, or salt-intolerant, species are limited to freshwater habitats, euryhaline organisms are capable of surviving in marine environments (Pan et al., 2013; Dieter and Cameron, 2014; Roberto, 2016).

1.2.4. Abiotic Characteristics (Non-Living Components)

Abiotic environmental factors and biological interactions form the structure of biotic communities, which make up an ecosystem. The kind of substrate, water depth, nutrient levels, temperature, salinity, and flow are a few of the crucial abiotic environmental elements of aquatic ecosystems. Determining the relative significance of these factors without conducting extensive experiments can often be challenging. Complex feedback loops could exist. Sediment, for instance, can indicate whether aquatic plants are present, but aquatic plants can also accumulate sediment through peat and trap it. A water body's dissolved oxygen content is often the primary factor in identifying the types and quantity of organic life present. Fish require dissolved oxygen to survive, however different species have different tolerance levels for low oxygen; in severe situations, some fish even start gulping down air. Aerenchyma production is frequently required of plants, and leaf size and shape can also change. In contrast, many anaerobic bacterial species die when exposed to oxygen.

The concentration of nutrients has a major role in regulating the number of algae species. The dominant species among algae can be determined by the relative abundances of phosphorus and nitrogen. While algae are an essential food source for aquatic life, an overabundance of algae can lead to fish population declines as they decompose. When an identical overabundance of algae decomposes in coastal environments like the Gulf of Mexico, it creates a dead zone—a hypoxic area of the water. The types of species that can be found in a body of water are also influenced by its salinity. While many organisms in freshwater environments are salt-intolerant, those in marine ecosystems can tolerate salinity. An estuary or delta's salinity level has a significant influence on the kinds of wetlands that exist there (fresh, intermediate, or brackish), as well as the animals that live there. Upstream dam construction may lessen spring flooding and sediment accumulation, which could cause saltwater intrusion into coastal wetlands. Salt concentrations that are detrimental to freshwater organisms are frequently absorbed by freshwater used for irrigation.

1.2.5. Threats

An aquatic ecosystem reaches a point where its capacity to withstand stress is exceeded, at which point its health declines. Chemical, biological, or physical changes to the surroundings can all put stress on an aquatic ecosystem. Water flow, temperature, and light availability are examples of physical modifications. Toxins, materials that consume oxygen, and nutrients that stimulate biostimulatory processes can all undergo chemical changes. The introduction of exotic species and excessive commercial species harvesting are examples of biological alterations. Aquatic ecosystems are susceptible to undue stress due to human populations. The disturbance of current plant and animal distribution patterns resulting from anthropogenic climate change can be detrimental to aquatic ecosystems. Coral reefs, crustaceans, deep sea biodiversity, and other biotic elements of these ecosystems have all suffered as a result. Because they trade off biodiversity for their intended uses, man-made aquatic ecosystems like irrigation channels, aquaculture ponds, and ditches may also be detrimental to naturally occurring ecosystems. Ditches, for example, serve mainly as drainage systems, but their existence has a detrimental effect on wildlife.

Numerous instances of extreme stress leading to adverse outcomes exist. The environmental past of the Great Lakes in North America serves as an example of this issue, specifically showing how various stresses—like overfishing, water pollution, and invasive species—can compound to create a problem. Invasive species and pollution have led to a similar decline in the English Norfolk Broadlands. The Gulf of Mexico's Lake Pontchartrain serves as an example of the detrimental consequences of several stresses, such as the building of levees, the clearing of swamps, the introduction of alien species, and the intrusion of salt water.

1.2.6. Conservation Approaches for the Aquatic Ecosystems

Established in 1994, the United Nations Convention on the Law of the Sea (UNCLOS) is one of the most notable approaches, particularly for the preservation of aquatic ecosystems. Fundamentally, this law took into account the sustainable use of ocean water resources as well as the encouragement and support of biodiversity. Large international non-governmental organizations, research institutions, and scientific groups can cooperate under the auspices of the UN to assess biodiversity globally and identify and list hotspots as indicators of biodiversity. But, insufficient information availability shouldn't be seen as a barrier to conservation efforts to stop future losses in freshwater or marine biodiversity. Communities that operate on a larger scale shouldn't wait to start a process until they have all the necessary information. There is enough substantial and trustworthy data available about fishes, marine reptiles, and other taxa to conclude that aquatic biodiversity is in crisis and could degenerate if not addressed.

It has been suggested that methods for protecting aquatic resources should be in line with ecosystem conservation, as this can support the development of biotic resources and aid in the preservation of natural ecosystems and habitats (Figure 1.3). Regulations governing conservation management should cover all broad areas to successfully conserve aquatic biodiversity. The goal of restoration should be aquatic areas that have experienced habitat loss, degradation, or destruction due to human activity. The conservation of plant and animal species that are experiencing a decline in population and are in danger of going extinct should be the same goal. Aquatic biodiversity and water systems are interdependent and have a close relationship. The general consensus is that the species composition, diversity, and physical structure of aquatic ecosystems are influenced by the quality of the water and the habitat.

Within a body of water, areas designated as bio-reserves can be established, where fishing and other activities are prohibited to preserve the aquatic biodiversity by safeguarding the plants, animals, and their habitat. These bio-reserves can also be very helpful for tourism, education, and recreation. They may even be able to increase the amount of fish reserves by improving the conditions that lead to the decline in fish populations. These bio-reserves serve as biological parks, fisheries reserves, sanctuaries, and marine protected areas. Bioregional management balances the needs of the local economy, society, and conservation while also helping to control factors that affect aquatic biodiversity. This covers both smaller-scale biosphere zones and larger reserves. The Figure 1.3 shows a schematic of the regional area protection plan and the global biodiversity monitoring system network. The goal of the Red Sea Biodiversity collaborative project was to evaluate the marine biodiversity of the Red Sea's deep water and Saudi Arabian coast while also developing an information resource at Senckenberg and the King Abdul Aziz Marine Museum (KAUMM).

Figure 1.3. *Representation of a biodiversity monitoring system.*

Source: Beck-O'Brien M, Bringezu S. Biodiversity Monitoring in Long-Distance Food Supply Chains: Tools, Gaps and Needs to Meet Business Requirements and Sustainability Goals. Sustainability. 2021; 13(15):8536. https://doi.org/10.3390/su13158536.

An important part of maintaining the harmony and conserving aquatic diversity is watershed management. Because aquatic resources frequently traverse multiple political territories on their geographical route, they are an open and unprotected system. Resource management and conservation are made more difficult in these kinds of situations. However, since local and regional watershed communities began working on lake and stream protection initiatives, the situation has recently improved. The highly biodiverse Napo drainage depression of the Amazon basin, which is in Ecuador, has experienced numerous human threats. The freshwater vertebrate and invertebrate taxa were the main focus of the investigation. Approximately 481 freshwater vertebrate species, including fish, birds, mammals, reptiles, and amphibians, as well as over 54 invertebrate families, were recorded between 200 and 4500 meters above sea level in the basin. These compilation data were used to create species richness maps. The goal of the study was to use labor, money, human effort, specialized knowledge, and time to establish conservation and management programs for the large species diversity of invertebrates in the large tropical Napo basin.

In addition to reducing sedimentation in water systems, reforestation can regulate soil erosion in the catchment or drainage areas of water bodies. This may give aquatic fauna a better environment for survival. Thermal power plants, chemical factories, and other industrial facilities shouldn't be built next to aquatic systems because their waste can harm the ecosystem that supports aquatic ecosystems and ultimately lead to a decline in biodiversity. Global research organization World Resources Institute USA has highlighted that classifying a certain plant or animal as threatened or endangered can help to protect biodiversity. Specialized educational programs can also be used to support the restoration of aquatic resources. The goal of the USDA Forest Service's unique course, for instance, is to preserve and restore the river ecosystem and the species that are connected to it. A biodiversity conservation management program is also offered by many international universities in an effort to educate the next generation about the value of biodiversity. To preserve the variety of aquatic resources, industrial and urban wastewater discharge regulations must be taken into account.

The most approachable method of educating the public about the preservation of aquatic biodiversity is through raising awareness on social media. Events sponsored by prizes, volunteer promotions, and educational initiatives can all effectively accomplish this. The creation of marine protected areas, or MPAs, is becoming more and more well-liked as a tool for managing coral reef systems. Although their place in the marine ecosystem is still poorly understood, non-coral invertebrates also greatly contribute to the biodiversity of coral reef ecology. For a very long time, Japan has been designating areas as marine protected areas. Among them are: 1) Natural Parks and Natural Seashore Conservation Areas, which work to preserve the natural landscape; 2) Nature Conservation Areas, Wildlife Protection Areas, Natural Habitat Conservation Areas, and areas designated as natural monuments for the preservation of the environment or an organism's habitat; and 3) Protected Water Surface Zones, which work to conserve aquatic life and plant life, as well as coastal marine resource development areas and other specific areas designated by various organizations, including fishing associations and prefectural governments (Pan et al., 2013; Dieter and Cameron, 2014; Roberto, 2016).

1.3. AQUATIC HABITATS

Fish have to navigate a variety of interrelated stressors because aquatic habitats include some of the most dynamic and complex environments on Earth. Fish frequently struggle with changes in important environmental factors combined with biotic obstacles and human pressures. Human pressures are creating more and more stressors. Multi-stressor studies are essential to understanding how multivariate phenomena, such as climate change, affect fish physiology. The world's habitats that are primarily made of water are included in the aquatic biome. Freshwater habitats and marine habitats are the two main categories of aquatic ecosystems based on salinity. Freshwater habitats are types of aquatic environments that have less than one percent salinity. Rivers, lakes, streams, ponds, wetlands, bogs, and lagoons are among them. Aquatic environments with salt concentrations greater than 1% are referred to as marine habitats. Seas, coral reefs, and oceans are among them. There are certain environments where freshwater and saltwater coexist. These consist of salt marshes, mangroves, and mud flats. Many different kinds of animals, such as fish, amphibians, reptiles, mammals, birds, and invertebrates, are supported by aquatic ecosystems.

1.3.1. Freshwater Habitats

Freshwater includes water from subterranean streams, lakes, reservoirs, and rivers, among other sources. The oceans constantly lose water, which subsequently re-emerges as rain, snow, or ice on land. Our rivers, streams, lakes, and subterranean streams receive freshwater from the melting of ice and snow on the mountains. Rain that has evaporated from the sea fills mountain streams, forming rivers and lakes. Rainwater runs off into freshwater rivers, which empties back into the ocean. Many different kinds of animals can be found in lakes, rivers, and streams. Freshwater habitats primarily consist of two types: rivers and lakes. Rivers and streams frequently feed lakes. They may also be closed spaces inhabited by unique species that are unique to the earth. Large creatures that can withstand strong currents can often be found in rivers, along with species like crabs and birds that hunt fish in the water. Many different species, such as fish, amphibians, reptiles, insects, birds, and mammals, find home in freshwater rivers. Streams and rivers are home to an astounding variety of fish species. A wide range of wildlife can be found in freshwater lakes. Certain species live their whole lives in a single location. Some come over for a quick bite to eat and drink. Numerous species have evolved specifically to survive in certain lakes. Large mammals that come to lakes for water include deer, giraffes, zebras, and primates. Human activity has had a major impact on many freshwater habitats. Waste water and chemicals and pesticides contaminate the water. Both the animals that consume the aquatic life and the plants that live there may be impacted. The Everglades, an 11.5-million-acre wetland in southern Florida, is the largest freshwater habitat in the world. The Amazon River originates in the Andes Mountains of South America and travels 4,000 miles (6,400 km) to the Atlantic Ocean, passing through six nations, including Ecuador and Peru. The largest lake in the world is Lake Baikal, located in the Russian region of Siberia. One-fifth of the freshwater on Earth is found in this body of water in North Asia. Thus, freshwater habitats are essential ecosystems for our planet, regardless of whether you're a hungry turtle, a plant that consumes pollutants, or a thirsty human who also enjoys playing in the water!

- **Importance of Freshwater Habitats:** Freshwater ecosystems support cultural significance, economic growth, recreational opportunities, biodiversity, and human well-being. Numerous indigenous plants and animals depend on freshwater. Certain species are specific to their environments and are only found in New Zealand. Water is deeply connected to Tangata Whenua. Rivers, or Awa, are an integral part of whakapapa, or genealogy, and freshwater is what keeps taniwha alive and safeguards wāhi tapu, or sacred areas. Important resources like harakeke (flax), Mahinga kai (harvesting), and habitat for culturally significant species (e.g., eels or tuna).
- **Threats and Pressures:** Many pressures have an effect on freshwater ecosystems. These include intensification and alteration of land use, deforestation, drainage, decreased flows, pollution, sedimentation, enrichment of nutrients, and the spread of invasive species. Our freshwater species have suffered greatly as a result of these pressures, as they are susceptible to overharvesting, interbreeding, invasive species competition, habitat loss, and degradation (Pan et al., 2013; Dieter and Cameron, 2014; Roberto, 2016).

1.3.2. Oceans

The world's largest habitat is created by the oceans. Over 75% of the planet is made up of oceans, home to countless animal species. Coastal, inshore habitats that are found near land and open ocean habitats that encircle the entire planet are the two primary categories of ocean habitat. While there are animals in all of the oceans, a greater diversity of species can be found in the shallower, richer waters. The ocean floor is home to trenches, plains, mountains, valleys, volcanoes, and underwater continental shelves, all of which are as vast and varied as the land. Most species are found in warmer coastal waters around the world. Compared to the deep ocean, these regions have more food sources. Shallower areas are typically home to smaller aquatic animals. There are more hiding spots for them in coastal waters, and there are fewer large predators. Larger animals typically favor areas on continental shelves that are lower below the waves.

For many marine animals, plankton—microscopic plants and animals, fish eggs, and animals in their larval form—provide an abundant food source. This essential food source is consumed by everything from tiny fish and crustaceans to the blue whale, the largest animal on the planet. The two biggest challenges facing ocean habitats are pollution and overfishing. Numerous plant and animal species suffer greatly as a result of pollution from the land and air building up in the sea. Numerous species face extinction as a result of overfishing.

1.3.3. Coral Reefs

The earth's richest habitats are coral reefs. Numerous plant and animal species, such as fish, reptiles, invertebrates, echinoderms, and crustaceans, find a home in these coastal habitats. Coral reefs can be found in tropical and subtropical coastal areas with year-round warmth during the day and night. Hard coral reefs and soft coral reefs are the two primary types of coral reef habitats. Animals called soft corals swim through the water before settling. The hard coral shells that remain after a coral dies are known as hard corals, and they are what build reefs. The largest coral reefs can be found all over Australia, southeast Asia, the Caribbean, the South Pacific Ocean's coastal regions, and the southwest coast of Africa. Coral reefs are home to an amazing variety of animal species that can coexist peacefully with little competition for food because they are so rich in life and biodiversity. Coral reef animals come in a wide variety of sizes, shapes, and colors.

Invertebrates of the sea urchin, starfish, and crustacean family inhabit coral reefs. In coral reefs, sea snakes hunt eels and small fish. There are numerous fish species, including seahorses and eels. Sharks visit coral reefs frequently in quest of prey, but they do not permanently reside there. In addition, sea turtles frequently visit coral reefs in quest of food. Storms, pollution, and commercial fishing pose a threat to coastal wildlife and coral reefs. Coral reefs are destroyed during the dredging process, which entails dragging fishing nets across the ocean floor. Numerous animal species that live in coral reefs are in danger of going extinct. Coral reef ecosystems can also be severely damaged by sea storms like tsunamis.

1.3.4. Wetlands

All over the world, wetlands can be found, usually in more temperate climates where vegetation grows rapidly. These vast expanses of water, which are divided by tiny landmasses, are rich in flora. Bogs, fens, marshes, and swamps are examples of wetlands. Numerous wildlife species, such as fish, amphibians, birds, mammals, reptiles, and insects, have evolved particularly to live in wetland habitats. Wetlands and swamps are the two primary categories of shallow watery areas. Typically, swamps are found in wooded areas. Mangrove trees are examples of trees that can thrive in saltwater environments and need lots of room for their roots. Large rivers or estuaries that flood when river banks collapse from excessive rain are typically home to wetlands. One of the planet's richest habitats are mangrove swamps. Living both above and below the water's surface are many animal species. Numerous animal species found in mangrove forests are unique to our planet. Small fish, amphibians, and reptiles can find refuge and a means of entry and exit from the water thanks to the massive roots of mangrove trees. Bigger animals can eat a lot of fish. In wetland environments, large aquatic birds with long beaks, like herons, spear fish. Alligators, crocodiles, snapping turtles, and crabs can be found in salt-water swamps. Reptiles and amphibians live along the water's edge. Wetland habitats are home to many insects, both living and reproducing. supplying frogs and lizards with food. Pollution and deforestation pose the biggest threats to wetlands. Because of their unique environmental adaptations, animals living in wetland habitats are susceptible to toxins found in the air and water.

1.3.5. Islands

Islands are created when land separates from larger land masses or beneath the sea, and volcanoes erupt. They can be found everywhere in the world. Their isolation gives rise to distinct wildlife species, frequently distinct from those inhabiting mainland habitats. Certain island animal species have evolved entirely apart from their mainland counterparts. There are many different types of habitats on different islands, such as wetlands, forests, tundra, and deserts. Islands have small areas and few resources, making their ecosystems delicate and prone to disruption. Numerous species have been harmed by human activity and the introduction of new species to islands, leading to their extinction or becoming endangered. The loss of habitat or food sources is especially harmful to island animals because they have nowhere else to go. Only Madagascar is home to lemurs, Papua New Guinea is home to tree kangaroos, New Zealand is home to kiwis, and the Indonesian islands of Borneo and Sumatra are home to orangutans. These species have adapted to their remote habitats after being cut off from the mainland. Since there were no large predators on the islands for them to escape, the kiwi and kakapo birds have adapted to a life without wings. Their survival is in danger due to human-introduced predators. Orangutans suffer from both the exotic pet trade and widespread deforestation in southeast Asia (Pan et al., 2013; Dieter and Cameron, 2014; Roberto, 2016).

1.4. WATER POLLUTION AND AQUATIC BIODIVERSITY

Water pollution has a significant impact on people's environmental health worldwide and is currently regarded as one of the most significant global issues facing both developed and developing nations. Water quality, conservation, and distribution are global priorities, but improper wastewater management has led to serious issues in many nations and exacerbated the global water crisis. In the 2030 Agenda for Sustainable Development, which emphasized future policies to guarantee that control of water pollution will be elevated in national and international priorities, the importance of water quality was acknowledged. Human habitation, agriculture, and industry are the main known causes of water pollution. About 80% of municipal wastewater worldwide is dumped untreated into water streams, and industrial operations are accountable for the yearly release of millions of tons of pollutants (toxic sludge, heavy metals, and solvents).

Constant population expansion, a rise in economic migration, and climate change all contribute to the depletion of natural water resources, endangering aquatic ecosystems and aquatic systems in general. It has been observed recently that aquatic biodiversity suffers numerous harms and, as a result, is severely declining in many nations. The fact that nearly primitive ecosystems are among those that are threatened is the worst problem. Although aquatic ecosystems are among the richest habitats in terms of species diversity and abundance, freshwater systems experience twice as much biodiversity degradation as other ecosystems, according to the 2005 Millennium Ecosystem Assessment (MEA 5). As a result, they are less able to provide ecosystem services, which has detrimental effects on human health. Scientists consider the annual ratio of all extinct species—marine, avian, aquatic, and terrestrial—to be an indicator of the loss of biodiversity. It is more than 10 times higher than what they consider to be the upper bound that is acceptable. The primary cause of this predicament is the long-known fact that ecosystem services are provided for free, which has resulted in extensive damage and negative effects on livelihoods and public health in general. As a result, a healthy ecosystem is a fundamental human goal and one of the Sustainable Development Goals (SDGs), which were ratified in September 2015. Considered as a whole, freshwater ecosystems, which include lakes, ponds, streams, and wetlands, make up about 15% of the planet's surface. Due to the small systems' neglect, earlier estimates were off.

Approximately 4.2 million km^2 are occupied by 300 million natural lakes worldwide, with small lakes measuring less than 1 km^2 being the most prevalent. In addition, man-made lakes cover roughly 335,000 km^2. Additionally, the surface area and volume of artificial water bodies increase quickly. The total area covered by the universal network of water streams is roughly 500,000 km^2. Wetlands span a continental area of 12 to 15 million km^2 at the same time. More than 10% of all registered animals and one-third of all vertebrate species call inland water bodies home. The actual knowledge of the diversity of registered freshwater species varies greatly amongst organismal groups. The primary cause of the current decline in biodiversity is the loss of ecological function. Despite a great deal of research on the connections between biodiversity and ecosystem functions, there is still a big knowledge vacuum regarding the impact of biodiversity loss on ecosystems, particularly those found in freshwater environments. The term "freshwater" is used here to refer to any inland water system, whether fresh or salty, which includes wetlands, rivers, lakes, and estuaries. A vital element of any water system is biodiversity. The relationship between water pollution and biodiversity will be discussed in the review that follows, starting with the importance of biodiversity and how it relates to ecosystems, and moving on to thcir primary threats. It is necessary to assess how biodiversity affects ecosystems and communities to clarify what conservation issues biodiversity faces.

1.4.1. What Is Water Pollution?

Among the most significant problems facing humanity in the 21st century are those related to water quality. The health of humans and other relevant organisms is negatively impacted by aquatic pollution, which is regarded as a major issue for both freshwater and marine environments. Through alterations in water quality, pollution has a direct or indirect impact on fish immune systems. Aquatic pollutants originate from a variety of sources, including municipal sewage deposited in the River Nile, industrial effluents, and agricultural runoffs. These pollutants gradually contaminate water, rendering it unfit for human use. Many contaminants found in agricultural wastewater, such as pesticides and herbicides, have an adverse effect on rivers and the people who use them. Heavy metals present in industrial effluents can be extremely toxic and can combine with suspended solids in home wastewater to form muck. Additionally, water pollution in Egypt, particularly in El-Max Bay and Bahary in Alexandria, impacted fish biological responses, which in turn caused food oxidation damage and degradation of the surrounding environment. Additionally, certain immune genes from fish samples of seabass (*Dicentrarchus labrax*) and seabream (*Pagrus auratus*) were impacted by water pollution (Pan et al., 2013; Dieter and Cameron, 2014; Roberto, 2016).

1.4.2. Biodiversity Importance

It is well known that the term biodiversity encompasses a broad concept. The UN Convention on Biological Diversity defines biodiversity as "living species variations from sources that include terrestrial, marine, and various aquatic ecosystems as well as ecological groups to which they belong: including diversity among species and also ecosystems." Thus, biodiversity encompasses the entire spectrum of species as well as genetic and ecological variation. Most biotic ecosystem processes, like production and decay, are supported by it. Only about two million species have been described out of the estimated five to 30 million total species on Earth. The surrounding society benefits economically from the ecosystem's service to humans. The Millennium Ecosystem Assessment (MEA) summarizes the following roles that biodiversity plays in ecosystem services: supporting: biodiversity promotes ecosystems through compositional, structural, and practical diversity; regulatory: biodiversity affects production, constancy, as well as ecosystem resilience; cultural: biodiversity provides humans with benefits related to spirituality, aesthetics, and some recreational activities; providing a direct and indirect source of food, fresh water, etc.

In addition, biodiversity has some significant benefits that are incalculable and separate from any other advantages. It is important to properly conceptualize the relationship between human well-being (wealth, security, and health) and biodiversity (traits, species, and genes) as society works to shift toward more sustainable development paths. Data analysis from earlier research indicated that, although mostly as independent terms, the terms biodiversity, human well-being, and sustainable development are becoming more and more used in public discourse. Some researchers have proposed that biodiversity should be explicitly included in a framework suitable for sustainable development as one of the internal variables that both influence and are influenced by human well-being.

1.4.3. Water Pollution and Biodiversity

The majority of aquatic life is extremely sensitive to environmental changes and reacts in various ways to pollution. Death or migration to another habitat represents the most extreme reactions. Reproductive capacity reduction and the inhibition of certain enzyme systems essential for regular metabolism are two examples of lessening responses. The significance of macrobenthic and zooplankton components in the trophic dynamics of freshwater ecosystems has been established. By occupying an intermediate position in the food chain, these organisms not only influence aquatic productivity but also provide information about the state of the environment at a given time. Furthermore, the significance of their diversity has increased recently, particularly because of the ability of some species to indicate any decline in water quality brought on by pollution and eutrophication.

In terms of biodiversity, both fish and bird species have seen significant declines in the lake over the past few decades. With the possible exception of the northwest basin, where a dam has led to hypersalinity, a negative water balance, and a loss of species diversity, the most important factor may have been the decline in the salinity of the water. A great deal of eutrophication and water pollution in the southeast basin have led to the extinction of many aquatic species. Pollutants from the wastewater discharge have impacted the benthic fauna in a few areas of the lake. Recent studies have documented fish deformation, with various fish species exhibiting atypical shapes. This phenomenon may be linked to pollution stemming from the lake's elevated eutrophication as a result of increased exposure to industrial and household waste discharge, as well as predation.

Diverse aquatic contaminants significantly impacted biodiversity, manifesting in various ways, within the ecosystem. The aquatic life was affected by pollution containing heavy metals. According to a 1992 report, the metallurgical sector accounts for 50% of industrial waste, with weaving, dyeing, and spinning contributing roughly 30%. About 25% of Egypt's industrial establishments are located in Greater Cairo, accounting for 35% of the country's total industrial activity. Of these, 40% is associated with the dumping of heavy metals into waterways. A quarter of the total heavy metal load in water streams is exchanged by roughly 150 industrial facilities. Only about 10% of Egypt's total heavy metal loads are exchanged in natural streams, although Alexandria is home to roughly 175 facilities or 25% of the country's industrial activity. In certain commercial fish species that were taken from the Egyptian coast along the Mediterranean Sea, the effects of heavy metals (Zn, Cu, Cd, Pb, and Hg) were investigated. Regarding the age or size of fish collected, it was discovered that there are no appreciable variations in certain metal concentrations, such as lead.

Furthermore, one of the most harmful contaminants in freshwater or marine water streams is thought to be microbial contamination. In addition to multicellular parasites that are dangerous to all living things, livestock excrement contains zoonotic microorganisms. Bacteria like *Salmonella* spp., *Campylobacter* spp., and *Clostridium botulinum* are examples of livestock pathogens that have an impact on public health. together with *Escherichia coli*

O157:H7. Furthermore, among the significant pathogens, parasitic protozoa are taken into consideration, such as *Microsporidia* spp. *Cryptosporidium parvum* and *Giardia lamblia* have the potential to infect hundreds of thousands of organisms annually. Large amounts of domestic, sewage, and raw agricultural effluents contaminated with heavy metals are dumped into Egypt's Lake Qarun. Fish immune defense mechanisms are inhibited in lakes due to unfavorable conditions. fish to a variety of infections as a result. *Trichomonas alginolyticus*, *Aeromonas hydrophila*, and *Parasitic isopoda* sp. were found in Lake Qarun and had a negative effect on fish populations.

1.4.4. Main Threats to Biodiversity

The term "Anthropocene" refers to an unofficial geologic time unit that we live in. Geologically speaking, humans have the ability to significantly alter our surroundings. Regretfully, our environment is harmed by the majority of these significant changes. Even more concerning is the startling rate at which these changes are occurring. Environmental alterations and extinctions happened naturally and slowly before the Anthropocene era. But species are disappearing during the Anthropocene period more quickly than they ought to. The introduction of invasive species, pollution, unsustainable resource use, habitat loss and destruction, and global climate change are some of the main direct threats to an environment's biodiversity. The main factors contributing to the general loss of biodiversity are the continuous growth in the human population and the ensuing rise in resource demand. Because they are caused by a multitude of intricately linked factors, they are difficult to address.

Major threats to Global Aquatic Biodiversity can be categorized into the following categories:

- **Climate Change:** Climate change leads to erroneous cycles in the hydrological, biogeochemical, and atmosphere. These variations include variations in temperature, seasonal changes, variations in the carbon cycle, and variations in solar radiation. Rising temperatures—which have risen by 0.6°C over the previous centuries—serve as evidence of these changes. By examining ice cores and tree rings, researchers were able to pinpoint this rise in temperature. Oceanic coral reef ecosystems significantly declined as a result of this temperature shift. Sea levels rose as a result of the temperature increase, rapidly submerging coastal areas. Sea levels are thought to have risen by 0.1 to 0.2 meters during the past century. These alterations could directly harm species and their habitats, which would reduce biodiversity.

- **Water Pollution:** There are four types of contaminants that can contaminate water: chemical, biological, radioactive, and physical. These pollutants can come from a variety of places, including home sewage, mining operations, industrial effluents, and agricultural runoff. An estimated 485,000 deaths from diarrhea are attributed to water pollution, which contaminates drinking water. The majority of these fatalities occur in Asia and Africa. Improper disposal of untreated wastewater can lead to the introduction of pathogenic microorganisms into sources of drinking water. Nuclear waste from nuclear power plants has the potential to enter water bodies. The two main types of pollutants are point and non-point sources. Point source pollution originates in a single, easily recognized location. Pollution that is not from a single source comes from multiple sources. Because it is difficult to determine where these pollutants originate, they are more difficult to address.

"Any contaminant that enters the environment from an easily identified and confined place" is the definition of point source pollution given by the US Environmental Protection Agency. Point source pollutants are frequently produced by oil refineries, paper mills, and auto plants—particularly those that use water in their manufacturing processes. They frequently dump wastewater—the water used in the manufacturing process—into bodies of water. The ecosystem of the water bodies may be harmed by the hazardous chemical pollutants present in this wastewater. Municipal wastewater treatment facilities are another frequent source of point source pollution. Municipal wastewater treatment plants frequently dump their effluents into streams, introducing microorganisms and nutrients into the water. A source of drinking water would get contaminated if this waterway was connected to it. Those who reside in the homes that use this drinking water source run the risk of contracting diseases that are spread by contaminated water. The nutrients found in wastewater can promote the growth of algae in the water, which could result in algal blooms that could lower the quality of the water.

It is challenging to address non-point source pollution because it is hard to determine its precise source. Take this scenario into consideration: the public has access to multiple sizable land parcels for grazing. These areas of land may be frequented by a large number of grazing animals who use them for grazing and even for urinating. Rainfall from these land plots could produce runoff that contains nutrients and microbiological pathogens. Runoff may reach adjacent bodies of water that could be connected to sources of potable water. Microbial pathogens would now be present in this source of drinking water. In addition, nutrients like nitrates and phosphorus would contaminate the source of drinking water. These nutrients may promote the growth of algae and result in algal blooms in the source of drinking water. Concerned parties would struggle to determine what went wrong when they learned that the source of drinking water was compromised. mainly due to the difficulty in identifying the source of the pollution. Developed countries acknowledge that the pollution issue has escalated to epidemic levels. The worst thing about them is that they harm ecosystems as a whole and human health, which results in a decline in biodiversity. For this reason, many nations are working to find a solution to the pollution issue.

- **Overexploitation:** This relates to the original definition of biodiversity, which states that different species in an ecosystem have complex relationships. Other species that depend on a particular aquatic species may suffer from their declining population as a result of overfishing or overexploitation. Many fish species, including grouper, swordfish, tuna, cod, salmon, and bass, are unfortunately overfished. Many studies have documented the destruction of ecosystems brought about by overfishing certain aquatic species. An ecological imbalance resulted from the overfishing of cod and sand eels, as a study on their declining populations found that copepod populations elevated.
- **Invasion by Exotic Species:** Certain species have inadvertently been carried and transported by humans through activities like travel to areas where they are not native. This can occasionally result in the invasive species dominating the new habitat and causing ecological harm. Damage of this kind may result in heightened vulnerability to physical and chemical insults caused by pollution and/or climate change.
- **Habitat Degradation:** An organism considers a habitat to be its home when it finds the ideal environmental conditions there to sustain it. A habitat is primarily made up of food, water, shelter, and space. Aquatic habitats have been severely damaged by dynamiting, bottom trawling, shrimp farming, dredging, and poison fishing, among other human activities. Because of the physical harm done to the habitat, the lack of food, and the destruction of shelters, it is now more difficult for organisms to reproduce. In addition to the previously mentioned activities, the hulls, propellers, and anchors of boats and ships used for transportation have harmed habitats. Whether intentional or not, direct contact between divers and snorkelers and the delicate aquatic environments they inhabit is inevitable. This contact frequently causes harm to aquatic environments. Marine

habitats can be harmed by even the massive network of submerged fiber optic cables that supports our internet. particularly when old ones are being maintained or new ones are being installed. Species that depend on a destroyed habitat would either disappear from the planet or move elsewhere. In any case, there would be a significant negative impact on species richness, abundance, distribution, genetic variation, and inter-population dynamics, all of which would have an enormous effect on biodiversity.

- **Flow Modification:** River flow patterns, for example, can be drastically altered, with consequences for aquatic ecosystems' physical habitat, food sources, behavior, community composition, energy expenditure, and population dynamics. The biodiversity of a water body may suffer as a result of changed flow regimes, according to an increasing number of research articles. Within flowing bodies of water, changes to the water are fairly common. Places with high demand for food safety and water storage experience them more frequently and severely. Low water levels in many major rivers across the world could be related to flow modification. A decrease in biodiversity may result from this drop in water levels' effects on the river's aquatic ecosystem (Pan et al., 2013; Dieter and Cameron, 2014; Roberto, 2016).

1.4.5. The Main Threats to Biodiversity

Threats to the world's aquatic biodiversity fall into several categories, including i) climate change; ii) water pollution; iii) overexploitation; iv) invasion by exotic species; v) habitat degradation; and vi) flow modification. The changes to the hydrological, biogeochemical, and atmospheric cycles are referred to as climate change. The subtle changes in daily average temperatures, the length of the rainy season, the carbon cycle, the temperature at night, and solar radiation are examples of fluctuations that can have an impact on living things. Compared to previous centuries, the twentieth century saw an increase in temperature of roughly 0.6°C. The scientific data required to demonstrate the trend of rising temperatures was established through research on tree rings and ice cores. Certain oceanic coral reef ecosystems have experienced a decline due to temperature changes. The swift rise in sea levels, predicted to rise by 0–2 meters by the end of the 20th century, poses a threat to the coastal areas, as they could quickly submerge. This is regarded as disastrous for a variety of ecotone communities and certain species. Because different species can adapt to different climate changes, past climate changes have resulted in ecosystems with varying species composition.

Water biodiversity is thought to be seriously threatened by the physical, biological, chemical, and radioactive pollution that contaminates water bodies. These pollutants can originate from a variety of sources, including mining operations, industrial effluents, residential sewage, and agricultural runoff. Worldwide, but particularly in Asia and Africa, pollution is a major cause of illness and even death. Certain physical pollutants, like temperature changes, can lead to visual pollution in water. Nuclear power plants emit radioactive matter pollution, and various pathogenic pollutants are released by untreated sewage. Sewage and agricultural drains are the main sources of two different categories of water pollution: point sources and non-point sources. Water bodies face significant threats from a growing number of sources of pollution, including chemical pollution, despite some industrialized nations making progress in reducing their water pollution from domestic and industrial sources. It has been acknowledged that pollution issues are widespread. Marine vertebrates, including large vertebrates and predators like sharks and tuna, which have been observed to be declining, are being severely impacted by overexploitation, particularly overfishing. Target species overfishing at legal levels can also have a general negative impact on ecosystems, especially when it involves high biomass ratios or food webs. For instance, overfishing of sand eels combined with the range shift of the copepod *Calanus finmarchicus*, which is regarded as a major food source for sand eels, has resulted in overexploitation of sand eel and cod stocks in UK waters.

It has been discovered that the deliberate and widespread introduction of exotic species increases the chemical and physical effects of human activity on freshwater ecosystems. This is because, in the first place, exotic species typically invade ecosystems that have already been altered or degraded by human activity. Several instances of the drastic effects of exotics on native species have been identified, including the European crayfish pandemic, salmonids in Southern Hemisphere streams and lakes, and the Nile perch, *Lates niloticus*, in Lake Victoria. These effects are expected to increase rapidly. Furthermore, certain indirect effects stand out, such as the influence of terrestrial plants (*Tamarix* spp.: *Tamaricaceae*), which modify stream flows in Australia and North America as well as the water regime of riparian soils. A combination of reactive factors, such as changes in drainage basins or direct effects on the aquatic environment, are collectively referred to as "habitat degradation." For instance, clearing forests typically results in changes to surface runoff and elevated river sediment loads, which can cause habitat changes like shoreline erosion, habitat strangling, obstruction of river bottoms, or even worsening of floodplains. In moving bodies of water, changes in flow are common. Despite their differences in kind and intensity, they frequently act aggressively, particularly in areas with wildly fluctuating flow patterns. This means that people in these places have the greatest need for food safety and water storage. The amount of water that dams retain globally, approximately 10,000 km —five times the total volume of water in the world—explains why human activity has altered river flows so much. Due to extensive water abstraction, several of the longest rivers in the world are currently dry. Future studies must take into account the likely severe effects of flow modifications on river biota, including fish.

1.4.6. Biodiversity Loss and Ecological Communities

It is unclear whether a decline in biodiversity affects an ecological community's capacity to recycle vital nutrients, produce biomass, and obtain biologically significant resources. According to all data released since 2005, there are fewer genes, species, and active groups of organisms, which both reduce efficiency and turn these resources into biomass. Across different trophic fields, studied ecosystems, and groups of organisms, the various impacts of biodiversity are observed to be remarkably proportionate. Although some processes and ecosystems offer different opportunities for examining boundaries that limit the effects of biodiversity, there are general primary principles that govern how communities affect ecosystem functions.

1.4.7. Biodiversity and Stability of Ecosystem Functions and Processes

Many forms of stability exist, and there isn't any proof, at least not in theory, that biodiversity could make any of them stronger. All biomass at various stages of diversity, for example, is a group property that has higher temporal stability according to both theory and data. Five summaries provide a clear understanding of how diversity influences ecosystem functions over time. These summaries indicate that resource capture and biomass production are more stable in various communities. Statistics average, over-yielding, and compensatory drive are the methods through which diversity confers stability. When average biomass production rises by diversity faster than the standard deviation, stability is encouraged by overyielding. When random differences in the population sizes of different species lessen the variance of the total ecosystem variables, the statistical average is produced. Desynchronization in environmental response is the outcome of interactions and/or unique responses to environmental inconstancy among different life forms, which comprise compensatory dynamics. Any change increases in tandem with biodiversity loss because the impact of biodiversity on ecosystem processes is nonlinear and saturated. The most recent experimental research, which takes into account the relationships between ecosystem functioning and biodiversity, show that while losses in biodiversity have varying effects on ecosystem functions, the overall rate of change increases with greater losses. For many processes that require further study, there are no precise quantitative estimates regarding the true level of biodiversity that can change ecosystem functions.

1.4.8. Biodiversity Loss and Common Species

A decrease in the number of common species and changes in the patterns of species dominance are also examples of biodiversity loss, although the majority of earlier research studies on the subject concentrated on the effects of species extinctions. It can be argued that common species are the catalysts for all ecosystem processes, and any reduction in their numbers could have negative effects on the ecosystem's primary function. Dominant fish, for instance, are essential to freshwater ecosystems because they primarily connect benthic and pelagic components with their swift movement and adaptable feeding behaviors. Because they frequently cover more trophic levels by having lower species richness, dominant fishes are known to be less likely to be replaced by other functionally equivalent ones. There have been numerous instances of this process, whereby overfishing results in changes to the way the ecosystem functions. The reduction in migratory fish populations leads to many issues since they frequently transport materials over great distances without having functional substitutes identified. Salmon fish are a real example of how major marine-derived nutrients are transferred on large scales into freshwater and terrestrial habitats. During their spawning season, they also reseed nutrients on finer ranges.

1.4.9. Cost Efficiency Plans for Conservation

Freshwater conservation strategies should ideally incorporate a clear assessment of conservation management expenses to address the intensifying competition for various water resources. Priorities for freshwater conservation identified based on social, biological, and financial costs allow for a clear evaluation of the trade-offs between management and use, and they also increase the likelihood of political and decision-maker support. Additional ideas regarding conservation costs were provided in the literature when it came to the choice of conservation regions. These ideas included opportunity costs in addition to acquisition, management, and transition costs. Examining conservation costs for freshwater communities is particularly difficult because it involves both the idea of costs for protecting a specific area and related costs related to the water needed to sustain that area over time. This includes calculating the benefits and costs for the entire catchment. The environmental field flow evaluation has led to a significant variation of certain formal techniques, emphasizing alternative water allocations within the catchment. These approaches essentially calculate the trade-offs in flow requirements that would support human demands for water resource management while also sustaining biodiversity. Although it hasn't been done yet, full integration of such conservation planning strategies offers enormous potential. Various conventional decision-making instruments were employed in the conservation planning of land areas to facilitate the analysis of a desired outcome based on quantitative or qualitative biological, social, and economic expenses (Pan et al., 2013; Dieter and Cameron, 2014; Roberto, 2016).

REFERENCES

1. Amphibia Web. (2005). Amphibia Web species numbers. AmphibiaWeb: Information on amphibian biodiversity and conservation. Berkeley. http://amphibiaweb.org/ (accessed on 16 October 2024).
2. Appeltans, W., Ahyong, S. T., Anderson, G., Angel, M. V., Artois, T., Baily, N., et al. (2012). The magnitude of global marine species diversity. *Current Biology, 22*(21), 2189–2202.
3. Benstead, J. P., De Rham, P. H., Gatolliat, J. L., Gibon, F. M., Loiselle, P., Sartori, M., Sparks, J. S., & Stiassny, M. L. J. (2003). Conserving Madagascar's freshwater biodiversity. *Biological Science, 53*(11), 1101–1111.
4. Berumen, M. L., Hoey, A. S., Bass, W. H., Bouwmeester, J., Catania, D., Cochran, J. E. M., Khalil, M. T., Miyake, S., Mughal, M. R., Spaet, J. L. Y., & Saenz-Agudelo, P. (2013). The status of coral reef ecology research in the Red Sea. *Coral Reef, 32*(3), 737–748.
5. Curtis, T. P., & Sloan, W. T. (2004). Prokaryotic diversity and its limits: Microbial community structure in nature and implications for microbial ecology. *Current Opinion in Microbiology, 7*(3), 221–226.
6. Dieter, H., & Cameron, H. (2014). *Nature in the Balance: The Economics of Biodiversity*. Oxford University Press.
7. Laundbrg, G., Kottelat, M., Smith, G. R., Undberg, G., Kottelat, M., Smith, G. R., Stiassny, M. L. J., & Gill, A. C. (2000). So many fishes, so little time: An overview of recent ichthyological discovery in continental waters. *Annals of the Missouri Botanical Gardens, 87*(1), 26–62.
8. Myers, N., Mittermeier, R., Mittermeier, G. C., DaFonseca, G. A. B., & Kent, J. (2000). Biodiversity hotspots for conservation priorities. *Nature, 403*(6772), 853–858.
9. Pan, J., Marcoval, M. A., Bazzini, S. M., Vallina, M. V., & De Marco, S. G. (2013). Coastal marine biodiversity challenges and threats. In *Coastal Marine Biodiversity Challenges and Threats* (pp. 43–67). CRC Press.
10. Roberto, C. G. (2016). Freshwater biodiversity: A review of local and global threats. *International Journal of Environmental Studies, 73*(6), 887–904.
11. Shlesinger, Y., Goulet, T., & Loya, Y. (1998). Reproductive patterns of Scleractinia corals in the northern Red Sea. *Marine Biology, 132*(4), 691–701.
12. Strayer, D., Downing, J. A., Haag, W. R., King, T. L., Layer, J. B., Newton, T. J., & Nichols, S. J. (2004). Changing perspectives on pearly mussels, North America's most imperiled animals. *Biological Science, 54*(5), 429–439.

13. Torsvik, V., Ovreas, L., & Thingstad, T. F. (2002). Prokaryotic diversity: Magnitude, dynamics, and controlling factors. *Science, 296*(5570), 1064–1066.

Chapter **2**

The Role of Aquatic Biodiversity in Ecosystem Functioning

Contents

INTRODUCTION

Global aquatic ecosystem stability and function are greatly dependent on aquatic biodiversity. These ecosystems cover a wide range of habitats, including marine settings like oceans and coral reefs as well as freshwater lakes and rivers. These systems are home to a diverse range of species, from large apex predators like sharks to microscopic phytoplankton, all of which contribute in different ways to the dynamics of the ecosystem. The aquatic biodiversity's contribution to the cycling of nutrients is one of its main roles. Within aquatic ecosystems, various species play distinct roles in recycling nutrients and decomposing organic matter.

For instance, when fungi and bacteria break down organic matter, nutrients like phosphorus and nitrogen are released back into the water, where they are subsequently taken up by aquatic plants and algae, which are primary producers. The foundation of the food web is made up of these primary producers, who provide support to animals at higher trophic levels such as fish, amphibians, reptiles, birds, and mammals.

Aquatic biodiversity also improves the stability and resilience of ecosystems. Ecosystems can withstand disturbances like pollution events, the effects of climate change, and invasive species better when they have a diverse range of species. When a species' functions are redundant, it means that even in the event of a decline or extinction, other species may be able to carry out comparable tasks and preserve ecosystem services and processes.

Aquatic biodiversity aids in the control of water quality in addition to resilience and nutrient cycling. Animals such as filter feeders (e.g., mussels, and clams) aid in water clarification by eliminating impurities and particulate matter. As natural buffers, seagrass beds and mangrove forests stabilize shorelines and lessen coastal erosion. Known as the rainforests of the sea, coral reefs shield coastlines from storms and serve as a home for a wide variety of species. The economic and social value of aquatic biodiversity is substantial, even beyond its ecological roles. Diverse fish stocks are essential to the aquaculture and fishing industries for both food and livelihood. Diverse and pristine marine environments are ideal for tourism and leisure pursuits like diving, snorkeling, and fishing.

Coastal communities' cultural customs and practices are frequently closely linked to the biodiversity of the freshwater or marine environments in which they live. In summary, aquatic biodiversity plays a complex and vital role in the functioning of ecosystems. Diverse aquatic ecosystems are essential to maintaining life on Earth for a variety of reasons, including resilience, nutrient cycling, control over water quality, and socioeconomic benefits. Therefore, protecting and restoring aquatic biodiversity through conservation efforts is not only ecologically necessary but also vital to human welfare and the health of the planet.

2.1. ROLE OF AQUATIC BIODIVERSITY IN ECOSYSTEM FUNCTION

An ecosystem is a system made up of both biotic and abiotic elements that work as a whole. All living things are considered biotic components, while non-living things are considered abiotic components. Therefore, an ecological community made up of various populations of organisms coexisting in a specific habitat is defined by ecosystem science. An ecosystem is a geographical area where organisms, weather, and landscape combine to form a "bubble of life," according to natural sciences like ecology and geography. In biology, an ecosystem is essentially defined as the basic building block of nature. Similar to how cells serve as the structural and functional building blocks of life in a living thing, ecosystems are the basic building blocks of nature.

A community and its surroundings make up an ecosystem. Ecology can be viewed at the level of the individual, the population, the community, or the ecosystem. Ecology is the scientific study of interactions between populations or between organisms and the environment. The physiology, reproduction, and development of each individual organism are the main concerns of ecology at the individual level. Ecology primarily addresses the characteristics and different factors that impact a population at the population level. Ecology examines interactions between populations and community patterns at the community level. Ecology combines all of them at the ecosystem level to comprehend how the system functions as a whole. Therefore, energy flow and nutrient cycles would be of greater importance to an ecosystem ecology than the concerns of individual species.

The meaning and origin of the term "ecosystem" can be derived from the Ancient Greek words "o□κος" ("oîkos") for "house" and "σύστημα" ("sústēma") for "organized body." The botanist Roy Clapham first used the term in the early 1930s to refer to the biological and physical elements of an environment. But the idea was initially presented by British ecologist Arthur Tansley in a paper titled "The Use and Abuse of Vegetational Concepts."

An ecosystem is a dynamic process made up of a community of living things interacting as a system with the abiotic elements of their surroundings, such as air, water, and mineral soil. Energy flows and nutrient cycles bind these biotic and abiotic elements together. Ecosystems can be any size, but they typically cover a small, defined geographic area because they are characterized by the network of interactions between organisms and their surroundings. In an ecosystem with deer and lions, for example, the lion has an eating relationship with the deer in order for it to survive. The other animals and plants that inhabit the same or nearby areas are impacted by this relationship. For example, plants may begin to flourish more when lions eat more deer because there will be fewer deer for the plants to be eaten.

2.1.1. Ecosystem Structure

The structure of an ecosystem consists of two major components (Figure 2.1):

1. biotic components; and
2. abiotic components.

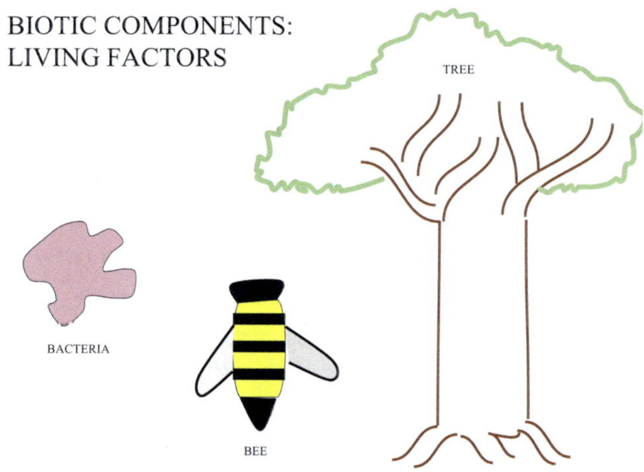

Figure 2.1. *Two components of the ecosystem: biotic and abiotic.*

Source: https://en.wikipedia.org/wiki/File:Biotic_Components_Living_Factors.svg.

2.1.1.1. Biotic Components

All living things are included in the biotic components. In general, there are two main categories of living organisms. They are known as prokaryotes and eukaryotes. Organelles (like a nucleus) that are membrane-bound are found inside the cells of eukaryotes. Conversely, prokaryotes are those that do not have membrane-bound organelles. Read this to learn more about the distinctions between these two groups. Protists, fungi, animals, and plants are eukaryotes. The prokaryotes are represented by bacteria and archaea. They all now have a "job" to perform within the ecosystem. For example, plants can capture light energy thanks to their chloroplasts.

Subsequently, they absorb water and carbon dioxide from their surroundings and convert them into sugar, a biomolecule that can be used to generate chemical energy (such as ATP). These organisms are known as producers because they can produce their own food through photosynthesis. After the producers come the primary consumers, which feed on the producers. These consumers, in turn, provide food for higher-level consumers, such as secondary and tertiary consumers. Animals serve as examples of consumers. Herbivores are animals that feed on plants, while carnivores are animals that feed on other animals. There are also omnivores, which consume both plants and animals.

2.1.1.2. Abiotic Components

All non-living objects, such as rocks, soil, minerals, water sources, and the surrounding atmosphere, are considered abiotic components. Abiotic elements play an ecological role, just like biotic elements do. Compounds and elements, for instance, are sources of nutrients. They are necessary for an organism's growth and metabolism. They give organisms a home or a place to live and grow, in addition to nutrients.

2.1.1.3. Interactions

Energy flows and nutrient cycles connect the biotic and abiotic components to one another and allow them to interact with one another as a system. For example, the photoautotrophs allow nutrients and energy to enter the system. These are the creatures, like plants and green algae, that are involved in photosynthesis. After that, the photoautotrophs are eaten by the heterotrophs, such as animals. This facilitates the flow of nutrients and energy throughout the body. These organisms' demise encourages the decomposers to break down. Through this process, the nutrients are returned to the environment so that the organisms can use them again.

The elements that are biotic and abiotic can also act as ecological environmental factors, influencing the ecosystem. The term "biotic factors" refers to the biotic elements whose biological activity has an effect on the ecosystem. The inanimate objects and the environmental elements of an ecosystem, such as pH, temperature, and climate, are considered abiotic factors.

The degree of predation in an ecosystem is an illustration of a biotic factor. Predation activity would probably rise in the event of a predator population increase. Thus, their prey's population density may be considerably reduced. The decline (or worse, the extinction) of these key species could also result in the decline (or worse, the extinction) of the organisms that depend on them if their prey is a key species, which means that another group of organisms depends on them for survival.

Regarding the abiotic component, they can control a species' population density or size. For example, unusually acidic precipitation with high hydrogen ion concentrations, known as acid rain, can have negative effects on the soil (e.g., leaching), in addition to the aquatic life and acid-sensitive plants. Abiotic elements include temperature, light, salinity, air, soil or substrate, and others besides pH (Chapman and Reiss, 2000; Smith, 2006; Meffe et al., 2013).

2.2. TYPES OF ECOSYSTEMS

The following are the different types of ecosystems based on the environment.

2.2.1. Aquatic Ecosystems

These can then be broken up into smaller ecosystems. For instance:

- **Pond Ecosystems**: These are usually relatively small water bodies. It includes various types of plants, amphibians and insects. Sometimes they include fish, but as these cannot move around as easily as amphibians and insects, it is less likely, and most of the time fish are artificially introduced to these environments by humans.
- **River Ecosystems**: As rivers always link to the sea, they are more likely to contain fish alongside the usual plants, amphibians and insects. These sorts of ecosystems can also include birds, because birds often hunt in and around water for small fish or insects.
- **Wetland**: It is a zone of flat lands that has groundwater of shallow depth and that ascend to the surface in determined periods, forming lagoons and marshes, until where they come to live hundreds of species. There are five classes of wetlands: marine, estuarine, lake, riparian and marshy.
- **Mangrove**: It is a grouping of semi-submerged trees that have been flooded with water, with high levels of salinity and therefore they develop and survive in coastal lands. The trees grow on long roots, which like stilts raise the trunks above the level of the waters. To reproduce, they quickly retain the seeds in the branches until they are about to develop. When the tide goes down they are able, within a few hours, to root and begin to grow before being again underwater.
- **Coral Reef**: It is one of the richest aquatic ecosystems of the planet, the product of the great amount of species that inhabit them (fish,

snails, corals and algae). The reef structure consists of large colonies of corals, accumulations of sediments and calcareous sands. They are found mainly in tropical regions and there are two types of coral, i.e., hard and soft.

According to different habitats of aquatic organisms, aquatic ecosystems are various kinds and divided into the following:

- **Benthic**: These are located at the bottom of aquatic ecosystems. In those that are not very deep, the main inhabitants are algae. In the deeper ones, the majority are consumers.
- **Nectonic**: These animals move freely, thanks to their means of locomotion can adapt to water currents.
- **Planktonic**: These living beings float in the terrestrial or marine water and are dragged by the water currents. They do not move by their own movements.
- **Neustonie**: These live floating on the surface of the water.

2.2.2. Terrestrial Ecosystems

These are areas where living things (plants, animals, etc.) inhabit and grow in the soil and atmosphere that envelop a specific terrestrial region. Ecological regions, also known as biomass, are larger ecosystems that include terrestrial ecosystems. Latitude, climate, temperature, and precipitation totals serve as the boundaries between these zones.

Different types of terrestrial habitats exist, such as deserts, grasslands, and forests, depending on the abiotic factors specific to each ecosystem:

- **Tropical Forests**: They usually have extremely dense ecosystems because there are so many different types of animals and they all live in a very small area. They have high biodiversity of plants and animals; are also one of the oldest ecosystems of the planet and are below 1200 meters in height; the temperature and light remain constant throughout the year.
- **Temperate Forest**: Those who have a good number of trees like mosses and ferns are for them. Temperate forests exist in regions where the climate changes significantly from summer to winter. Summers and winters are clearly defined and trees lose their leaves during the winter months.
- **Swamp**: Situated just before the arctic regions, a swamp is defined by evergreen conifers. The temperature is below zero degrees for almost half a year, the rest of the months, is full of migratory birds and insects.

- **Tundra**: It has an extremely cold climate. The ground remains frozen for much of the year. Its rainfall is very low, so it reduces the growth of living organisms. There are no large trees, only small plants (mosses, lichens and other tree species) remain present in this region.
- **Desert**: Found in regions that receive annual rainfall of less than 25%. They occupy about 17% of all the land on our planet. Due to the high temperatures, low availability of water and intense sunlight, the fauna and flora are scarce and underdeveloped.
- **Savanna**: Tropical meadows are seasonally dry and have few individual trees. They support a large number of predators and herbivores.
- **Grassland**: The temperate grasslands are completely devoid of large shrubs and trees. Grasslands could be categorized as mixed grass, tall grass and grassy meadows. Grasslands support a wide variety of life and can have very complex and involved ecosystems.
- **Mountain**: Offers a dispersed and diverse matrix of habitats where a large number of animals and plants can be found. The harsh environmental conditions that normally prevail at higher altitudes, the alpine vegetation can only survive. The animals that live there have thick fur coats to prevent cold and hibernation in the winter months.
- **Deserts**: Quite the opposite of tundra in many ways, but still harsh, more animals live in the extreme heat.
- **Savannas**: These differ from deserts because of the amount of rain that they get each year. Whereas deserts get only a tiny amount of precipitation every tea, savannas tend to be a bit wetter which is better for supporting more life.
- **Forests**: There are many different types of forests all over the world including deciduous forests and coniferous forests. These can support a lot of life and can have very complex ecosystems.

2.2.3. Ocean Ecosystems

Ocean ecosystems are made up of a wide variety of environments and living things that work together to support the planet's overall health and functionality. Oceans, which make up more than 70% of the planet's surface, are vital for maintaining biodiversity, controlling climate, and giving marine life and human societies access to resources and services. Phytoplankton, the microscopic organisms that form the foundation of the marine food web, are at the center of ocean ecosystems. Because they produce oxygen and sequester carbon dioxide through photosynthesis, phytoplankton are essential to the atmosphere's composition and the global carbon cycle. They serve as the main food source for zooplankton, which supports animals at higher trophic levels like fish, seabirds, and marine mammals. From the brightly lit surface waters to the deepest ocean trenches, where rare species have adapted to harsh conditions and complete darkness, there is an interconnected web of life.

Some of the most productive and biodiverse environments on Earth are found in coastal ocean ecosystems, which include coral reefs, salt marshes, estuaries, and mangrove forests. Estuaries are vital habitats for migrating birds and act as nurseries for young fish. In addition to providing habitat for a variety of marine life, including crustaceans and endangered species like manatees and sea turtles, mangrove forests and salt marshes also serve as barriers against storm surges and coastal erosion. Known as the "rainforests of the sea," coral reefs are hotspots of biodiversity that support millions of people globally by giving them access to food, shelter, and a means of subsistence.

Although they occupy less than 1% of the ocean floor, coral reefs are home to about 25% of all marine life. Large areas of water in the open ocean are home to a variety of species that have evolved to withstand varying temperatures, depths, and nutrient levels. Because of the productivity of phytoplankton, surface waters are teeming with life, supporting fish stocks that are significant to the economy, such as sardines and tuna. Unique species that have adapted to live in the cold and darkness, such as giant squid, deep-sea fish, and bioluminescent organisms, can be found in deeper waters.

Deepwater hydrothermal vents and cold seeps sustain ecosystems powered by chemosynthesis, the process by which bacteria transform substances into energy. Ocean ecosystems are seriously threatened by human activity, which includes pollution, habitat destruction, overfishing, and climate change. In addition to harming marine life and deteriorating habitats, pollution from plastics, chemicals, and nutrient runoff can deplete fish stocks and upset marine food webs. Through ocean warming, acidification, and sea level rise, climate change intensifies these threats, affecting marine biodiversity and ecosystem resilience.

To reduce these risks and maintain the wellbeing and productivity of marine environments, conservation and sustainable management of ocean ecosystems are essential. The implementation of marine protected areas, the encouragement of sustainable fishing methods, the reduction of plastic pollution, and the mitigation of climate change through international efforts to lower greenhouse gas emissions are examples of conservation measures. Ocean ecosystems are dynamic, networked systems that sustain a vast array of species and offer priceless ecological services. The health of human societies, marine life, and the earth itself depends on maintaining and restoring these ecosystems.

It is the earth's largest salt water ecosystem that contains millions of species. There are different types of ocean ecosystems:

- **Shallow Water**: Some tiny fish and coral only live in the shallow waters close to land.
- **Deep Water**: Big and even gigantic creatures can live deep in the waters of the oceans. Some of the strongest creatures in the world live right at the bottom of the sea.
- **Warm Water**: Warmer waters, such as those of the Pacific Ocean, contain some of the most impressive and intricate ecosystems in the world.
- **Cold Water**: Less diverse, cold waters still support relatively complex ecosystems. Plankton usually form the base of the food chain, followed by small fish that are either eaten by bigger fish or by other creatures such as seals or penguins.

2.2.4. Ecosystems According to the Degree of Human Intervention

Ecosystems can be categorized based on the level of human intervention; they can be as natural or pristine as possible, with little to no human influence, or as highly modified or artificial ecosystems heavily impacted by human activity. Ecosystems that are pristine, or natural, are those that have not been significantly impacted by humans and have retained their natural biodiversity and ecological processes.

These ecosystems include deep ocean habitats, untouched wetlands, isolated forests, and wilderness regions. They frequently show a great diversity of species and intricate relationships between different organisms, which represent long-term evolutionary processes. Maintaining genetic diversity among species, scientific research, and conservation initiatives all depend on pristine ecosystems. Ecosystems that are semi-natural have a moderate human influence but still contain a lot of natural elements.

Managed forests, traditional farming landscapes, and certain freshwater ecosystems with low pollution levels are a few examples. These ecosystems continue to support a variety of native species and ecological functions even though human activities like controlled burning and selective logging may have changed them. Intentional human modifications to achieve particular goals while preserving ecological integrity are what constitute semi-artificial ecosystems.

Urban parks, gardens, and artificial wetlands intended for wastewater treatment are a few examples. These ecosystems are maintained to improve biodiversity, offer areas for recreation, or rebuild damaged habitats.

Although they frequently need constant human intervention for upkeep and management, they can support a variety of wildlife and ecosystem services. Human activity significantly alters artificial ecosystems for commercial or social gains, frequently at the expense of ecological processes and natural biodiversity.

Natural habitats are replaced or fragmented in heavily altered ecosystems, such as urban areas, industrial zones, and intensive agricultural landscapes. These ecosystems might be unable to sustain as many native species and might need a lot of inputs like pesticides, fertilizers, and irrigation.

The categorization of ecosystems according to the extent of human intervention draws attention to the variety of ways in which humans affect natural settings and emphasizes the significance of conservation and sustainable management strategies in preserving biodiversity and ecosystem services.

Global ecosystem resilience and long-term health depend on striking a balance between human needs and ecological integrity:

- **Natural Ecosystems**: Man has not intervened in their formation, such as forests, lakes, and deserts.
- **Artificial Ecosystems**: Man actively participates in its formation, such as dams, parks, and gardens.

2.2.5. Trophic Structure, Ecological Energetic and Productivity

In an ecosystem, organisms are arranged in tiers called tropic structures, wherein organisms that have similar functions and food sources are represented at each level. The movement of energy from one trophic level to the next is also shown in trophic structure diagrams. We are able to determine the estuary's productivity by arranging it into a trophic structure. In essence, productivity is the estuary's capacity to produce organic matter. A productive estuary is sustainable—that is, has a high survival rate, little to no invasive species, and organisms that continuously carry out life processes.

This trophic structure adopts a bottom-up perspective when examining the aquatic ecosystem. The organisms classified as primary producers, or bottom tier, have the highest energy efficiency, while top predators, or top tier, have the lowest energy efficiency. While top fish or predators need a variety of organisms to function, they are less energy efficient than primary producers because they generate their own food. An alternative way to put it is that predators require a lot more energy than phytoplankton does.

2.2.5.1. Ecological Energetics

Ecological energetics describes the energy flow in an ecosystem. Energy is the capacity for exertion. The sun emits radiation in addition to light. On Earth, some of these radiations land. Light radiation strikes an object. The object warms up after absorbing light. As a result, heat energy is created from light energy. Heat is absorbed by land and sea.

It causes regions of the earth to get hot or cold. It creates clouds and air movement. The sun is the Earth's energy source. The green plants take up a tiny portion of the radiant energy. They convert this energy, which is glucose, into chemical energy. The plants provide this chemical energy to all other living things. Heat is the result of significant energy loss. During photosynthesis, very little of the solar radiation is used.

The ecosystem's living elements receive their energy from this stored energy. Red and blue light are absorbed by green plants. Nevertheless, only 1% to 5% of the energy that is absorbed is transformed into energy for food. An organism quickly loses the energy it used once in the form of heat. Energy flows in a single direction at all times. The laws of thermodynamics govern how energy moves through an ecosystem.

It has been estimated that about 57% of the solar energy is absorbed in the atmosphere. It is scattered in space. About 36% is absorbed by water or land. About 8% falls on plants. 80 to 85% is absorbed by green plants and the rest of it is reflected. About 50% of the absorbed energy is used in photosynthesis, i.e., it is stored in the form of potential energy of organic compounds of green plants.

2.2.5.2. Ecological Productivity

The term "ecological productivity" describes how plants initially fix solar energy and then use that energy to feed on dead biomass as detritivores, plant-eating herbivores, and animal-eating carnivores. A food web is this complex of energy fixation and utilization. The productivity of green plants is referred to as primary productivity by ecologists.

Net primary productivity is less than gross primary productivity because it accounts for energy losses necessary to maintain plant respiration. Gross primary productivity is the total amount of energy fixed by plants. Vegetation biomass increases with time if the net primary productivity of green plants in an ecosystem is positive. Herbivorous animals are covered by gross and net secondary productivities, whereas carnivores are covered by tertiary productivities.

Ecological productivity within food webs is characterized by a pyramidal structure. Generally speaking, plants make up over 90% of the food web's total productivity, followed by herbivores for the most part and carnivores for less than 1%. Eventually, decomposer organisms consume any dead plant or animal biomass; if ecological conditions prevent this process from happening effectively, the dead biomass will accumulate as peat or other nonliving organic matter.

Because of differences in the availabilities of solar radiation, water, and nutrients, the world's ecosystems differ greatly in the amount of productivity that they sustain. Deserts, tundra, and the deep ocean are the least productive ecosystems, typically having an energy fixation of less than 0.5×10^3 kilocalories per square meter per year (thousands of kcal/m^2/yr; it takes one calorie to raise the temperature of one gram of water by 34°F [1°C] under standard conditions, and there are 1,000 calories in a kcal). Grasslands, montane and boreal forests, waters of the continental shelf, and rough agriculture typically have productivities of $0.5–3.0 \times 10^3$ kcal/m^2/yr. Moist forests, moist prairies, shallow lakes, and typical agricultural systems have productivities of $3–10 \times 10^3$ kcal/m^2/yr. The most productive ecosystems are fertile estuaries and marshes, coral reefs, terrestrial vegetation on moist alluvial deposits, and intensive agriculture, which can have productivities of $10–25 \times 10^3$ kcal/m^2/yr.

2.2.5.3. Types of Productivity

Productivity refers to the rate of formation of biomass in the ecosystem. It can also be referred to as the energy accumulated in the plants by photosynthesis. There are two types of productivity, namely:

1. Primary Productivity
2. Secondary Productivity

2.2.5.3.1. Primary Productivity

Primary Productivity refers to the generation of biomass from autotrophic organisms such as plants. Photosynthesis is the primary tool for the creation of organic material from inorganic compounds such as carbon dioxide and water. Primary productivity can be divided into two aspects:

- Gross primary productivity
- Net primary productivity
- **Gross Primary Productivity:** The solar energy trapped by the

photosynthetic organism is called gross primary productivity. All the organic matter produced falls under gross primary productivity. This depends upon the photosynthetic activity and environmental factors.

- **Net Primary Productivity:** This is estimated by the gross productivity minus energy lost in respiration.

NPP = GPP – Energy lost by respiration

It the net energy stored in the plants. This energy serves as food for the animals that feed on plants. It is measured as the amount of organic matter produced in a community in a given time. Annually, over 170 billion tons of net primary productivity occurs over the entire biosphere.

2.2.5.3.2. Secondary Productivity

Animals, who are heterotrophs, have an impact on secondary productivity. It is the build-up of energy at the level of the consumer. Unlike primary productivity, it is constantly shifting from one organism to another. The movement of organic materials between different trophic levels is what causes this process to happen. It is also known as the rate at which heterotrophic biomass is increasing. Numerous protists, fungi, animals, and bacteria are among the organisms that have an impact on secondary production.

2.2.5.4. Unit of Productivity

Typically, productivity is expressed in units of mass per unit volume (or surface) per unit time (Chapman and Reiss, 2000; Smith, 2006; Meffe et al., 2013).

2.3. ENERGY FLOW IN ECOSYSTEMS

The chemical energy of food is the main source of energy required by all living organisms. This energy is transmitted to different trophic levels along the food chain. This energy flow is based on two different laws of thermodynamics:

- First law of thermodynamics states that energy can neither be created nor destroyed, it can only change from one form to another.
- Second law of thermodynamics states that as energy is transferred more and more of it is wasted.

One of the main things that allows for the survival of so many different kinds in the ecosystem is the flow of energy. Solar energy is the main energy source for nearly all living things on Earth. It's funny to learn that less than half of the sun's effective radiation reaches Earth. Effective radiation is defined as radiation that plants can use to perform photosynthesis. The earth's atmosphere typically reflects back into space the majority of the solar radiation that strikes it. The term PAR, or photosynthetically active radiation, refers to this effective radiation. Only about 2–10% of the energy that is present in photosynthetically active radiation is utilized by plants for the process of photosynthesis, compared to the 40–50% that we receive overall.

Since plants are the ecosystem's producers and all other organisms rely on them either directly or indirectly for survival, this percentage of PAR therefore supports life on Earth. The food web and food chain facilitates the flow of energy. Plants, which are the ecosystem's producers, use their chloroplasts to absorb sunlight, and during the process of photosynthesis, some of that energy is converted into chemical energy.

When herbivores eat plants as food and transform the chemical energy stored in plant products into kinetic energy, or heat, energy degradation takes place. This energy is stored in a variety of organic products made by the plants and transferred to the primary consumers in the food chain.

The secondary consumers came next. Additional degradation will take place when first-order carnivores—also known as secondary consumers—eat these herbivores. Lastly, energy will be lost once more when tertiary consumers eat the carnivores.

As a result, the flow of energy is unidirectional. Entities occupy a specific position in the food chain, referred to as their trophic level, which is determined by the source of their nourishment or food. The first trophic level is occupied by producers, followed by herbivores (primary consumers) in the second and carnivores (secondary consumers) in the third (Refer to Figure 2.2).

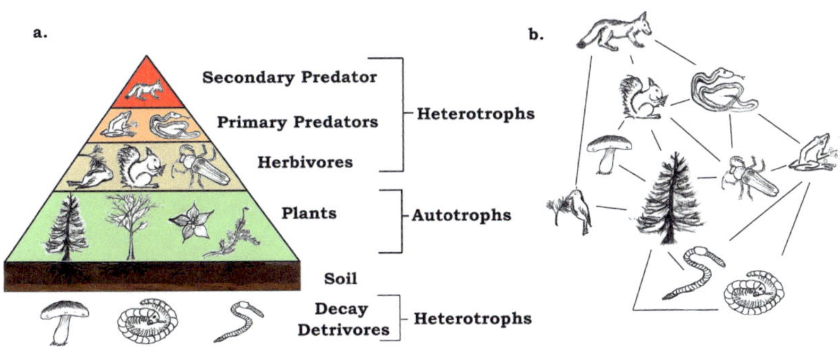

Figure 2.2. *Diagrammatic representation of trophic levels in an ecosystem.*

Source: By Thompsma – Own work, CC BY 3.0, https://commons.wikimedia.org/w/index.php?curid=15367675.

Any dead organism is transformed into detritus, or dead biomass, which provides decomposers with energy. For their energy needs, organisms at every trophic level rely on those at the lower trophic level. The standing crop is the specific mass of living material that each trophic level has at any given time. The biomass, or mass of living things in a given area, is used to measure standing crops. A species' biomass is reported as either fresh weight or dry weight. Furthermore, the energy flow in a food chain abides by the 10% law. This law states that 90% of energy is lost to evaporation and only 10% is transferred from one trophic level to the next. This is depicted as an energy pyramid in the following Figure 2.3, which provides a clear explanation.

2.3.1. Trophic Level

The producers and consumers in the ecosystem can be arranged into different feeding groups and are known as trophic level or the feeding level.

1. The producers (plants) represent the first trophic level.
2. Herbivores (primary consumers) present the second trophic level.
3. Primary carnivores (secondary consumers) represent the third trophic level
4. Top carnivores (tertiary consumers) represent the last level.

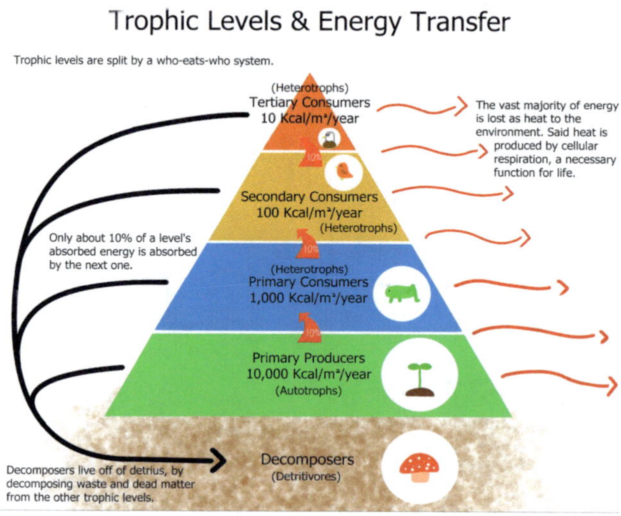

Figure 2.3. *Energy flow through different trophic levels.*

Source; By WDiment – Own work, CC BY-SA 4.0, https://commons.wikimedia.org/w/index.php?curid=105570493.

2.3.2. Food Chain

The term "food chain" refers to the process by which food energy moves from the producers through a succession of organisms (herbivores, carnivores, and decomposers) through repeated eating and being eaten. Grazing food chains and detritus food chains are the two main categories of food chains found in nature.

There are basically three different types of food chains in the ecosystem, namely:

i. **Grazing Food Chains**: which starts from the green plants that make food for herbivores and herbivores in turn for the carnivores. Ecosystems with such type of food chain are directly dependent on an influx of solar radiation.

This type of chain thus depends on autotrophic energy capture and the movement of this captured energy to herbivores. Most of the ecosystems in nature follow this type of food chain.

A simple grazing food chain (GFC) is depicted below:

The phytoplanktons →zooplanktons →Fish sequence

or

the grasses →rabbit →Fox sequences are examples, of grazing food chains.

ii. **Detritus Food Chains**: start from the dead organic matter to the detrivore organisms which in turn make food for protozoan to carnivores, etc.

Organic detritus is the starting point of the detritus food chain (DFC). It is composed primarily of bacteria and fungi, which are heterotrophic decomposers. They break down detritus, or dead organic matter, to obtain the energy and nutrients they need. Another name for them is saprotrophs (sapro: to break down). Digestive enzymes secreted by decomposers break down waste and dead materials into simple inorganic materials that are then absorbed by them. Within an aquatic ecosystem, GFC serves as the main energy transfer pathway.

In contrast, a much higher proportion of energy moves through the detritus food chain in a terrestrial ecosystem than through the GFC. At some levels, the grazing food chain and the detritus food chain may be related. In a natural ecosystem, certain animals such as crows and cockroaches consume some of the DFC organisms. are those that eat everything.

iii. **Parasitic Food Chain:** In this type of food chain, large organisms either the producer or the consumer is exploited and therefore the food passes to the smaller organisms.

A parasitic food chain is also an auxiliary food chain. It begins with the host and usually ends in the parasite.

2.3.3. Food Web

In nature, simple food chains are extremely uncommon because an organism can get nourishment from multiple trophic levels. As a result, the different food chains in an ecosystem are linked to one another to create a network known as the food web. A food chain tracks just one food pathway; a food web depicts all potential energy and nutrient transfers among the organisms in an ecosystem. Food webs are crucial to an ecosystem's ability to remain stable.

2.3.4. Ecological Pyramids

The trophic levels in an ecosystem are represented graphically by ecological pyramids. Every pyramid has producers, or the first trophic level, at its base and tertiary, or top level, consumers at its apex. The three ecological pyramids that are typically examined are the energy pyramid (c), the biomass pyramid (b), and the number pyramid (a) (Figure 2.4).

- **Pyramid of Number:** The number of organisms in each trophic level is regarded as a level in this kind of ecological pyramid. With a few exceptions, such as the detritus food chain, where numerous organisms feed on a single dead plant or animal, the number pyramid is typically upright.
- **Pyramid of Biomass:** Each level of this specific ecological pyramid accounts for the quantity of biomass generated by each trophic level. With the exception of oceans, where a large number of zooplanktons rely on a comparatively small number of phytoplanktons, the biomass pyramid is likewise upright.
- **Pyramid of Energy:** Because energy flows in food chains only in one direction, the energy pyramid is the only kind of ecological pyramid that is always upright. Additionally, a certain amount of energy is lost to the environment with each increase in trophic level.

Nevertheless, ecological pyramids have certain drawbacks, such as the inability to account for species that are members of two or more trophic levels. It does not account for a food web; instead, it assumes a simple food chain, which is something that almost never occurs in nature. Moreover, despite their critical role in the ecosystem, saprophytes are ignored in ecological pyramids.

- **Ecosystem Homeostasis:** The equilibrium, or balance, of the organisms within an ecosystem is known as ecosystem homeostasis. This indicates that the ecosystem's species populations are largely stable. These populations will fluctuate over time, but in the near run, they ought to follow cycles that center around an average value (Chapman and Reiss, 2000; Smith, 2006; Meffe et al., 2013).

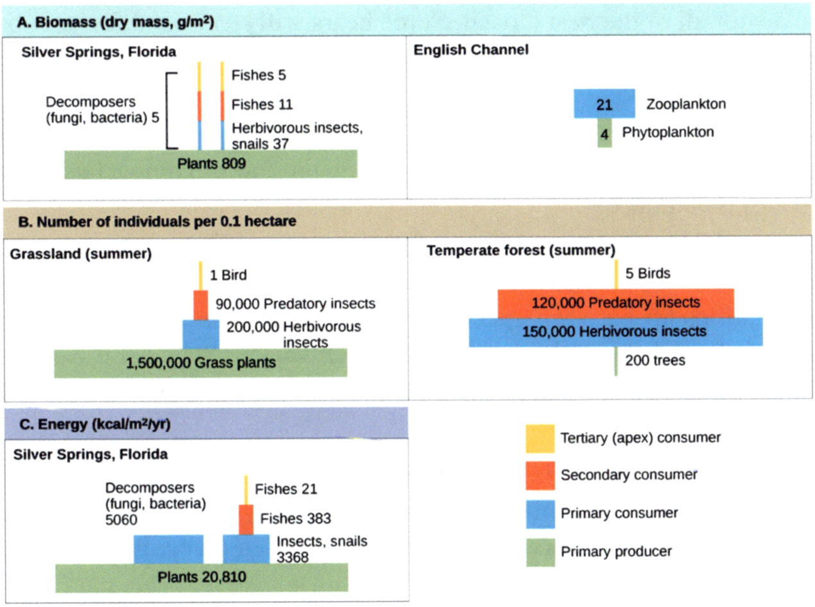

Figure 2.4. *Ecological pyramids.*

Source: https://courses.lumenlearning.com/wm-nmbiology2/chapter/energy-flow-through-ecosystems/.

2.4. ECOSYSTEM STABILITY

The ability of an ecosystem to remain in a steady state, or equilibrium, even after a stressor or disturbance has occurred is known as ecosystem stability. An ecosystem must have systems in place that enable it to recover from disturbances and return to its initial state in order to be deemed stable.

An ecosystem's ability to endure in its current state is referred to as its stability. Natural ecosystems are frequently extremely sensitive to changes, like the influx or extirpation of a species. When population sizes of a given species remain within a sustainable range, an ecosystem is considered to be in equilibrium and is considered healthy (not too many of a certain species dead or alive). Consider the case of bears (who only eat fish) and salmon (who are only hunted by bears):

1. If the bear population grows particularly large one year, the total population will require more fish to feed it. This will make the salmon population shrink.
2. Over time, if there are fewer salmon, there will not be enough food for all of the bears to eat. Some bears will starve and fewer cubs will be able to prosper, leading to a smaller overall population the next year.
3. As more time passes and the bear population gets smaller, the salmon population will start to increase again due to having fewer natural predators.

In a perfectly stable ecosystem, this cycle can continue indefinitely. Of course, natural ecosystems are far more complex than this simple example, but the relationship between species applies to complex ecosystems as well. Because of all of the interdependence between various species, it can be very difficult to understand the implications of small changes to an ecosystem in the real world because

1. The relationships between species can be incredibly complex and
2. Actual experimentation in real-world ecosystems can be catastrophic by the time observable changes have taken place; it's often too late to stop their effects.

Ecosystem stability is often divided into two components: resistance and resilience.

2.4.1. Resistance

The ability of an ecosystem to remain at equilibrium in spite of disturbances is called resistance. The capacity of an ecosystem or community to continue operating normally in the face of possible disruption is known as resistance. Greater resistance is implied by a slight alteration in species abundance caused by a disturbance as opposed to a larger alteration. When a system eventually adapts to changes in the ecosystem, resistance to ecological change is typically examined in its absence.

These studies frequently demonstrate that when there is little ecological redundancy and no other species in the ecosystem can take up the ecological role of important species, change happens more quickly. The recent eradication by humans of important species, like top predators, can significantly reduce redundancy and, as a result, change an ecosystem's capacity to withstand ecological change, especially in marine systems.

Example

1. **Drought-Resistant Soils Are Those Rich in Organic Matter and High in Biodiversity:** Rich in organic matter, healthy soils are those that have been nourished by agro-ecological fertilizers (such as compost, animal dung, green manures, etc.), are better at retaining water and less prone to erosion. Numerous scientific studies demonstrate that the most crucial component in helping soils become more drought-resistant and better equipped to deal with decreasing and increasingly unpredictable rainfall is organic matter. In addition to allowing more water to be held in the soil for longer periods of time and facilitating infiltration during heavy rains, organic matter increases the amount of pore space in the soil. This increases the total amount of water that can be captured. Therefore, compared to soil that is deficient in organic matter, soil that is rich in organic matter requires less water to grow crops. Additionally, the presence of organic matter increases the activity of fungi, earthworms, and microorganisms, which reduces the density and compaction of the soil and improves its physical characteristics for water retention. Because of all these qualities, soils high in organic matter are more drought-resistant, which improves the farm's overall water-use efficiency as well as that of the crop.
2. **Invasion Resistance:** Reduces in native taxonomic richness were associated with higher invading species survival and percentage cover in many marine environments. This implies that the integrity of the native species pool enhances invasion resistance, just like in terrestrial plant ecosystems. Diverse systems, for instance, make greater use of resources like available space.

Reducing the richness of native taxa in benthic (sea floor) communities that were artificially assembled was associated with higher invading species survival and percent cover. The only resource available to invaders was open space, and communities were protected from invasion by increased temporal stability when species richness was higher. It is also anticipated that a high level of biodiversity will increase community resilience by generating insurance through functional redundancy. Despite the paucity of research on the impacts of biodiversity on marine ecosystems, what is known indicates that marine systems may have invasion resistance mechanisms not unlike those of terrestrial systems.

2.4.2. Resilience

How readily an ecosystem returns to equilibrium after being disturbed is called resilience. The ability to bounce back after a disturbance is called resilience. A resilient community or ecosystem may be completely disrupted by disturbance but quickly return to its former state. Ecosystem resilience refers to the capacity of an ecosystem to recover from disturbance or withstand ongoing pressures.

It gauges an ecosystem's capacity to withstand shock without disintegrating into a different state governed by distinct processes. Resilience is a dynamic system of disturbance and recovery rather than a single optimal ecological state.

The boundaries of natural variability are known as thresholds, and they are crossed when an ecosystem shifts to a new, alternative state following disturbance or invasion and is unable to return to its previous state through natural processes.

Example

Coral reefs and other tropical marine ecosystems are frequently disrupted by a variety of human activities as well as hazards like cyclones, outbreaks of crown-of-thorns starfish, and freshwater influxes. These occurrences frequently harm, strain, or even destroy ecosystem components. A resilient ecosystem will be able to fully recover from these disruptions and return to its pre-impact state of biodiversity and health given enough time. In a similar vein, an ecosystem that is resilient might be able to withstand the stresses brought on by these disruptions without showing any signs of deterioration.

2.4.2.1. Factors That Affect Resilience

Understanding and evaluating ecosystem resilience is difficult due to the multitude of factors that can impact it. The innate biology and ecology of an ecosystem's constituent species or habitats, the state of these individual components, the type, intensity, and duration of the impacts, and the extent to which potential impacts have been eliminated or minimized all affect an ecosystem's capacity to absorb or recover from impacts and the rate at which it does so. Populations of species or habitats can frequently absorb or recover from impacts if all of these conditions are met, allowing the ecosystem to continue operating. If there are restrictions, the ecosystem's ability to absorb effects without altering will be less than ideal, which will cause recovery to take much longer or possibly fail (Chapman and Reiss, 2000; Smith, 2006; Meffe et al., 2013).

2.5. ECOSYSTEM FUNCTION AND VALUES

Ecosystem functions, or what the ecosystem does, are the physical, chemical, and biological processes or characteristics that support an ecosystem's ability to sustain itself. The provision of habitat for wildlife, the cycling of carbon, or the nutrient trapping are a few instances of ecosystem functions. Thus, the processes or functions that take place within ecosystems, like wetlands, forests, or estuaries, can be used to describe them.

2.5.1. Ecosystem Services

The positive effects that ecosystem functions have on the environment or on humans are known as ecosystem services. Ecosystem services include things like harvesting plants or animals, maintaining the food chain, and providing clean water or beautiful scenery. An ecosystem needs to interact with humans, or at the very least, be appreciated by them, to provide services to humans. Ecosystem services are valuable to society, but their functions are value-neutral.

2.5.2. Some Factors that Complicate Ecosystem Management Decisions

Because different forms of market failure are linked to the environment and natural resources, decisions regarding ecosystem management are made more difficult. When markets fail to fully account for the social costs and benefits of a good, market failures result. For instance, the cost of burning gasoline imposes pollution costs on society that are not fully reflected in the price of gasoline. Market failures associated with ecosystems include the following: (i) a large number of ecosystem services are public goods; (ii) externalities impact a large number of ecosystem services; and (iii) property rights associated with ecosystems and their services are frequently ill-defined. Ecosystem services are frequently considered public goods, meaning that anyone can enjoy them without detracting from the enjoyment of others.

An aesthetic perspective, for instance, is a pure public benefit. It doesn't matter how many people appreciate the view—others can too. Certain services might be considered quasi-public goods, meaning that their use may reduce the enjoyment of others to a certain extent. For instance, anyone may be able to access a public recreation area. But, a crowded place can make people feel less satisfied with it. Public goods have a problem in that, despite their value, no one is motivated to pay to keep them maintained. Therefore, in order to produce the most beneficial quantity, collective action is necessary. Externalities are uncompensated side effects of human activity that can have an impact on ecosystem services. For instance, there is a negative externality for the people downstream if runoff from agricultural land pollutes a stream. The issue with negative externalities is that they usually result in no compensation for the harm they cause to the people or ecosystems they affect. Ultimately, in the absence of a clear definition, property rights pertaining to natural resources may be exploited excessively, as there would be no motivation to preserve them. Uncontrolled fisheries, for instance, are a resource that is available to anybody who wishes to harvest fish. Since there is no single owner of the resource, unrestricted access may eventually result in significant declines in fish populations due to overfishing. Ecosystem valuation measures the costs of market failures to society in terms of lost economic benefits, which can assist resource managers in mitigating their effects. The costs to society can then be used to hold those accountable or as a gauge for how worthwhile it is to take steps to lessen or eliminate negative environmental effects. For instance, by lessening the crowding in the crowded public recreation area, the benefits to the general public could be increased. This could be achieved by enlarging the space or reducing the amount of guests. The higher economic benefits of lessening crowding can be compared to the costs of putting various options into practice. If agricultural runoff pollutes a stream, the benefits of cleaning it up can be weighed against the costs of taking steps to lessen the runoff, or the amount of fines or taxes that should be imposed on the offenders. The advantages of lessening overfishing in open-access fisheries can be weighed against the costs of regulations or the financial impact on the commercial fishing sector if access is restricted.

2.5.3. Ecosystem Values

Ecosystem values quantify the importance and value of ecosystem services to human well-being. By calculating the price people are willing to pay to maintain or improve ecosystem services, economists are able to determine the value of these services to individuals. This isn't always the case, though, for many reasons. Most importantly, many ecosystem services, such as a day spent watching wildlife or enjoying an ocean view, are not traded in markets, although some ecosystem services, like fish or lumber, are. Thus, many ecosystem services are provided without direct payment from humans. In addition, people's willingness to pay might not be evident because they are unfamiliar with buying such goods. However, this does not imply that ecosystems or the services they provide are worthless or that their worth cannot be expressed in monetary terms. Ecosystem services do not always need to be traded on open markets to be valued in US dollars. What is needed is an estimate of the amount of purchasing power (in dollars) that users are willing to forfeit to utilize the ecosystem services or the amount that users would have to pay to forfeit it if forced to make a decision that is comparable to one they would make in a market.

2.5.4. Types of Values

Ecosystem values are divided into various categories by economists. Use values and non-use, or passive use values, are the two primary categories. Non-use values are those that aren't connected to the actual use of an ecosystem or its services, or even the option to use them. In contrast, use values are determined by how the environment is actually used. The value obtained from using an item or service—such as hiking, fishing, bird watching, or hunting—is thus referred to as use value. Indirect uses may also be included in use values.

For instance, visitors to an Alaskan wilderness area benefit directly from its direct use. Indirect use values could be obtained by others from watching a television program about the region and its wildlife. Indirect use values can also be obtained by people from an input that goes into making something else that they use directly. For instance, recreational anglers who catch the fish that eat the lower-ranking aquatic organisms receive indirect use values from them.

The value that people attach to the possibility of enjoying something in the future, even if they aren't using it right now, is known as option value. It is a kind of use value as a result. For instance, someone might be willing to pay to preserve the Alaskan wilderness area in order to keep the option of visiting it at some point in the future. Comparably, bequest value is the importance people attach to the knowledge that something will be available for enjoyment to future generations. Therefore, the willingness of people to pay to protect the environment for future generations is a key indicator of bequest value. To ensure that future generations can enjoy the Alaskan wilderness area, for instance, someone might be willing to pay a price to preserve it. Values that are unrelated to the actual use of an item or service, or even the option to use one, are known as non-use values, sometimes known as passive use values. The non-use value that people attach to the knowledge that something exists, even if they never see it or use it, is known as existence value. For instance, someone may be willing to pay to preserve the Alaskan wilderness area just because they appreciate its existence, even if they never plan to visit or even want to. It is obvious that a single individual can gain multiple benefits from the same ecosystem. Therefore, the total economic value of a good or service is the sum of its relevant use and non-use values (Chapman and Reiss, 2000; Smith, 2006; Meffe et al., 2013).

2.6. NUTRIENT CYCLING IN ECOSYSTEMS

Every element on Earth is continually recycled. The constituents of all living things, including oxygen, carbon, nitrogen, phosphorus, and sulfur, are major elements. The movement of these kinds of chemicals and elements between living things and their physical surroundings is referred to as a biogeochemical cycle. Chemicals that organisms consume are reabsorbed by the soil, air, and water through processes like respiration, excretion, and decomposition. These processes occur throughout the food chain. An element frequently forms compounds with other elements as it progresses through this cycle as a result of natural reactions in the atmosphere, hydrosphere, or lithosphere as well as metabolic processes in living tissues. The biogeochemical cycle refers to this type of periodic material exchange between living things and their non-living surroundings.

Nutrient cycles are of two types:

- Gaseous; and
- Sedimentary.

The storehouse for the gaseous nutrient cycle (e.g., the atmosphere, nitrogen, carbon cycle), and for the sedimentary cycle (e.g., phosphorus and sulfur cycle), the reservoir is found in the crust of the Earth. The surroundings, for example, soil, humidity, temperature, pH, and so forth. control the rate at which nutrients are released into the atmosphere. The reservoir's purpose is to make up for the deficit brought on by an imbalance in the rates of influx and outflow.

2.6.1. Carbon Cycle

Through the process of photosynthesis, carbon is converted from carbohydrates into carbon dioxide, which is then absorbed by living things. The producers then transfer these organic materials (food) to the consumers (herbivores and carnivores). Eventually, this carbon is released back into the surrounding medium through respiration or the animals' and plants' natural decomposition. When fossil fuels are burned, carbon is also recycled (Figure 2.5).

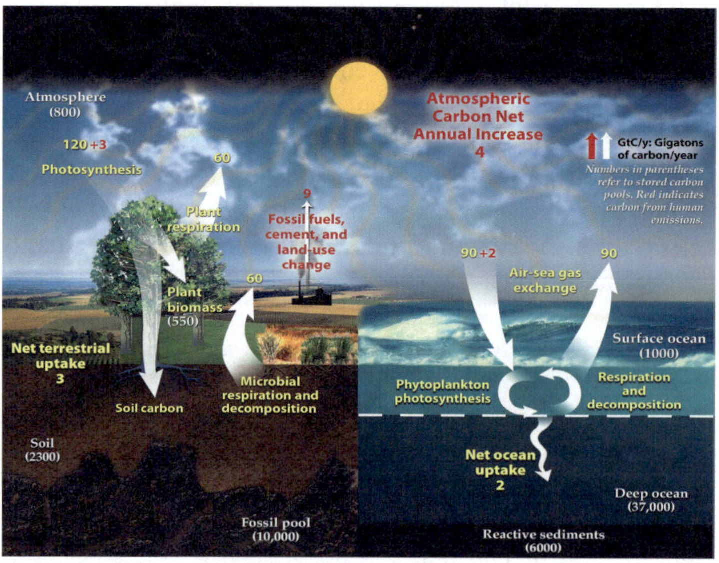

Figure 2.5 Carbon cycle.

Source: https://en.wikipedia.org/wiki/File:Carbon_cycle.jpg.

2.6.2. Nitrogen Cycle

As a necessary component of protein, nitrogen is needed by all living things, including humans. Although almost 79% of the nitrogen in our atmosphere is present, most living things are unable to directly utilize it. Similar to carbon dioxide, nitrogen undergoes a cycle of transitioning from gaseous to solid and back again through the actions of a diverse range of organisms. All living things depend critically on the cycling of nitrogen. The following lists the five primary processes that are necessary for the nitrogen cycle (Figure 2.6).

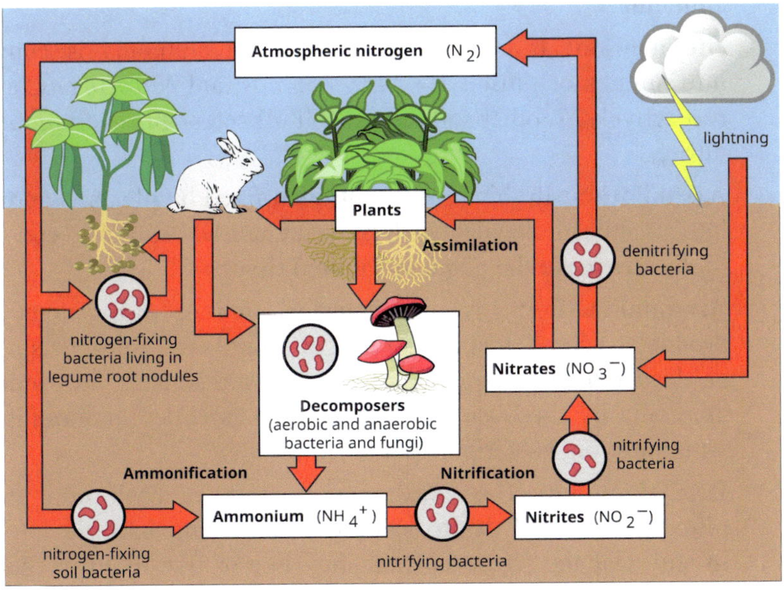

Figure 2.6. *Nitrogen cycle.*

Source: https://cdn-ccnli.nitrocdn.com/hPbgvskpJmIfLTeYLYGdGpfRfGKbrYYq/assets/images/optimized/rev-c209f7f/upload.wikimedia.org/wikipedia/commons/thumb/b/bd/Nitrogen_Cycle_2.svg/1280px-Nitrogen_Cycle_2.svg.png.

2.6.2.1. Nitrogen Fixation

This process involves the conversion of gaseous nitrogen into Ammonia, a form in which it can be used by plants. Atmospheric nitrogen can be fixed by the following three methods:

- **Atmospheric Fixation**: Lightening, combustion and volcanic activity help in the fixation of nitrogen.

- **Industrial Fixation**: At high temperatures (400°C) and high pressure (200 atm.), molecular nitrogen is broken into atomic nitrogen which then combines with hydrogen to form ammonia.
- **Bacterial Fixation**: There are two types of bacteria: symbiotic bacteria, such as *Rhizobium* found in the root nodules of leguminous plants, and free-living or symbiotic bacteria, including *Nostoc, Azotobacter*, and *Cyanobacteria*. These bacteria are capable of combining atmospheric or dissolved nitrogen with hydrogen to form ammonia.
- **Nitrification:** It is a process by which ammonia is converted into nitrates or nitrites by *Nitrosomonas* and *Nitrococcus* bacteria respectively. Another soil bacteria Nitrobacter can covert nitrate into nitrite.
- **Assimilation:** In this process nitrogen fixed by plants is converted into organic molecules such as proteins, DNA, RNA, etc. These molecules make the plant and animal tissue.
- **Ammonification:** Living organisms produce nitrogenous waste products such as urea and uric acid. These waste products as well as dead remains of organisms are converted back into inorganic ammonia by the bacteria. This process is called ammonification. Ammonifying bacteria help in this process.
- **Denitrification:** The conversion of nitrates back into gaseous nitrogen is called denitrification. Denitrifying bacteria live deep in soil near the water table as they like to live in an oxygen-free medium. Denitrification is the reverse of nitrogen fixation.

2.6.3. Water Cycle

Life requires water. Without water, no organism can survive. precipitation, such as slush dew, rain, or snow. is the only water source on the planet. The term "water cycle" refers to the continuous movement of water in the biosphere as a result of direct evaporation and evapotranspiration, which return water vapor that was taken from the earth's atmosphere (hydrological cycle). Earth is a watery planet in the solar system, with water covering roughly two-thirds of its surface. Nevertheless, only a very tiny portion of this is accessible to plants and animals. The distribution of water on Earth's surface is not uniform. The majority of water on Earth, approximately 95% of it, is chemically bonded to rocks and does not cycle. Approximately 97.3% of the remaining 5% is found in the ocean, and 21.1% is made up of polar ice caps. Therefore, the amount of fresh water that is present is only 0.6 percent in the form of soil, ground, and atmospheric water vapor. The water cycle is fueled by two factors: (i) solar radiation; and (ii) gravity. The water cycle involves two primary processes: precipitation and evaporation. The sun's heat energy causes water in lakes, ponds, rivers, and streams to evaporate. These two processes alternate with one another. Additionally, plants transpire vast volumes of water. When water evaporates in the atmosphere, it condenses into clouds that move with the wind. In the mountains, clouds collide with the chilly air.

On average, 84% of the water is lost from the surface of the through oceans by evaporation. While 77% is gained by it from precipitation. Water run-off from lands through rivers to oceans makes up 7% which balances the evaporation deficit of the ocean. On land, evaporation is 16% and precipitation is 23% (Figure 2.7).

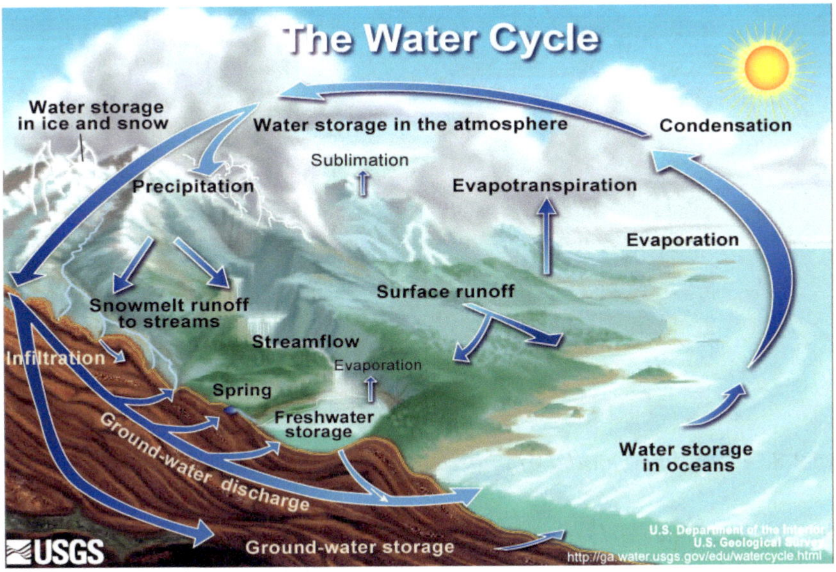

Figure 2.7. Water cycle.

Source: By John M. Even / USGS – USGS – http://ga.water.usgs.gov/edu/watercycleprint.html (English Wikipedia, original upload 27 April 2005 by Brian0918 en:Image:Water cycle.png), Public Domain, https://commons.wikimedia.org/w/index.php?curid=313135.

2.6.4. Phosphorus Cycle

Nucleic acids, biological membranes, and cellular energy transfer systems are all heavily dependent on phosphorus. This element is also required in large amounts by many animals for the synthesis of teeth, bones, and shells.

Phosphates, or natural phosphates, are found in rock, which serves as a natural source of phosphorus. Tiny amounts of these phosphates dissolve in soil solution as rocks weather, and plant roots absorb them (FigURE 2.8). This element is obtained by herbivores and other animals from plants. Phosphate-solubilizing bacteria break down waste materials and dead organisms to release phosphorus. There is no respiratory release of phosphorus into the atmosphere, in contrast to the carbon cycle. The other two primary and significant distinctions between the carbon and phosphorus cycles are as follows: (i) the amount of phosphorus that is incorporated into the atmosphere by rainfall is significantly less than that of carbon; and (ii) there is very little phosphorus gas exchange between organisms and their surroundings (Chapman and Reiss, 2000; Smith, 2006; Meffe et al., 2013).

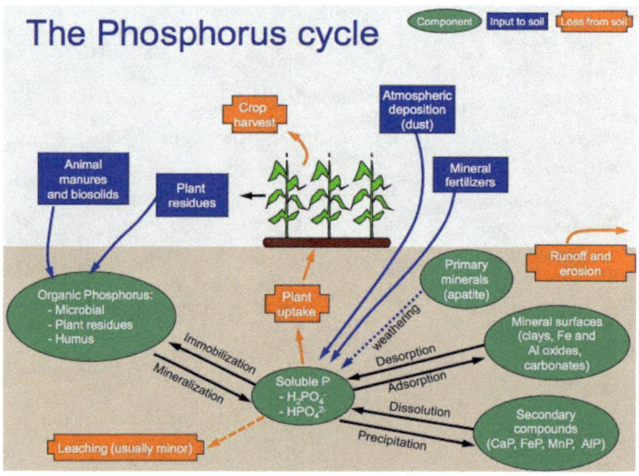

Figure 2.8. *A simplified model of phosphorus cycling in a terrestrial ecosystem.*

Source: By Welcome1To1The1Jungle at English Wikipedia, CC BY 3.0, https://commons.wikimedia.org/w/index.php?curid=48498089.

2.7. PRINCIPLES OF ECOSYSTEM MANAGEMENT AND NATURE CONSERVATION

The goal of ecosystem management is to preserve important ecological services and replenish natural resources while satisfying the demands of present and future generations on a socioeconomic, political, and cultural level. Ecosystem management is a method of managing natural resources with the dual goals of addressing socioeconomic, political, and cultural demands and ensuring the long-term sustainability and persistence of an ecosystem's services and function.

Although sustainable ecosystem management techniques have been used by indigenous communities for millennia, the term "ecosystem management" first appeared in print in the 1990s as a result of increased awareness of the interdependence of ecosystems, human dependence on them, and their influence (e.g., perturbation, ecological robustness). Ecosystem management, which builds on conventional natural resource management, incorporates institutional, socioeconomic, and ecological knowledge and priorities through the involvement of a wide range of stakeholders.

Ecosystem management is a comprehensive, flexible approach to assessing and attaining resilience and sustainability, in contrast to command and control approaches to natural resource management, which frequently result in reductions in ecological resilience. Because of this, implementation varies depending on the context and can include landscape-scale conservation, strategic management, and adaptive management, among other approaches.

The term "ecosystem management" was formalized in 1992 by F. Dale Robertson, the then Chief of the U.S. Forest Service. Robertson stated, "By ecosystem management, we mean an ecological approach… [that] must blend the needs of people and environmental values in such a way that the National Forests and Grasslands represent diverse, healthy, productive and sustainable ecosystems." A variety of additional definitions of ecosystem management exist, although definitions of this concept are typically vague. *For example*, Robert T. Lackey emphasizes that ecosystem management is informed by ecological and social factors, motivated by societal benefits, and implemented over a specific timeframe and area. F. Stuart Chapin and co-authors highlight that ecosystem management is guided by ecological science to ensure the long-term sustainability of ecosystem services, while Norman Christensen and coauthors underscore that it is motivated by defined goals, employs adaptive practices, and accounts for the complexities of ecological systems. Peter Brussard et al., suggest ecosystem management balances preserving ecosystem health while sustaining human needs.

Ecosystem management is still a contentious and vague concept in natural resource management, partly because some of its tenets are based on disputed scientific and policy claims.

Understanding a large portion of the controversy surrounding ecosystem management requires an understanding of these claims. Professional managers of natural resources, who usually work out of government agencies and trade associations, frequently use the argument that ecosystem management is merely a continuation of earlier management strategies to avoid having to address contentious statements.

2.7.1. Principles of Ecosystem Management

The long-term sustainability of ecosystems' capacity to produce goods and services is a basic tenet of ecosystem management since intergenerational sustainability is a requirement for management, not an afterthought. The ideal system management objectives should be well-defined and made public, with regard to the future paths and characteristics of the system under management. A thorough ecological understanding of the system, encompassing ecological dynamics, connectedness, and the context in which the system is embedded, is another crucial prerequisite. It's also critical to comprehend how humans fit into ecosystems and how to apply adaptive management. Ecosystem management is applicable in both intensively managed ecosystems (e.g., wilderness areas) and as part of a plan for wilderness conservation (e.g., agroecosystem and forestry near the natural world).

Core principles and common themes of ecosystem management:

1. **Systems Thinking:** Management has a holistic perspective, instead of focusing on a particular level of biological hierarchy in an ecosystem (e.g., only conserving a specific species; only preserving ecosystem functioning).
2. **Ecological Boundaries:** Ecological boundaries are clearly and formally defined, and management is place-based and may require working across political or administrative boundaries.
3. **Ecological Integrity:** Management is focused on maintaining or reintroducing native biological diversity, and on preserving natural disturbance regimes and other key processes that sustain resilience.
4. **Data Collection:** Broad ecological research and data collection is needed to inform effective management (e.g., species diversity, habitat types, disturbance regimes, etc.).
5. **Monitoring:** The impacts of management methods are tracked, allowing for their outcomes to be evaluated and modified, if needed.
6. **Adaptive Management:** Management is an iterative process in which methods are continuously reevaluated as new scientific knowledge is gained.
7. **Interagency Cooperation:** As ecological boundaries often cross administrative boundaries, management requires cooperation among a range of agencies and private stakeholders.
8. **Organizational Change:** Successful implementation of management requires shifts in the structure and operation of land management agencies.

9. **Humans and Nature:** Nature and people are intrinsically linked, and humans shape, and are shaped by ecological processes.
10. **Values:** Humans play a key role in guiding management goals, which reflect a stage in the continuing evolution of social values and priorities.

2.7.2. Nature Conservation

One of the many cutting-edge strategies for mitigating climate change is conservation ecology, also known as nature conservation. We can guarantee that future generations will likewise have a healthy planet to live on and grow on by contributing to the conservation of our natural resources.

A movement known as "nature conservation" aims to prevent the extinction of wildlife and natural resources while also reducing the rate of climate change and its other negative effects. Nature conservation works to stop further destruction, lessen habitat loss, create protected areas, and guarantee that there is sufficient biological diversity within the planet's ecosystems.

2.7.2.1. Purpose of Nature Conservation

Conservation efforts can support sustainability, improve ecosystem services, and keep the environment healthier. By keeping an eye on the health and longevity of our planet, conservation can help preserve natural resources for use by future generations. There are various ways to conserve nature, and all of them are meant to be beneficial to all living things on the planet.

2.7.2.2. Six Nature Conservation Methods

Conservationists and local communities can all have a hand in conserving our natural world.

Below are a few methods of nature conservation:

- **Planting Trees**: Each year, approximately 15 billion trees are cut down. Trees absorb the carbon dioxide in the atmosphere, converting it into breathable oxygen.
- **Using Alternative Energy Resources**: Switching to solar power or geothermal energy can help reduce our reliance on fossil fuels, cutting down on the amount of greenhouse gases released into the atmosphere, while also reducing the need for mining and drilling.
- **Establishing Protected Areas**: Establishing protected areas (like national parks and nature reserves) can help protect biodiversity from

harmful land use and human activities, keeping wildlife habitats and animal food resources safe.

- **Protecting Biodiversity**: Biodiversity ensures a healthy overall natural environment. Conservation efforts aimed at biodiversity can be done on a local level (e.g., by encouraging native plant growth) or through legislation, like the United States Endangered Species Act of 1973. Federal decision-making can help keep animals that are on the brink of species extinction from completely dying out.
- **Hunting Restrictions**: Wildlife conservation programs establish laws limiting the time, areas, or animal allotment for hunting and fisheries, which can help reduce the amount of wildlife that are killed every year.
- **Proper Planting**: Making sure you know where your plant seeds come from can help prevent the spread of invasive species, which can devastate indigenous crops, and threatening biodiversity (Chapman and Reiss, 2000; Smith, 2006; Meffe et al., 2013).

REFERENCES

1. Barot, S., Ugolini, A., & Brikci, F. B. (2007). Nutrient cycling efficiency explains the long-term effect of ecosystem engineers on primary production. *Functional Ecology, 21*(1), 1–10. https://doi.org/10.1111/j.1365-2435.2006.01225.x.
2. Basu, R. N. (2004). *A Compendium of Terms in Ecology and Environment.* Naya Udyog.
3. Berkes, F., Colding, J., & Folke, C. (2000). Rediscovery of traditional ecological knowledge as adaptive management. *Ecological Applications, 10*(5), 1251–1262. https://doi.org/10.1890/1051-0761(2000)010[1251:ROTEKA]2.0.CO;2.
4. Chapman, R. L., & Reiss, M. J. (2000). *Ecology: Principles & Application.* Cambridge University Press.
5. Dash, M. C. (2001). *Fundamentals of Ecology* (2nd ed.). Tata McGraw-Hill Company.
6. Gregorich, E. G., et al. (2001). *Soil and Environmental Science Dictionary.* Canadian Society of Soil Science. CRC Press LLC.
7. Joshi, P. C., & Joshi, N. (2009). *A Textbook of Ecology and Environment.* Himalaya Publishing House.
8. Kellert, S. R., Mehta, J. N., Ebbin, S. A., & Lichtenfeld, L. L. (2000). Community natural resource management: Promise, rhetoric, and reality. *Society and Natural Resources, 13*(8), 705–715. https://doi.org/10.1080/089419200750035575.
9. Kormondy, E. J. (2002). *Concepts of Ecology* (4th ed.). Pearson Education.
10. Lavelle, P., Dugdale, R., Scholes, R., Berhe, A. A., Carpenter, E., Codispoti, L., et al. (2005). Nutrient cycling. In *Millennium Ecosystem Assessment: Objectives, Focus, and Approach*. Island Press.
11. Meffe, G., Nielsen, L., Knight, R., & Schenborn, D. (2013). *Ecosystem Management: Adaptive, Community-Based Conservation*. Island Press.
12. Montes, F., & Cañellas, I. (2006). Modelling coarse woody debris dynamics in even-aged Scots pine forests. *Forest Ecology and Management, 221*(1–3), 220–232. https://doi.org/10.1016/j.foreco.2005.10.019.
13. Ochoa-Hueso, R., Delgado-Baquerizo, M., King, P. T. A., Benham, M., Arca, V., & Power, S. A. (2019). Ecosystem type and resource quality are more important than global change drivers in regulating early stages of litter decomposition. *Soil Biology and Biochemistry, 129*, 144–152. https://doi.org/10.1016/j.soilbio.2018.11.009.

14. Odum, E. P., & Barret, G. W. (2005). *Fundamentals of Ecology* (5th ed.). Thomson Brooks Cole.
15. Podolinsky, S. (2004). Socialism and the unity of physical forces. *Organization & Environment, 17*(1), 61–75. https://doi.org/10.1177/1086026603262092.
16. Romero, J., Lee, K., Pérez, M., Mateo, M. A., & Alcoverro, T. (2007). Nutrient dynamics in seagrass ecosystems. In Larkum, A. W. D., Orth, R. J., & Duarte, C. M. (Eds.), *Seagrasses: Biology, Ecology and Conservation* (pp. 227–270). Springer. https://doi.org/10.1007/978-1-4020-2983-7.
17. Smith, T. M., & Smith, R. L. (2006). *Elements of Ecology* (6th ed.). Pearson Education.
18. Sutherland, W. J. (2003). Evidence-based conservation. *Conservation in Practice, 4*(3), 39–42. https://doi.org/10.1111/j.1526-4629.2003.tb00068.x.
19. Weiner, D. R. (2000). *Models of Nature: Ecology, Conservation and Cultural Revolution in Soviet Russia*. University of Pittsburgh Press.

Chapter **3**

Threats to Aquatic Biodiversity

Contents

INTRODUCTION

Numerous and intricate threats to aquatic biodiversity present serious obstacles to the stability and well-being of aquatic ecosystems around the world. The destruction of habitat, which is frequently caused by human activities like agriculture, deforestation, and coastal development, is one of the main threats. Many aquatic species depend on vital habitats like wetlands, coral reefs, and mangroves for their survival, all of which are lost as a result of these activities. Pollution, which taints water bodies with dangerous materials like pesticides, heavy metals, and plastic waste, is another serious threat. Aquatic life is particularly vulnerable to the destructive effects of pollution, which can destroy many species by upsetting their reproductive cycles and causing illnesses. A type of pollution known as eutrophication, which is brought on by an overabundance of nutrients from agricultural runoff, also results in toxic algal blooms that lower the water's oxygen content and produce dead zones where marine life cannot survive.

An additional serious danger to aquatic biodiversity is overfishing. Depletion of fish stocks and non-target species bycatch have the potential to upend food webs and bring down entire marine ecosystems. Fish populations are not the only marine creatures impacted by overfishing; other marine organisms that depend on fish for food are also impacted. More people are realizing that one of the biggest threats to aquatic biodiversity is climate change. The distribution and behavior of aquatic species are impacted by rising temperatures, ocean acidification, and shifting weather patterns. For instance, coral reefs are extremely sensitive to temperature fluctuations and are currently undergoing widespread bleaching events as a result of global warming. In a similar vein, different aquatic species' migratory patterns and breeding cycles can be upset by fluctuations in ocean currents and temperature.

Aquatic biodiversity can potentially be threatened by invasive species that humans have unintentionally or purposely brought into the area. These alien species have the power to displace native species from their habitats, outcompete them for resources, and bring diseases. Invasive species can have detrimental effects, changing ecosystem functions and causing declines in native biodiversity. A multifaceted strategy involving efficient conservation tactics, pollution prevention measures, sustainable fishing methods, and international efforts to slow down climate change is needed to address these threats. The rich diversity of life in our aquatic ecosystems can only be preserved by limiting pollution, promoting policies that support biodiversity, and protecting and restoring aquatic habitats.

3.1. WATER POLLUTION

One of the biggest problems facing humanity in the 21st century is water quality. Both freshwater and marine environments are negatively impacted by aquatic pollution, which is also harmful to other relevant organisms and human health. By altering the water's quality, pollution can have a direct or indirect impact on fish immune systems. Aquatic pollutants come from a variety of sources, including municipal sewage deposited in the River Nile, industrial effluents, and agricultural runoffs. These pollutants gradually contaminate water, rendering it unfit for human use. Many pollutants from pesticides and herbicides found in agricultural wastewater have an adverse effect on rivers and the people who use them. High levels of toxicity are present in industrial effluents, including heavy metals that can combine with suspended solids in home wastewater to form muck. Additionally, fish biological responses were impacted by water pollution in Egypt, particularly in El-Max Bay and Bahary, which ultimately resulted in food oxidation damage and degradation of the environment. Additionally, certain immune genes from fish samples of seabass (*Dcentrarchus labrax*) and seabream (*Pagrus auratus*) were impacted by water pollution.

3.1.1. Types and Sources of Water Pollutants

Numerous elements, such as pathogenic microorganisms, putrescible organic waste, fertilizers and plant nutrients, heat, petroleum (oil), hazardous chemicals, sediments, and radioactive materials, can contaminate water bodies. The various kinds of water pollution are discussed below.

There are two types of sources of water pollution: dispersed and point. A point source is a pipe or channel, like the ones used to empty a city sewer system or an industrial facility. A wide, unconstrained region that allows a variety of pollutants to enter a body of water, like runoff from an agricultural area, is known as a dispersed (or nonpoint) source. Because the contaminated water has been gathered and transported to a single location where it can be treated, point sources of water pollution are simpler to manage than dispersed sources. Dispersed sources of pollution are challenging to manage, and even with significant advancements in the construction of contemporary sewage treatment facilities, a significant portion of water pollution issues still stem from these sources.

3.1.1.1. Domestic Sewage

The main source of pathogens—microorganisms that cause disease—and putrescible organic materials is domestic sewage. All sewage from cities and towns is likely to contain pathogens of some kind because pathogens are expelled in feces, potentially posing a direct threat to public health. The kind of threat putrescible organic matter poses to the quality of water is distinct. The dissolved oxygen content of the water decreases as organic materials break down organically in the sewage due to the action of bacteria and other microorganisms. Since fish and other aquatic life depend on high oxygen levels to survive, this puts the quality of lakes and streams in jeopardy. Furthermore, active pharmaceutical ingredients that can harm aquatic organisms and possibly contribute to antibiotic resistance are frequently found in domestic sewage. While pathogens and organics are reduced in wastewater by sewage-treatment processes, they are not entirely removed (Chapman, 1992; Fischer and Lindenmayer, 2007; Ashok, 2015) (Figure 3.1).

Figure 3.1. *Sewage.*

Source: By SuSanA Secretariat – https://www.flickr.com/photos/gtzeco-san/6619766435/, CC BY 2.0, https://commons.wikimedia.org/w/index.php?curid=36511998.

Another important source of plant nutrients, primarily phosphates and nitrates, is domestic sewage. The overabundance of phosphates and nitrates in water encourages the growth of algae, which can occasionally result in abnormally dense and quick growths known as "algal blooms." Because microorganisms use oxygen to digest algae during the decomposition process, the amount of oxygen dissolved in the water decreases when the algae die. The organic wastes are subsequently broken down by anaerobic organisms, which do not need oxygen to survive. This process releases gases like methane and hydrogen sulfide, which are toxic to aerobic (oxygen-requiring) life forms. The process known as eutrophication is the way a lake goes from being clean and clear, with a balanced aquatic community and a relatively low concentration of dissolved nutrients, to being nutrient-rich and algal-filled, and finally becoming oxygen-deficient and waste-filled. The process of eutrophication is slow, natural, and unavoidable. However, it can cause a body of water to age and die sooner than it should if it is accelerated by human activity and water pollution (a phenomenon known as cultural eutrophication).

3.1.1.2. Solid Waste

One of the main causes of water pollution is the incorrect disposal of solid waste. Garbage, trash, electronic waste, construction and demolition waste, and rubbish are all considered forms of solid waste and are produced by various activities such as residential, commercial, institutional, and industrial ones. The issue is particularly severe in developing nations since they might not have the resources or the infrastructure necessary to properly dispose of solid waste, or they might not have the regulations in place to prevent improper disposal. Certain locations purposefully dump solid waste into bodies of water.

Figure 3.2. *Plastic bottles and other garbage in a lake.*

Source: Public Domain, https://commons.wikimedia.org/w/index.php?curid=16396.

If trash or other debris is brought to bodies of water by animals, the wind, or precipitation, it can also turn from land pollution into water pollution. Significant levels of pollution from solid waste in inland waterways have the potential to eventually reach the ocean. Pollution from solid waste is unsightly, detrimental to aquatic ecosystems, and directly harmful to wildlife. Toxic or hazardous waste can be found in many solid wastes, including plastics and electronic waste, which decompose and release dangerous chemicals into the water (Figure 3.2).

Plastic pollution is a growing concern for aquatic environments. The ocean receives the majority of the plastic waste produced on land because it is located downstream of almost every place on Earth. Every year, millions of tons of debris, mostly improperly discarded plastic litter, find their way into the world's oceans. Waves and UV light can break down plastic pollution into smaller particles known as microplastics, which are not biodegradable and have a length of less than 5 mm (0.02 inches). Primary microplastics, such as microbeads in personal care products and plastic fibers in synthetic textiles (e.g., nylon), also enter the environment directly, through any of various channels—for example, from wastewater treatment systems, from household laundry, or from unintentional spills during manufacturing or transport. Sadly, every aquatic organism examined in many freshwater and marine studies has been found to contain microplastics.

Figure 3.3. *Breakdown of plastic pollution in ocean waters. Sunlight and seawater embrittle plastic, and the eventual breakdown of larger objects into microplastics makes plastic available to zooplankton and other small marine animals.*

Source: By United States Fish and Wildlife Service. – Public Domain Images, United States Fish and Wildlife Service., Public Domain, https://commons.wikimedia.org/w/index.php?curid=24859656.

These microscopic plastics have been detected in seafood and are thought to have made their way up the marine food chains from zooplankton and tiny fish to large marine predators. Additionally, microplastics have been found in drinking water. Their impact on health is uncertain (Figure 3.3).

3.1.1.3. Toxic Waste

Poisonous, radioactive, explosive, carcinogenic (caused to cause cancer), mutagenic (caused to cause damage to chromosomes), teratogenic (caused to cause birth defects), and bioaccumulative (that is, increasing in concentration at the higher ends of food chains) waste are all classified as toxic. Toxic chemical sources include surface runoff containing pesticides used on suburban lawns and agricultural areas (chlordane, dieldrin, heptachlor) and improperly disposed wastewater from industrial plants and chemical process facilities (lead, mercury, chromium).

3.1.1.4. Sediment

Debris (e.g., silt) that comes from building projects or soil erosion can enter water bodies through surface runoff. Suspended silt disrupts a body of water's natural equilibrium and prevents sunlight from penetrating. Additionally, it can interfere with fish and other life forms' reproductive cycles. When it settles out of suspension, it can suffocate organisms that live on the bottom.

3.1.1.5. Thermal Pollution

Heat causes the water's ability to hold dissolved oxygen in solution to decrease, and it speeds up fish metabolism, which is why it's regarded as a water pollutant. Important game fish species (e.g., trout) are not able to survive in water with very little dissolved oxygen. Cooling water discharged from power plants into rivers is a common practice that generates a lot of heat; the water may be up to 15°C (27°F) warmer than naturally occurring water. Another example of thermal pollution is the rise in water temperatures brought on by global warming.

3.1.1.6. Petroleum (oil) Pollution

Petroleum (oil) pollution occurs when oil from roads and parking lots is carried in surface runoff into water bodies. Accidental oil spills are also a source of oil pollution—as in the devastating spills from the tanker *Exxon Valdez* (which released more than 260,000 barrels in Alaska's Prince William Sound in 1989) and from the Deepwater Horizon oil rig (which released more than 4 million barrels of oil into the Gulf of Mexico in 2010). Oil slicks eventually move toward shore, harming aquatic life and damaging recreation areas (Chapman, 1992; Fischer and Lindenmayer, 2007; Ashok, 2015) (Figure 3.4).

Figure 3.4. *Oil pollution.*

Source: By US Gov NOAA – US Gov NOAA, Public Domain, https://commons.wikimedia.org/w/index.php?curid=10905.

3.2. HABITAT FRAGMENTATION AND ITS EFFECT ON AQUATIC BIODIVERSITY

Environments are dynamic sites where organisms breed, live, and eat. Aquatic species rely on their unique habitat for survival, but one of the main threats to biodiversity loss is fragmentation. When habitat degradation results in a species becoming threatened or endangered, it is frequently caused by habitat fragmentation. Therefore, protecting the natural ecosystems that will increase aquatic diversity and lower extinction rates is urgently needed.

3.2.1. Habitats

The term "habitat" describes the spaces where a specific organism can survive and reproduce. The environment in which an animal lives, including all biotic and abiotic resources that impact its survival and procreation, is referred to as its habitat by wildlife ecologists. The scientific underpinning of wildlife conservation and management of wildlife species and their habitats is wildlife ecology. One of the main ideas in wildlife ecology is habitat. For an organism to survive, it needs a specific type of habitat. Some have very specific needs, while others can tolerate a lot of variation. There is a positive correlation between species diversity and habitat area.

Habitats can be either aquatic or terrestrial. Forests, grasslands, wetlands, and deserts are examples of terrestrial habitats. These broad biomes encompass more specialized environments with different types of soil, altitudes, temperatures, and vegetation. Freshwater and marine environments are examples of aquatic habitats. Marshes, rivers, lakes, ponds, and streams are examples of freshwater habitats. Estuaries, salt marshes, the coast, the intertidal zone, reefs, bays, the open sea, the sea bed, deep water zones, and shallow water zones are examples of marine habitats.

Numerous spatial scales are used to study habitats, from microhabitats to macrohabitats. Microhabitats comprise elements needed by an individual to carry out a specific activity, such as reproduction and foraging. While macrohabitats consist of all the elements needed to sustain a population of a specific organism. The degree of connectivity between different types of habitats and the dispersal mechanisms that organisms use within them differ. For instance, because the medium is fluid in aquatic systems, organisms can spread passively. However, compared to highly connected systems like the open ocean, the movement of microbes within a more solid matrix like soil may be more constrained (Hanson et al., 2017).

Strong selection pressures encourage individuals to select high-quality habitats for a particular activity, such as foraging, breeding, displaying, etc., because habitat quality affects the fitness of the organisms. The contribution of resources to individual survival and reproduction as well as population persistence is known as habitat quality. The behavioral and physiological mechanisms through which organisms select certain locations to carry out particular activities are known as habitat selection processes.

3.2.2. Habitat Fragmentation

A habitat is a place where living things coexist with their inanimate counterparts. It describes how the habitat is divided up into smaller, more isolated pieces. It causes discontinuities or fragmentation in an organism's preferred environment by causing habitat loss and habitat configuration changes. According to Connor and McCoy (2017), it is currently one of the biggest threats to both terrestrial and aquatic biodiversity.

Aquatic habitat fragmentation results in population fragmentation and ecosystem degradation, which have an impact on species abundance and the reduction of aquatic biodiversity worldwide. It causes habitat degradation and loss, which have an impact on biodiversity. Populations dispersed across large geographic areas are vulnerable to extinction for many species if the networks of patches are not sufficiently linked by dispersal. The expansion of agriculture in tandem with climate change is one example of an anthropogenic disturbance that has caused habitat loss and fragmentation worldwide. In order to enhance efforts to conserve species in the future, we must improve our capacity to manage landscapes. Usually found in formerly forested areas, a fragmented landscape is defined by a sharp contrast between vegetation patches and their surrounding matrix. It has been demonstrated that the fragmentation of tropical rainforests has had negative effects on old-growth tropical rainforests that are still intact. These effects include modifications to the dynamics, structure, and function of the ecosystems. For example, long-term increases in tree mortality near the boundaries of habitat patches result in canopy gaps, which alter the species composition and structure of the forest. The biodiversity of these rainforests has suffered greatly as a result of this fragmentation. One million species from Tropical Rainforests are thought to have already gone extinct as a result of land-use changes during the past 100 years. Numerous animals exhibit extreme sensitivity to the size of the forest fragments that make up their home range. Small habitat fragments lose species at a very high rate once they become isolated. It has been demonstrated that local extinctions of birds and butterflies (extirpations) happen more quickly in small (1–10 ha) than in large (100 ha) fragments of TRF. The drying out of forest edges makes it easier for fires to spread, including lightning strikes, which would normally go out in the moist interior of tropical rainforests (Flaspophler et al., 2001).

Natural habitats eventually disappear as human land use takes over an area. The ability of species to adapt to changing environments is harmed by habitat fragmentation, which also poses a threat to the size of species populations. Because it obstructs gene flow from one generation to the next in small populations, this occurs at the genetic level. Genetic recombination and mutations have a higher chance of surviving in the species gene pool in larger populations, but in smaller populations, mutations can result in the expression of deadly genes that drive a species to extinction and hinder the process of natural selection. A component of this is the phenomenon known as genetic drift, which results in changes in a population's genetic makeup because of chance or unpredictability rather than natural selection. Over time, it causes variations in allele frequencies. Genetic drift causes some traits to gradually become more common while others become less common in small populations; however, these changes are not driven by natural selection, but rather by random events. Small populations with genetic uniformity are more susceptible to high infection rates and the quick spread of diseases. It also lessens a population's capacity to adapt and respond to changes in the environment. Reduced gene flow also lessens the likelihood that local adaptations will proliferate across the entire species. In all ecosystems, the location and extent of an individual's movement are determined by landscape characteristics and geomorphological limitations, which have an impact on local community structure and species coexistence. When the aquatic environment degrades to the point where it cannot support life, this is known as aquatic habitat destruction and loss. Both natural and human activities have an impact on this process (Thibault Datry, 2017).

3.2.3. Aquatic Habitat Fragmentation

Aquatic habitats that are diverse and healthy support populations of fish and other aquatic organisms that are also healthy and diverse. Natural materials found in aquatic habitats include rocks, coral, gravel, sand, and mud; vegetation types include macrophytes, snags, seaweeds, seagrasses, mangroves, and saltmarsh; riffles, pools, and reefs; and the ecosystem as a whole, which includes wetlands, floodplains, streams, estuaries, lakes, and beaches. The requirements for habitat vary among species and can alter over the course of a fish's life. The proper water conditions (freshwater, estuary, or saltwater), temperature, quantity (depth and flow), and quality are all necessary for an aquatic ecosystem. In freshwater, estuarine, and marine systems, among other terrestrial and aquatic ecosystems, fragmentation is essential (e.g., The aquatic ecosystem is seriously threatened by seas, coral reefs, and seagrass meadows (Ledger et al., 2011).

Freshwater fragmentation has been accelerated by human activity, especially because of the overuse of water by expanding populations. Additionally, because of reduced rainfall brought on by climate change in many regions, there is a risk that widespread habitat loss and fragmentation will result. Droughts cause river flows to decrease, which changes the amount of wetted habitat (depth and width of water) and its structure. They also raise water temperatures, lower dissolved oxygen levels, and change the availability of nutrients (Dahm et al., 2003). The fragmentation caused by this erratic drought can have disastrous consequences for aquatic food webs.

Oceans, coral reefs, and seagrass meadows are examples of marine systems that are susceptible to fragmentation. The open ocean, for instance, may seem to be rather homogeneous, but it actually contains distinct vertical and horizontal regions that are divided by physicochemical barriers like frontal systems and pycnoclines. These regions are more permeable to larger organisms (e.g., fish (both anadromous and catadromous) as opposed to the smaller species.

Excessive fishing causes habitat loss and fragmentation of coral reefs, and rising sea temperatures are causing an increasing frequency of coral bleaching. Reef fish and other motile invertebrates are becoming less abundant and diverse as a result of the loss of structural complexity in these broken coral environments. Seagrass meadows are one of the other marine systems; they are distinctive, fruitful, and varied ecosystems. They are impacted by urchin grazing, eutrophication, severe weather, fragmentation brought on by dredging and boating, and wasting disease. The fragmentation of seagrass meadows can cause rapid declines in species diversity and abundance by affecting epifaunal diversity and abundance. Seldom used marine environments that are impacted by fragmentation are sea ice, salt marshes, and kelp forests. Fish biomass and abundance in kelp forests are decreased by habitat loss. Key grazers, or snails, undergo behavioral changes as a result of the salt marsh die-off as they seek cover from blue crab predation. Declines in the mating success and searching efficiency of polar bears and other top predators, as well as modifications to the structure of phototrophic communities and the relative abundance of dominant marine taxa, are the outcomes of increased fragmentation of sea ice habitats (Chapman, 1992; Fischer and Lindenmayer, 2007; Ashok, 2015).

3.2.4. Effect of Aquatic Fragmentation

- **Effect on Fish Food:** Aquatic plants such as seagrasses and macroalgae (seaweed), plankton (microscopic animals and plants), mosquito larvae, dragonflies, water beetles, boatmen, shrimp, mayflies, and other fish are among the food sources for fish. Fish have specific needs when it comes to habitat. Therefore, although habitat changes may not directly impact fish, they may have an impact on their food source. Thus, the fish are unable to survive in that environment.
- **Places of Fish Shelter:** Fish and other aquatic animals require cover from predators as well as from harsh environmental elements like floodwaters and direct sunlight. Fish have places to hide in freshwater environments such as patches of vegetation, spaces behind rocks, and the areas around submerged logs and branches (snags). Fish in estuaries are protected from predators by seagrass beds and mangrove forests. Fish that have these shelters removed from their natural habitat are at risk of being eaten by predators. Moreover, submerged plants oxygenate water, absorb nutrients, and contribute to the reduction of pollution, while marginal plants offer crucial habitats for both vertebrates and invertebrates.
- **Fish Movement:** The majority of fish swim over long distances to get from one place to another. Fish movement can be obstructed by habitat fragmentation resulting from the construction of barriers like road crossings, weirs, dams, or floodgates. Fish populations in the area may suffer greatly as a result.

- **Places to Breed:** Fish tend to procreate in the swiftly moving waters of rivers and streams. Fish are unable to reproduce due to obstructions in the running water. Rising water temperatures and longer days are the catalysts for breeding in many freshwater fish species. Fish reproduction may be hindered by actions that change a stream's natural water temperature, such as the summertime release of cold water from dam bottoms (Gehrke et al., 2002). Moreover, the building of dams in a river's path changes the river's flowing water into pond-like stagnation. Fish that are designed to reproduce in moving water do not reproduce in still water.
- **Fecundity:** The type of habitat that fish choose to lay their eggs in is very specific. For example, certain indigenous fish deposit their eggs in watery plants or hollow logs. Some use gravel to build their nests, such as freshwater catfish. Fish breeding grounds can be overrun by anthropogenic interference in most cases.

3.2.5. Causes of Aquatic Habitat Fragmentation

The causes may be anthropogenic or environmental.

i. **Anthropogenic Factors:** Human activities for agriculture, like construction of dams, bridges, roads, housing developments, etc., urbanization, population explosion, harvesting natural resources – for industrial production, etc.

ii. **Environmental Factors:** Climate change, floods, droughts, volcanoes, etc.

3.2.5.1. Construction of Dams on Rivers (Barriers)

Dams have altered the hydrology, morphology, and function of numerous freshwater ecosystems, leading to significant changes in the abundance and distribution of species, including extinctions. The flowing water of a river becomes the stagnant water of an artificial lake when dams are built across its course. Although dams and reservoirs can boost energy output, they can also alter and fragment river systems, endangering fisheries in the process. The fish species that live in freshwater ecosystems in Gabon, an extremely biodiverse country in Africa, have been impacted by the proposed construction of 38 hydropower dams.

3.2.5.2. Construction of Roads and Bridges

Transportation infrastructure creates a barrier to many species in terms of migration and dispersal, which leads to habitat loss and fragmentation.

Moreover, traffic related to infrastructure raises the risk of fauna mortality, which exacerbates the fragmentation effect (Friessell and Trombulak, 2000).

3.3. CLIMATE CHANGE

One of the biggest worldwide issues of our day is climate change, which has profound effects on aquatic systems as well as other natural ecosystems. The earth is warming, precipitation patterns are changing, and extreme weather events are happening more frequently. These changes are affecting the species that live in aquatic habitats and upsetting the ecological processes that support them. This paper explores the ways that aquatic biodiversity is threatened by climate change, looking at the mechanisms, evidence, and effects in a range of aquatic ecosystems, from freshwater lakes and rivers to coastal and oceanic settings.

3.3.1. The Science of Climate Change and Its Impacts on Aquatic Systems

Climate change is driven primarily by the increase in greenhouse gas concentrations in the atmosphere, primarily carbon dioxide (CO_2), methane (CH_4), and nitrous oxide (N_2O). These gases trap heat, leading to a rise in global temperatures, melting ice caps and glaciers, and altering weather patterns. For aquatic ecosystems, these changes manifest in several ways: rising water temperatures, shifts in precipitation and runoff patterns, sea level rise, and changes in the frequency and intensity of extreme weather events such as storms and droughts.

- **Temperature Rise and Its Effects:** The warming of the water is one of the most obvious effects of climate change on aquatic systems. Elevated temperatures have the potential to cause thermal stratification in bodies of water, thereby impacting the nutrient and oxygen mixing that is essential for the survival of numerous aquatic organisms. For example, thermal stratification in lakes and reservoirs can result in anoxic (oxygen-depleted) conditions at deeper layers, which can kill fish and cause the sediments to release harmful substances like hydrogen sulfide.
- **Changes in Precipitation and Runoff:** Freshwater ecosystems are greatly impacted by changes in precipitation patterns, including variations in the frequency and intensity of rainfall. Water quality and the wellbeing of aquatic organisms can be negatively impacted by increased rainfall and flooding, which can cause habitat destruction, erosion, and sedimentation in rivers and streams. On the other hand,

lower water levels brought on by droughts and decreased snowpack can concentrate pollutants and increase the stress on aquatic life.

- **Sea Level Rise and Coastal Impacts:** Ecosystems along the coast and in estuaries are seriously threatened by rising sea levels, which are caused by the melting of polar ice and the thermal expansion of seawater. When saltwater seeps into freshwater environments, it upsets the equilibrium of salinity, which impacts the species distribution and ecosystem processes. Moreover, storm surge frequency and intensity increase with rising sea levels, causing habitat loss, coastal erosion, and the deterioration of important habitats like seagrass beds, salt marshes, and mangroves.

3.3.2. Impacts on Freshwater Ecosystems

Freshwater ecosystems, such as wetlands, lakes, and rivers, are especially susceptible to the consequences of climate change. These ecosystems are crucial to biodiversity because they support a wide range of species and vital ecological processes like carbon storage, flood control, and water filtration (Chapman, 1992; Fischer and Lindenmayer, 2007; Ashok, 2015).

- **Altered Flow Regimes:** River and stream flow regimes are changing as a result of modifications to precipitation patterns and snowmelt timing. The amount of water available for use by humans and wildlife has been impacted in many areas by lower river flows during the crucial summer months due to early and decreased snowmelt and snowfall. Increased precipitation intensity in some places can cause flash floods, which can suffocate aquatic habitats and lower water quality by upsetting ecosystems and adding to the sediment load in streams.
- **Temperature Effects on Species Distribution:** Freshwater species are becoming more widely distributed as a result of rising temperatures; they frequently migrate to higher latitudes or altitudes in search of cooler climates. This change may result in a shift in the composition of the community, with certain species thriving and others facing extinction or decline. For instance, many species of cold-water fish, like salmon and trout, are migrating upstream or to higher altitudes where there is still adequate habitat. This change may lessen genetic diversity and make populations more susceptible to illnesses and other stresses.
- **Increased Frequency of Harmful Algal Blooms (HABs):** Harmful algal blooms (HABs) are becoming more common in freshwater

systems due to rising water temperatures and increased nutrient runoff from urban and agricultural areas. Toxins that are dangerous to humans, livestock, and aquatic life can be produced by these blooms. For instance, the cyanobacterium *Microcystis aeruginosa* produces toxins that can harm a person's liver or cause neurological problems in animals. With continued climate change, it is anticipated that HAB frequency and severity will rise, posing serious risks to ecosystem health and water quality.

3.3.3. Impacts on Marine and Coastal Ecosystems

Climate change has an equal impact on marine and coastal ecosystems, affecting everything from deep-sea habitats to mangroves and coral reefs. Because they serve as many marine species' feeding, breeding, and nursery grounds, these ecosystems are essential to biodiversity.

- **Ocean Warming and Coral Bleaching:** Ocean warming, which causes coral bleaching, is one of the most well-researched effects of climate change on marine ecosystems. Corals are extremely sensitive to temperature changes; even a small rise in water temperature can force them to drive out the zooxanthellae, which are symbiotic algae that live inside their tissues. As a result, the corals lose their vibrant color and essential source of energy. Extended bleaching episodes impair corals' resilience to disease and decrease their ability to reproduce. The world's largest coral reef system, the Great Barrier Reef, has seen numerous bleaching events that have significantly reduced the amount of coral cover and biodiversity.
- **Ocean Acidification:** The marine pH is falling as a result of ocean acidification, a process brought on by the oceans' absorption of excess CO_2. Ocean organisms, especially those with calcium carbonate shells or skeletons like corals, mollusks, and certain plankton species, may be severely impacted by this shift in ocean chemistry. These organismal capacity to create and maintain their calcium carbonate structures may be hampered by ocean acidification, which would lower their chances of surviving and procreating. Since many species depend on these organisms for both food and habitat, this has an impact on the entire marine food web.
- **Sea Level Rise and Coastal Erosion:** Coastal ecosystems are directly threatened by sea level rise, which could result in the loss of vital habitats like seagrass beds, salt marshes, and mangroves. In addition to providing vital functions like carbon sequestration and water filtration, these habitats also guard against storm surges and

coastal erosion. These habitats' tendency to migrate landward as sea levels rise is frequently impeded by human infrastructure, which results in habitat loss and a decline in biodiversity. Furthermore, the frequency of storms and storm surges has increased, which worsens coastal erosion and further deteriorates these essential ecosystems.

3.3.4. Impacts on Species and Ecosystem Interactions

In addition to having an impact on individual species, climate change is also changing how species interact with one another and how ecosystems are structured. These alterations may have a domino effect, upsetting pollination networks, food webs, and ecosystem services.

- **Changes in Species Distribution and Migration Patterns:** Many species are changing their range and migratory patterns in response to changes in habitat quality and temperature. In search of appropriate temperatures, for instance, a large number of marine species are migrating poleward or to deeper waters. The survival and reproductive success of these species may be impacted by this shift, which may interrupt well-established migratory routes and breeding grounds. Many species' chances of surviving are also decreased by the fact that shifting conditions may cause inconsistencies in the timing of migration, breeding, and the availability of food sources.
- **Altered Food Web Dynamics:** Both in freshwater and marine environments, the structure and dynamics of food webs are changing due to climate change. For instance, in freshwater lakes, the base of the food web, phytoplankton communities, can be impacted by variations in temperature and nutrient availability. These modifications may have an impact on herbivores, predators, and top predators as they move up the food chain. Changes in phytoplankton communities and species migration can cause disturbances to food web interactions in marine systems, resulting in changes to the abundance and composition of species. The health of marine ecosystems and the availability of food for commercial fisheries may be significantly impacted by this.

3.3.5. Adaptation and Mitigation Strategies

Adaptation and mitigation strategies must be used in tandem to address the effects of climate change on aquatic biodiversity. To prevent future warming, these tactics seek to decrease greenhouse gas emissions and increase the resilience of ecosystems.

- **Protected Areas and Habitat Restoration:** The creation and efficient administration of protected areas is an essential tactic in the preservation of aquatic biodiversity. These regions can aid in the restoration of damaged habitats and serve as havens for species impacted by climate change. Furthermore, restoring wetlands, mangroves, and seagrass beds are examples of habitat restoration projects that can strengthen an ecosystem's resistance to the effects of climate change. These ecosystems are essential for storing carbon, protecting against storm surges, and offering habitat to a wide variety of species.
- **Sustainable Fisheries Management:** Maintaining fish populations and the means of subsistence for communities that depend on fishing requires adapting fisheries management techniques to changing circumstances. This entails putting policies in place like flexible quota systems, fishing prohibitions during mating seasons, and habitat preservation measures. Furthermore, encouraging sustainable aquaculture methods can help conserve biodiversity and lessen the strain on wild fish populations.
- **Reducing Greenhouse Gas Emissions:** Limiting the effects of climate change on aquatic biodiversity requires reducing greenhouse gas emissions, which can only be achieved through mitigation. To cut CO_2 and methane emissions, this entails switching to renewable energy sources, enhancing energy efficiency, and encouraging sustainable land-use practices. Achieving global emission reduction targets requires international cooperation and policies like the Paris Agreement that support climate action (Chapman, 1992; Fischer and Lindenmayer, 2007; Ashok, 2015).

3.4. INVASIVE SPECIES

The biodiversity of Earth has been greatly impacted by humans, and many of these effects are widespread. The two biggest threats to species loss in lakes and streams are changes in land use and exotic invasive species. By long-distance trade, purposeful introduction of certain species, and hitchhiking of others, humans have proven especially successful in eradicating biogeographic barriers. Many freshwater species have been relocated as a result. Many freshwater species have become invasive, and some have had detrimental effects on the environment and the economy, even though the majority of introduced species fail to establish and spread. Terrestrial and marine ecosystems are less diverse per unit area than freshwater ecosystems. Additionally, freshwater ecosystems actively participate in the cycling of nutrients and water, which results in products and services for human societies. Invasive species from a broad range of taxonomic groups have also significantly altered freshwater ecosystems. Thus, it is essential to comprehend the mechanisms underlying the entry, dissemination, and eventual effects of invasive species in these environments.

Many significant syntheses that are pertinent to the discussion of invasive species in aquatic environments include reviews of the following topics: functional homogenization, invader traits and impacts, assessments of invasion potential, epidemiology and control, and climate change. This paper aims to investigate and examine the little-researched relationships between aquatic invasive species (AIS) and anthropogenic and natural environmental changes in freshwater ecosystems, as well as the difficulties these interactions pose for humans in the 21st century. Specifically, we look at the effects of invasive species, how humans help aquatic invasive species spread faster than they would otherwise, the traits of successful invaders, how vulnerable communities are to invasion and how this can lead to an "invasional meltdown," the role that microbial invasions and their hosts play in the interaction between invasion and other human impacts, and living with invasive species as a permanent fixture of the environment.

3.4.1. Impacts of Invasive Species

Numerous known and possible effects on community structure and ecosystem function have been linked to biological invasions of freshwater ecosystems. As demonstrated by various aquatic organisms and ecosystem types, the establishment rates of AIS are likewise high. The current rate of invasions far outpaces that of the past events that occurred over geologic time scales, although biotic exchanges and paleo-invasions contributed to biodiversity as we know it. The effects of invasions are undoubtedly cause for concern given the high level of biodiversity and the susceptibility of freshwater communities to biotic exchange. Even if an invasion is a delayed reaction to earlier climatic events or appears to have little impact on the overall function of the ecosystem, the change itself is frequently viewed as undesirable.

Many AIS have negatively impacted the environment and human interests, even though some have had negligible or no effects at all. Since disturbances to these levels have other unanticipated consequences, we concentrate here on the effects of AIS on communities and ecosystems. For instance, invasive species can quickly alter entire ecosystems when they act as ecosystem engineers. The transition of the Hudson River from a pelagic-based food web to a littoral-based food web after the zebra mussel invasion serves as an example of this scenario. In areas like Lake Erie and the Hudson River, where zebra mussels are highly dense, the majority of the algae can be filtered out of the water column, leaving little food for pelagic zooplankton and increasing light for vascular plants. The increased growth in turn strengthens the structure of their habitats among many other consequences. The impact of invasive plants in Australia, where high densities of African Poaceae changed stream hydrology by trapping sediments in the channel, is another illustration of how an ecosystem can change. Nutrient dynamics are then greatly impacted by changes in the hydrology. In general, the effects of AIS on ecosystems remain largely unexplored and present a challenge for future studies.

It is commonly known that invasive species alter freshwater food webs. A lot of focus has been on investigating the effects of fish on communities since game fish have been widely introduced by fisheries agencies and predatory fish frequently regulate the community structure of lakes and streams. Additionally, invasive plants have an impact on communities. For instance, *Myriophyllum spicatum*, the Eurasian water milfoil, creates dense surface mats that stifle light and diminish the variety of native aquatic plants. The reduced plant community provides macroinvertebrates and higher-order consumers with a less abundant food source. Similarly, studies carried out in ponds demonstrated that in contrast to macrophyte multi-species treatments, dense monotypic stands of hydrilla (*Hydrilla verticillata*) altered the taxonomic composition of invertebrate assemblages; however, the invertebrate taxonomic richness of both treatments was the same. With changes in the phylogenetic structure of bacterial communities and metabolic profiles, as hydrilla density rises, hydrilla probably has an effect on microbes as well.

3.4.2. Invasibility: Susceptibility of Aquatic Communities to Invasion

The inability of non-native species to colonize a new ecosystem is frequently explained by biotic resistance in conjunction with abiotic environmental variables. Elton (1958) put forth the biotic resistance hypothesis, which holds that the primary factor influencing the success of an invasion is the species diversity of the recipient community. Basically, competition, predation, and parasitism are the reasons behind a native community's resistance to the invasion of non-native species. In general, the more native species there are, the more intense these negative interactions become. Therefore, diversity is frequently used as a stand-in for native resistance in studies examining the function of biotic resistance. A meta-analysis, however, revealed that while this type of biotic resistance hinders invader establishment and performance, it is not a perfect barrier against plant invasions.

In freshwater ecosystems, the function of biotic resistance to invasions has been extensively studied. Since native species become more diverse and densely populated, macrophytes' growth is inhibited, demonstrating the significance of competition in biotic resistance. Beavers' effects on one species of invasive macrophyte in wetlands provide evidence that herbivory also contributes to a decrease in the abundance of invasive macrophytes. Invasive species can also be impacted by native predators. In rivers, zebra mussels are more likely to die due to predators like crabs, while in lakes, the abundance of non-native crayfish is reduced by native fish. Apple snail invasions in rivers are resisted by fish, turtles, and crayfish, among other predators.

Parasitism also lowers invasion success, though it is less studied. In a Canadian river, for instance, an oomycete parasite decreased the number of an invasive amphipod species and permitted coexistence with a different native species. The majority of these studies bolster the idea that biotic resistance plays a part in keeping invasive species in check and even driving them out of their new environments. If an invader is prevented from reaching the critical density required to sustain a viable population, this decline might be significant. Furthermore, by raising the threshold for community collapse, a higher native diversity can strengthen a freshwater community's resistance to invasion.

The spatial scale affects biotic resistance. At coarser scales (plant patches or entire ecosystems), invasion success is typically positively correlated with diversity but negatively correlated with native diversity on finer scales. The "invasion paradox" refers to this pattern, which can be explained by the greater significance of biotic interactions (e.g., competition, and predation) at small sizes as well as abiotic factors (e.g., coarse scales, such as habitat heterogeneity and resource availability. Positive relationships between invasive establishment and native macrophyte diversity have been demonstrated at coarse scales in both natural lakes and reservoirs, as an example from field studies of aquatic communities. However, studies carried out at finer spatial scales revealed the opposite pattern, with invasive species growth rates negatively correlated with native species diversity. Aquatic ecologists should conduct more studies that cover multiple scales at the same time because different factors probably affect invasions at different spatial scales.

3.4.2.1. Invasional Meltdown

"Community-level phenomenon in which the net effect of facilitations would lead to an increasing rate of establishment of introduced species and/or an accelerating impact" is the definition of invasive meltdown (IM). The term "meltdown" refers to the idea that invasions will eventually have a positive feedback loop that increases the total number of successful invaders. The IM hypothesis predicts an increase in species richness unless native species are replaced by invaders. Thus, it would seem that the biotic resistance hypothesis—which links increased species richness to improved resistance to invasion—and the IM hypothesis are at odds. IM was one of the few theories with strong empirical backing in a recent review of the tenets of invasion biology. Nevertheless, descriptive case studies of facilitation between non-native species make up the majority of this support. The IM prediction, however, that prior invasions lead to higher rates of subsequent invasions and the impact of invasive species, has not yet been proved beyond a reasonable doubt. Simberloff's conclusion holds true for freshwater ecosystems as well. A genuine meltdown due to increasing invasion rates and/or impacts over time is not supported by the data, despite the fact that numerous observational and experimental studies have shown facilitation between non-native species.

The Laurentian Great Lakes, with a well-documented history of multiple invasions and their effects, are among the best cases for the potential occurrence of IM in freshwater ecosystems. Results consistent with the IM model showed that invasions increased over time and that both direct and indirect facilitation between non-native species was more common than negative interactions in these lakes. However, the primary basis for this conclusion is correlative observations from field surveys, which could be skewed by additional variables. For instance, over time, there have been more opportunities for new invasions due to increased propagule pressure resulting from increased vector traffic. Indeed, De Vanna et al. (2011) have cast doubt on the conclusion regarding the existence of a true IM in the Great Lakes. They claimed that a better explanation for what is happening in those lakes is the indiscriminate facilitation of both native and non-native species by the ecosystem engineer Dreissena (zebra mussels and quagga mussels).

Future research to demonstrate IM would benefit from carefully designed experiments involving multiple species. Such experiments require a test of two questions. First, are invasive species facilitating each other? And, second, does facilitation between non-natives propagate in such a way as to impact the community level and enhance invasion rates? These kinds of experiments require careful attention to controls and understanding what exactly are the response variables (e.g., invasion rate) and treatments (e.g., presence or absence of prior invader). To conclude about a true meltdown, both the individual performance of non-natives and their impacts would have to be greater when they interact than when they are alone.

3.4.2.2. The Role of Reservoirs

Nowadays, reservoirs make up a sizable share of freshwater ecosystems, and as hydroelectric development picks up speed in South America and Asia, so does their representation. Invaders use reservoirs as stepping stones to establish themselves in newly created watersheds. The purpose of reservoirs is to serve as a source of propagules near uninvaded water bodies while also changing the habitat and increasing its susceptibility to invasions. According to recent research, reservoirs are in fact more prone to invasion than natural systems. A comparison of reservoirs and natural lakes in northern Wisconsin later confirmed the high invasibility of reservoirs, which was initially observed in dammed mountain streams. Many AIS, including rusty crayfish, rainbow smelt (*Osmerus mordax*), zebra mussels, Eurasian water-milfoil, and spiny water fleas (*Bythotrephes longimanus*), are much more likely to occur in reservoirs than in natural lentic systems, according to the latter study. This effect was robust even when potentially confounding environmental factors were taken into account. The present study has unambiguously demonstrated the function of reservoirs in enhancing connectivity by diminishing the separation from other aquatic ecosystems and expediting the proliferation of invasive species. However, it is still unclear how much of a role biotic resistance, disturbance, or habitat degradation play in reservoirs.

Reservoirs typically have fewer species and are young, early successional systems, making them less resistant to biotic invasion. A study conducted in Mediterranean reservoirs showed that native fish were primarily impacted by abiotic factors, while the richness of non-native fish was negatively correlated with the richness of the native fish community, highlighting the significance of biotic resistance. The impact of biotic resistance on the invasiveness of artificial systems has also been illustrated through the use of zooplankton. For example, seeding newly built tanks with native species eggs reduced the establishment success of invasive zooplankton one year later, compared to tanks without a native assemblage, in a mesocosm experiment (interpreted as a model of a reservoir in its earlier phase).

Abiotic factors may also be crucial in invasive species' colonization of reservoirs. These abiotic variables include deteriorating water quality and more fluctuations in water levels brought on by hydropower and flood control regulations, which decrease macrophytes and produce less-than-ideal habitats for native species. The riffle-pool structure that distinguishes the streams that mainstem reservoirs impound is absent from them. Additionally, because dams limit fish migration opportunities, reservoirs lack the flood pulses that large rivers and their floodplains are known for. A great deal of shallow water reservoirs are highly vulnerable to eutrophication and sediment resuspension. Invasive species are more likely to reside in habitats that have been chemically and physically degraded. Reservoirs foster an environment that is more favorable for invasive species that can withstand high temperatures as well as wildlife that can survive in damaged ecosystems. As a result, reservoirs can facilitate the establishment of invasive species through habitat degradation and permit the selection of taxa with functional characteristics distinct from those of the native assemblage.

Freshwater ecosystems are significantly impacted by the wide-ranging environmental changes brought about by the construction of reservoirs. Because reservoirs remove biogeographic barriers and alter habitat, they play a major role in the global homogenization of freshwater faunas as cosmopolitan lake taxa replace endemic riverine species. Reductions in native species in reservoirs due to invasion have been documented for several assemblages. In light of this, it's critical to comprehend how changes in taxonomy and homogenization translate into adjustments in the functional groupings' relative abundance and changes at the ecosystem level. There is a functional difference between invaders and native assemblages; they are not a randomly assembled group of species. Further disruption of the food web, a decline in resistance to anthropogenic disturbance, and an increase in invasiveness may ensue from the ensuing functional homogenization. To comprehend the function of reservoirs in the functional homogenization of freshwater faunas, natural systems with similar sizes and locations are required (Chapman, 1992; Fischer and Lindenmayer, 2007; Ashok, 2015).

3.5. OVERFISHING AND UNSUSTAINABLE FISHING PRACTICES

The global aquatic biodiversity is seriously threatened by overfishing and irresponsible fishing methods, which has a substantial impact on food security, marine ecosystem health, and the livelihoods of millions of people worldwide. When fish are taken more quickly than they can reproduce, this is known as overfishing, and it depletes fish stocks and disturbs marine ecosystems. A variety of actions that endanger marine life and habitats are included in unsustainable fishing practices, such as the use of harmful fishing gear, bycatch, and illegal, unreported, and unregulated (IUU) fishing. In order to preserve the long-term health of marine ecosystems, this paper examines the causes, effects, and remedies of overfishing and unsustainable fishing methods. It emphasizes the critical need for sustainable management and conservation initiatives.

3.5.1. Overfishing

The excessive and unsustainable removal of fish and other aquatic species from our freshwater and oceanic environments is referred to as overfishing. Put more simply, it happens when fish populations decline as a result of more fish being caught than can spawn offspring. The unsustainable exploitation of marine resources at a rate higher than their natural rate of replenishment is known as overfishing. In addition to fish, other marine animals harmed by this destructive practice include dolphins, sharks, and turtles, which frequently end up as accidental bycatch in fishing operations (Figure 3.5).

Figure 3.5. *Overfishing is the excessive and unsustainable extraction of fish and other aquatic species from oceans.*

Source: By C. Ortiz Rojas – http://www.photolib.noaa.gov/htmls/fish2172.htm, Public Domain, https://commons.wikimedia.org/w/index.php?curid=5920890.

3.5.2. Causes of Overfishing and Unsustainable Fishing Practices

Overfishing and unsustainable fishing methods are caused by many variables, from weak governance and enforcement to social and economic forces.

- **Economic and Social Drivers:** The last few decades have seen a sharp rise in the demand for seafood due to factors like population growth, rising incomes, and a growing understanding of the health benefits of eating fish. The demand for fish has resulted in significant strain on fish stocks, especially on highly sought-after species. Furthermore, fishing is a significant source of income for millions of people worldwide, especially in coastal areas. Overfishing and the use of unsustainable practices can result from the economic incentives to catch as many fish as possible, which are frequently caused by poverty and a lack of alternative livelihoods.

- **Advances in Fishing Technology:** The efficiency and capacity of fishing operations have increased due to technological advancements in fishing gear and vessels, which have allowed fishermen to catch more fish over larger areas and at greater depths. With the help of advanced fish-finding and navigational technology, fishing vessels of today can locate and catch fish more precisely. Fisheries productivity has increased as a result of these developments, but they have also led to overfishing and habitat degradation in marine environments.
- **Inadequate Management and Enforcement:** Ensuring the sustainability of fish stocks requires effective management of fisheries. On the other hand, insufficient management and enforcement of regulations plague numerous fisheries globally. Weak governance structures, a lack of scientific data on fish populations, and a lack of funding for enforcement and monitoring can all contribute to this. Political and commercial interests occasionally work against the implementation of sustainable fishing methods, which can result in overfishing and habitat degradation.

3.5.3. Ecological Consequences of Overfishing and Unsustainable Fishing Practices

The ecological impacts of overfishing and unsustainable fishing practices are profound and far-reaching, affecting marine ecosystems, biodiversity, and the resilience of ocean environments.

- **Depletion of Fish Stocks:** Fish stock depletion is the most obvious effect of overfishing. Due to overfishing, the populations of many commercially significant fish species, including Atlantic cod, bluefin tuna, and several shark species, have drastically declined. When fish populations are unable to reproduce quickly enough to maintain their numbers, depletion can result in the collapse of fisheries. The demise of these species may have a ripple effect on other species and the ecosystem's general health throughout the marine food web.
- **Disruption of Marine Food Webs:** Because overfishing removes important species that are vital to the ecosystem, it can upset marine food webs. For instance, the eradication of top predators like sharks and big predatory fish may cause prey populations to rise, which may have detrimental effects on species and habitats lower in the food chain. The structure and function of marine ecosystems may change as a result of this disturbance, making them less resilient to alterations in the environment and other stresses.

- **Habitat Destruction:** Marine habitats are seriously harmed by destructive fishing techniques like dynamite and bottom trawling. Coral reefs, sponge gardens, and seagrass beds are examples of benthic habitats that can be destroyed by bottom trawling, a practice that involves dragging large nets along the seafloor. Numerous marine species depend on these habitats for their survival because they offer food, cover, and breeding grounds. Ecosystem services may deteriorate and biodiversity may disappear as a result of the destruction of these habitats.
- **Bycatch and Its Impacts:** A serious problem with many fishing methods is bycatch, which causes accidental capture and death of non-target species like young fish, marine mammals, sea turtles, and seabirds. Significant ecological effects can be caused by bycatch, especially for species that are already endangered or vulnerable. For instance, the decline of several sea turtle populations has been attributed to the bycatch of sea turtles in longline and shrimp trawl fisheries. Furthermore, the capture of juvenile fish worsens the overfishing issue by lowering the future reproductive potential of fish stocks (Figure 3.6).

Figure 3.6. *Bycatch.*

Source: By National Observer Program, NMFS/SEFSC – http://www.st.nmfs.noaa.gov/observer-home/about-nop/activities/index, Public Domain, https://commons.wikimedia.org/w/index.php?curid=26033905.

3.5.4. Economic and Social Consequences

The economic and social impacts of overfishing and unsustainable fishing practices are significant, affecting the livelihoods of fishing communities, food security, and the global economy.

- **Impact on Fishing Communities:** Fishing is a major source of income for millions of people worldwide. Fishing communities may suffer greatly as a result of declining fish stocks and deteriorating marine habitats, which could result in reduced income, unstable food supplies, and worsening poverty. Due to their frequent lack of resources and alternate sources of income, small-scale and artisanal fishermen are especially vulnerable to the decline in fish stocks. The difficulties these communities face may be made worse by the disappearance of customary fishing grounds and the competition from large industrial fishing fleets.
- **Food Security:** For billions of people worldwide, especially those in developing nations, fish is a vital source of protein and other essential nutrients. Overfishing-related fish stock declines can jeopardize food security by lowering the availability and cost of fish for human consumption. This may have detrimental effects on diet and health, especially for populations that are more susceptible and depend primarily on fish for protein. Furthermore, declining fish stocks may put more strain on remaining food supplies, which would worsen the situation of food insecurity.
- **The Economic Losses:** Overfishing and irresponsible fishing methods result in significant financial losses. For fishing communities and nations, the collapse of their fisheries may result in a loss of income, jobs, and economic stability. Furthermore, there may be substantial costs associated with controlling and reducing the effects of overfishing, such as restoring habitat, reducing bycatch, and enforcing laws. For the fishing industry to remain profitable and for the people who depend on it to have a place to live, long-term fisheries sustainability is essential (Jackson et al., 2001).

REFERENCES

1. Ahmed, M. H., El Leithy, B. M., Donia, N. S., et al. (2006). Monitoring the historical changes of Manzala Lake ecosystems during the last three decades using multidate satellite images. *EcolLaw*, 120–133.
2. Ashok, K. V. (2015). Values and need of biodiversity conservation. *Bioherald: An International Journal of Biodiversity and Environment, 5*(1, 2), 77–79.
3. Ashok, K. V. (2016). Biodiversity: Its different levels and values. *International Journal on Environmental Sciences, 7*(2), 143–145.
4. Attrill, M. J., & Power, M. (2002). Climatic influence on a marine fish assemblage. *Nature, 417*(6886), 275–278. https://doi.org/10.1038/417275a.
5. Babcock Hollowed, A., Hare, S. R., & Wooster, W. S. (2001). Pacific basin climate variability and patterns of northeast Pacific marine fish production. *Progress in Oceanography, 49*, 257–282.
6. Benton, T. G., Vickery, J. A., & Wilson, J. D. (2003). Farmland biodiversity: Is habitat heterogeneity the key? *Trends in Ecology & Evolution, 18*(4), 182–188.
7. Bishai, H. M., & Khalil, M. T. (1997). *Freshwater Fishes of Egypt*. National Biodiversity Unit.
8. Boulenger, G. A. (1907). *Zoology of Egypt: The Fish of the Nile*. Hugh Press.
9. Brooks, T. M., Pimm, S. L., & Oyugi, J. O. (1999). Time lag between deforestation and bird extinction in tropical forest fragments. *Conservation Biology, 13*, 1140–1150.
10. Bryant, D., Burke, L., McManus, J., & Spalding, M. (1998). *Reefs at Risk: A Map-Based Indicator of Threats to the World's Coral Reefs*. World Resources Institute.
11. Chapman, D. (1992). *Water Quality Assessments: A Guide to the Use of Biota, Sediments and Water in Environmental Monitoring*.
12. Church, J. A., Gregory, J. M., Huybrechts, P., Kuhn, M., Lambeck, K., Nhuan, M. T., Qin, D., & Woodworth, P. L. (2001). Changes in sea level. In *Climate Change 2001* (pp. 1–20).
13. Doligez, B., & Boulinier, T. (2008). Habitat selection and habitat suitability preferences. In *Encyclopedia of Ecology* (pp. 1810–1830).
14. Fahrig, L. (2003). Effects of habitat fragmentation on biodiversity. *Annual Review of Ecology, Evolution, and Systematics, 34*, 487–515.
15. Fahrig, L. (2019). Habitat fragmentation: A long and tangled tale. *Global Ecology and Biogeography, 28*(1), 33–41.

16. Fischer, J., & Lindenmayer, D. B. (2007). Landscape modification and habitat fragmentation: A synthesis. *Global Ecology and Biogeography, 16*, 265–280.
17. Fishar, M. R. A., & Williams, W. P. (2006). A feasibility study to monitor the macroinvertebrate diversity of the River Nile using three sampling methods. *Hydrobiologia, 556*, 137–147.
18. Fishar, M. R. A., Kamel, E. G., & Wissa, J. B. (2003). Effect of discharged water from Shoubra El-Khima electric power station into the River Nile (Egypt) on the aquatic annelids. *Journal of the Egyptian Academy for Environmental Development, 4*, 83–100.
19. Flaspohler, D. J., Temple, S. A., & Rosenfield, R. N. (2001). Effects of forest edges on ovenbird demography in a managed forest landscape. *Conservation Biology, 15*, 173–183.
20. Gattuso, J. P., Frankignoulle, M., Bourge, I., Romaine, S., & Buddemeier, R. W. (1998). Effect of calcium carbonate saturation of seawater on coral calcification. *Global Planetary Change, 18*, 37–47.
21. Guirado, M., Pino, J., & Roda, F. (2006). Understorey plant species richness and composition in metropolitan forest archipelagos: Effect of forest size, adjacent land use, and distance to the edge. *Global Ecology and Biogeography, 15*, 50–62.
22. Harrison, P. A., Berry, P. M., Simpson, G., et al. (2014). Linkages between biodiversity attributes and ecosystem services: A systematic review. *Ecosystem Services, 9*, 191–203.
23. Hassan, M. M. (2008). *Ecological Studies on Zooplankton and Macrobenthos of Lake Edku, Egypt.*
24. Intergovernmental Panel on Climate Change (IPCC). (2001). *Summary for Policymakers: A Report of Working Group I of the Intergovernmental Panel on Climate Change*. Cambridge University Press.
25. Ismail, A., & Hettiarachchi, H. (2017). Environmental damage caused by wastewater discharge into Lake Manzala in Egypt. *American Journal of Bioscience and Bioengineering, 5*(6), 141–150.
26. Jackson, J. B. C., Kirby, M. X., Berger, W. H., Bjorndal, K. A., Botsford, L. W., Bourque, B. J., Bradbury, R. H., Cooke, R., Erlandson, J., Estes, J. A., Hughes, T. P., Kidwell, S., Lange, C. B., Lenihan, H. S., Pandolfi, J. M., Peterson, C. H., Steneck, R. S., & Tegner, M. J. (2001). Historical overfishing and the recent collapse of coastal ecosystems. *Science, 293*, 629–638.

27. Khan, R. A. (2003). *Faunal Diversity of Zooplankton in Freshwater Wetlands of Southeastern West Bengal*. Zoological Survey.
28. Klyashtorin, L. B. (1998). Long-term climate change and main commercial fish production in the Atlantic and Pacific. *Fisheries Research, 37*, 115–125.
29. Kumar, A., & Verma, A. K. (2017). Biodiversity loss and its ecological impact in India. *International Journal on Biological Sciences, 8*(2), 156–160.
30. Mace, G. M., Norris, K., & Fitter, A. H. (2012). Biodiversity and ecosystem services: A multilayered relationship. *Trends in Ecology & Evolution, 27*(1), 19–26.
31. Malmqvist, B., & Rundle, S. (2002). Threats to the running water ecosystems of the world. *Environmental Conservation, 29*(2), 134–153.
32. Rahel, F. J. (2002). Homogenization of freshwater faunas. *Annual Review of Ecology and Systematics, 33*, 291–315.
33. Rashad, H. M., & Abdel Azeem, A. M. (2010). Lake Manzala, Egypt: A bibliography. *Assiut University Journal of Botany, 39*(1), 253–289.
34. Revenga, C., Campbell, I., Abell, R., et al. (2005). Prospects for monitoring freshwater ecosystems towards the 2010 targets. *Philosophical Transactions of the Royal Society B, 360*(1454), 397–413.
35. Seddon, N., Mace, G. M., Naeem, S., et al. (2016). Biodiversity in the Anthropocene: Prospects and policy. *Proceedings of the Royal Society B, 283*(1844), 2016–2094.
36. Vorosmarty, C. J., McIntyre, P. B., Gessner, M. O., Dudgeon, D., Prusevich, A., Green, P., Glidden, S. E., Bunn, S. E., Sullivan, C. A., Reidy Liermann, C., & Davies, P. M. (2010). Global threats to human water security and river biodiversity. *Nature, 467*, 555–561.
37. Xie, Z., Xiao, H., Tang, X., et al. (2008). Interactions between red tide microalgae and herbivorous zooplankton: Effects of two bloom-forming species on the rotifer *Brachionus plicatilis* (O. F. Muller). *Hydrobiologia, 600*(1), 237–245.

Chapter **4**

Climate Change and Aquatic Biodiversity

Contents

INTRODUCTION

Climate change significantly impacts aquatic biodiversity, posing a profound threat to marine and freshwater ecosystems. As global temperatures rise, the intricate balance of these ecosystems is disrupted, leading to widespread consequences for aquatic life and the human communities that depend on them.

One of the most evident effects of climate change on aquatic biodiversity is the warming of water bodies. Increased temperatures alter the habitat ranges of many species, forcing them to migrate to cooler areas, often leading to overcrowding and competition for resources in these new regions. Species unable to migrate or adapt face a heightened risk of extinction. For instance, coral reefs, which are highly sensitive to temperature changes, experience bleaching events that devastate their populations and the myriad species that rely on them for habitat. Additionally, climate change contributes to ocean acidification, caused by the increased absorption of carbon dioxide by seawater. This process lowers the pH levels of the ocean, adversely affecting calcifying organisms such as corals, mollusks, and certain plankton species. The weakening of these organisms has cascading effects throughout the food web, disrupting predator-prey relationships and altering ecosystem dynamics.

Freshwater systems are not immune to these changes. Altered precipitation patterns and increased evaporation rates lead to changes in water availability, affecting the habitats of freshwater species. The increased frequency and intensity of extreme weather events, such as storms and floods, can also result in habitat destruction and increased pollution, further stressing aquatic ecosystems.

Human communities that depend on aquatic ecosystems for food, economic activities, and cultural practices are directly impacted by these changes. The decline in fish populations, for example, threatens food security and livelihoods, particularly in regions heavily reliant on fishing.

In conclusion, climate change poses a multifaceted threat to aquatic biodiversity, disrupting habitats, altering species distributions, and threatening the delicate balance of marine and freshwater ecosystems. Addressing these challenges requires global cooperation and concerted efforts to mitigate climate change and protect the resilience of aquatic ecosystems.

4.1. CLIMATE CHANGE AND ITS IMPACTS ON AQUATIC ECOSYSTEM

Human changes to ecosystems have a significant impact on the diversity of life on Earth (Baillie et al., 2004). Climate change continuously modifies biodiversity. These days, human activity is adding a new kind of climate variability to this natural variability, which threatens to hasten the loss of biodiversity that is already under stress from other human stresses. Around 70% of the earth's surface is submerged under water. Aquatic ecosystem abundance and distribution are already shifting due to climate change. The currents that travel across the surface of the earth will alter in response to even small variations in the temperature of the water. An aquatic ecosystem can be broadly classified into two types: freshwater ecosystems and marine ecosystems.

4.1.1. Climate Change Within the Ocean

The increase in greenhouse gases within the earth's atmosphere is set to change three fundamental variables:

- **Reduced Total Carbonate Alkalinity:** Seawater's overall carbonate alkalinity will drop as atmospheric CO_2 levels rise (Gattuso et al., 1998; Kleypas et al., 1999). It is anticipated that this specific variable will significantly alter the ocean's acidity and carbonate ion pool. By 2050, the aragonite saturation state in the tropics will have decreased by 30% due to a doubling of carbon dioxide concentrations in the atmosphere.
- **Increased Sea Level:** Sea level variations have had a significant effect on the distribution and abundance of terrestrial and marine biodiversity. As global temperature rises due to climate change, sea levels will rise. This happens as a result of glacier melting, alterations in ice sheet distribution, and thermal expansion of ocean water. Sea level rise is predicted to be between 9 and 29 cm over the next 40 years, or between 28 and 29 cm by 2090 (Church et al., 2001; IPPC, 2001). Sea level rise, according to Nichols et al. (1999), may result in the loss of up to 22% of the world's coastal west lands by the year 2080. By the end of the 21st century, this percentage is probably going to increase to 70% of the world's coastal wetlands lost due to human activity alone.
- **Sea Temperature Increase:** The notable rise in temperature has not been dispersed equally. The marine environment is influenced by sea temperature. Changes in global temperatures can have a direct impact on the rates and directions of ocean water movement because of their direct effects on the density of seawater.

4.1.2. Deep Sea Biodiversity

The deep sea is becoming more widely acknowledged as a significant biodiversity reservoir. More species are thought to exist in the deep seabed than in any other marine habitat. Pollution, shipping, military activity, and climate change are some of the factors threatening the marine ecosystem and biodiversity, but fishing is currently the biggest threat.

Bottom trawling poses the biggest risk to deep-sea biodiversity. There is greater harm done to seamounts and the cold-water corals they support by this kind of high seas fishing. Numerous commercial bottom-dwelling fish species call these habitats home.

4.1.3. Fish Populations

Coastal fishing provides essential livelihoods for numerous millions of individuals worldwide. Various researchers are emphasizing the significant overfishing of these fisheries and the subsequent decrease in fish populations as the primary cause of ecological transformation during the last two centuries (Jackson et al., 2001). New findings suggest that oceanic and climate fluctuations could have a leading influence on fish populations (Klyashtorein, 1998; Babcock Hollowed et al., 2001; Attrill and Power, 2002). The relationship between climate fluctuations and fish populations might be intricate. In certain instances, minor shifts could impact conditions and bring about critical alterations in the fish species' life cycles. Generally, the most extensive impacts of climate change are observed in primary and secondary production within marine ecosystems.

4.1.4. Coral Reef

Scleractinian corals form a framework that dominates ecosystems in tropical intertidal and subtidal areas. Over the past 20 years, they have experienced significant changes, many of which have been attributed to stressors such as climate change (Bryant et al., 1998). Large populations of fish, birds, turtles, and marine mammals are supported by these ecosystems' rich and complex food chains, despite the absence of external nutrients. Carbonate reef formation becomes more challenging at higher latitudes due to a decrease in light, temperature, and seawater's carbonate alkalinity in a pole-ward direction. Climate change has already had a significant impact on coral reefs. Over the past 30 years, there has been a noticeable increase in major coral reef disturbances, which are unquestionably associated with times when sea temperatures have risen above average.

4.1.4.1. Freshwater Ecosystem

The following broad categories include the threats to freshwater fauna: pollution, invasive species spread, habitat loss and degradation, hydrological modifications, and nutrient enrichment. The risks that already exist are compounded by a changing climate and rising UV radiation levels. Rapid changes in land use, habitat modification, and climate change are thought to pose a particularly significant threat to aquatic ecosystems.

- **Importance of Freshwater Ecosystems:** Corals that lose their cells quickly experience coral bleaching. Colonies that have been bleached turn from brown to white, frequently revealing stunning host pigments. Reef-building coral communities are likely to have significant effects on marine biodiversity if these vital symbionts are lost. Reef-building corals that lose them may face mortality rates of up to 90%. Corals provide the vital framework that many other species rely on for survival. Coral communities that support fish that rely on them for food, shelter, or settlement care may undergo significant changes, which could have a significant impact on marine biodiversity. Corals provide the vital framework that many other species rely on for survival. Fish that rely on corals for food, shelter, or settlement remedies run the risk of going extinct or seeing drastic fluctuations in population. Numerous other organisms are at risk as well.

4.1.5. Climate Change and the Hydrologic Cycle

It is challenging to forecast changes in the hydrologic cycle because freshwater ecosystems are inherently sensitive to them. A warmer climate will cause more water to evaporate from water surfaces and more transpiration from plants, intensifying the water cycle. Lake ecosystems will be directly impacted by future climate change due to increased temperatures and modifications to the hydrologic cycle.

4.1.6. Biological Impacts

The biodiversity of rivers and streams is negatively impacted by rapid climate change in many ways. Many taxonomic levels may experience extinction as a result of climate change. At the species level, ecologically highly restricted geographic ranges make species more susceptible to extinction worldwide. This is especially true for fish, which are at risk of going extinct worldwide when there are regional variations in the proportionate occurrence of specialized species.

The quantity and quality of aquatic ecosystems are currently seriously threatened by human demands. Future climate change and the various human stressors on aquatic ecosystems will interact. The current shifts in biodiversity are still primarily caused by habitat modification caused by humans. The earth's biodiversity is susceptible to even minute variations in its climate. Every man has multiple needs, but only when these needs are satisfied can we create a sustainable biodiversity.

There is nobody that we can directly hold accountable for the current national food crisis. Hunger, poverty, corruption, social evils, etc., are all diseases that belong to a single, massive, vicious cycle. The time has come for grassroots activism and public dialogue. Good governance and political will must be translated into concrete reform initiatives. To assist the poor in our nation, we must step forward and cooperate.

The planning and management of biological resources to ensure their widespread use and steady supply while preserving their quality, value, and diversity is known as biodiversity conservation. With careful planning and management, extinction must be prevented. It is imperative that we increase our efforts to slow down the loss of biodiversity and put long-term safeguards in place to protect this priceless resource. We must consider the investments we will make in the future, including whether to fund the distribution of humankind or its future. Enforcing the law alone cannot lead to the conservation of biodiversity. We must have an internal love for the earth and all its inhabitants. Biodiversity is life, is the motto for the International Year of Biodiversity. Life itself is biodiversity. Remember that biodiversity serves as nature's insurance against calamities.

4.2. EFFECT OF CLIMATE CHANGE ON AQUATIC ECOSYSTEM AND PRODUCTION OF FISHERIES

Climate change, food security, and their intricate interactions have gained international attention during the past few decades. Our quick population growth has ruined the natural world and contaminated the surrounding waters. Every day, the amount of greenhouse gases in the atmosphere rises. As a result, we are threatened by both global warming and other weather-related phenomena like floods, droughts, and cyclones. Changes in the earth's climate can happen all over the planet or just in one particular area. However, it has an impact on all ecosystems, including aquatic ones. Because the average temperature of the air and water is changing at the same time, aquatic organisms are particularly vulnerable to climate change. The primary causes of climate change in the aquatic system are rising sea levels and temperatures, altered monsoon patterns, extreme weather events, and water stress, which have an impact on aquatic life, including fish stocks, both directly and indirectly. It directly affects an organism's growth pattern and physiological behavior, which reduces the organism's ability to reproduce and ultimately results in death. Indirectly, it might change aquatic ecosystems' composition, productivity, structure, and function. In the end, all of these effects lead to a decline in fish production. It disrupts fish farmers' financial situation and severely damages their ability to make a living. This chapter will address how fish farmers' livelihoods and fish production are impacted by climate change and how it might be lessened with the right policies.

4.2.1. Causes of Climate Change

The terms "climate forcing" or "forcing mechanisms" refer to the variables that have the potential to alter the atmospheric system or climatic regime. Thus, two categories of forcing mechanisms exist: i.e., two types of forcing mechanisms: internal and external. Natural climatic system processes like thermohaline circulation, among others, are internal forcing mechanisms. There are two other categories of external forcing mechanisms: natural mechanisms such as variations in solar output, volcanic eruptions, etc., and anthropogenic mechanisms such as the emission of greenhouse gases and various other pollutants. The climate is changing as a result of all these processes. However, there is a wealth of evidence suggesting that human activity is the primary cause of this terrible condition. These are described below.

- **Fossil Fuel Burning:** Fossil fuel burning is one of the most important sources of climate change. As fossil fuels contain carbon for many years, they can release back CO_2 into the air. This is one of the direct causes of carbon emission in the air, which can cause all sorts of environmental problems including global warming.
- **Livestock Farming:** Through livestock farming, methane (CH_4) gas is emitted into the atmosphere. As we know, CH_4 is a greenhouse gas, so capable of trapping a huge amount of heat from the sun. In that way, they can contribute to global warming in broad sense.
- **Aerosols:** Aerosols also represent a big problem for the climate today. Aerosols are very small naturally occurring particles in the atmosphere. Previously the number of aerosols in the atmosphere was much less, but now the level is increasing.
- **Use of Fertilizers:** The use of fertilizers in both agricultural and aquacultural farmland can increase the availability of food sources to us. To meet the growing demand for food, the use of fertilizers has increased rapidly. Fertilizer contains a huge amount of nitrous oxide, which is responsible for a steady increase in the earth's surface temperature.

4.2.2. Changes in Aquatic Ecosystem due to Climate Change

4.2.2.1. Temperature

Since all aquatic life, including fish and aquatic invertebrates, is poikilothermic, its body temperature fluctuates with the temperature of its surroundings. They are therefore extremely sensitive to changes in the temperature in their living environment outside. These organisms will migrate to a location where their internal system permits them to regain internal homeostasis when the temperature of the external environment rises above their threshold. We call this process "behavioral thermoregulation." This will cause a swift migration to the water body's cooler regions. The aquatic life is able to migrate from semi-enclosed areas and shallow coastal waters to deeper, colder waters. It would take a conservative stance despite the detrimental effects of these phenomena, such as the destruction of coral reefs and an increase in ocean acidification. The migratory phenomenon alone has the ability to cut the tropical regions' maximum catch potential by 40%.

One of the main effects of climate change is temperature rise, which has a significant impact on hiring practices. Because of the negative effects of temperature enhancement, some stocks may develop an intolerance for sustainable fishing efforts because they perceive them as overfishing [6]. Fish that live in water with higher temperatures will not grow as large or reach their maximum size because the increased temperature will speed up their metabolism. Fish species extinction would be observed locally, particularly in freshwater and diadromous species. Terrestrial species exhibit a higher overall migration rate (15–37%) than marine species due to their greater potential for migration.

Due to coral reef bleaching brought on by the rising temperatures, reef fisheries would be severely impacted.

4.2.2.2. Primary Productivity

The availability of nutrients in the water body is determined by temperature and light levels, which in turn impacts primary productivity. Because of the variability of the climate, less precipitation would result in less runoff from the land, starving mangroves and wetlands and harming the local fisheries. In some other locations, heightened precipitation from extreme weather events, such as flooding, caused a sharp rise in the water body's nutrient level, leading to eutrophication and the washout of fertilizer, which allowed dangerous algal blooms, or red tides, to form in the water. The majority of small-scale fisheries are found in lower latitudes, which are most affected by climate change and see a decline in the sector's primary productivity.

4.2.3. Impact of Climate Change on Fish Production and Ecosystem

4.2.3.1. Aquaculture System

Aquaculture and fishing are heavily reliant on the interplay of different elements, such as the ocean's environment and the planet's temperature. Thus, alterations to wind patterns, rainfall, sea level, ocean acidity, and air and sea surface temperature patterns will have a negative impact on aquaculture and fisheries.

4.2.3.1.1. Marine System

The production of marine fish is significantly impacted by climate change. Numerous alterations in the ocean are noted as a result of changing climatic conditions, such as elevated temperatures, melting of polar ice, rising sea levels, modifications to the ocean current system, and acidification of seawater. The Indian Ocean's temperature will rise by 1–3°C during the next few decades. Plankton is the species that will be impacted by these conditions the first. In the marine ecosystem, it serves as the foundation of the food chain. Seabirds, fish, and corals are among the other species that will be impacted concurrently. Marine organisms such as oysters, shrimp, and corals would not be able to form their outer covering or shell through the process of calcification due to increased ocean acidification. As a result, the development of cracks in the marine food chain has an impact on the entire marine food web.

4.2.3.1.2. Freshwater Systems

Freshwater ecosystems are particularly vulnerable to climate change. The lake's trophic status, depth, and size all affect how vulnerable this system is to climate change. Positive effects were seen for warm-water species and negative effects for cold-water species, according to Field et al. The freshwater lake system is witnessing changes in their forms and distribution, and in certain instances, their disappearance, as a result of the severe impacts of climate change. These are the characteristics of the dynamics shift in runoff, evaporation, and precipitation. Long-term increases in fish production are encouraged by climate change because it causes the production rate of invertebrate prey to increase logarithmically with temperature. For every 10°C increase in temperature, the rates increase by two to four times. On the other hand, the composition of prey species will shift as a result of climate change. The long-term improvement of fish production may be negatively impacted by this modification. Due to timing mismatch, climate change will soon result in a decrease in fish production.

Determining a species' resilience to climate change depends critically on its ability to migrate. This is especially true for freshwater species.

4.2.3.1.3. Coral Reef

A significant source of revenue for many developing nations is the coral reef. For over half of all marine species, coral serves as a home. However, the ecosystem's coral reefs are currently in grave danger. The primary causes are rising temperatures, increased acidity, etc.

Climate change-related impact on the coral reef can be based on three different time scales.

- **Years**: Coral bleaching which increased in recent years and results in the degradation of reefs.
- **A few Decades Later**: Acidification increased and carbonate structures degenerate.
- **Multi-Decades:** Weakened the structural integrity of the reefs which caused large-scale composition shifts.

One of the most resilient ecosystems, the coral reef is unable to recover from both mild chronic stressors and acute ones. However, Hughes et al, claim that the reef ecosystem cannot withstand both acute and chronic stress.

The ocean's pH drops as acidity rises, causing a decrease in aragonite saturation that can interfere with coral calcification. The world's ocean's increased acidity is a serious threat to coral reefs over the long run, but it is unclear how coral growth will affect the rising acidity. In deep cold water corals, aragonite saturates to a depth of 90–150 meters. These deep-cold water corals clearly show the effects of acidification.

Reef fish biodiversity suffers if corals are reduced as a result of negative effects from climate change. Grandcourt and Cesar claim that bleaching events and climate change have a negative impact on coastal fisheries. We can draw the conclusion that animals that rely on coral reefs for their food and habitat suffer long-term effects when these reefs are destroyed.

4.2.3.2. Global Marine Biodiversity

Climate change is a significant factor that influences both the historical and future distribution of biodiversity. Aquatic organism species richness is strongly influenced by environmental factors. Marine species may alter as a result of ocean warming, particularly in terms of their depth and latitudinal ranges. Larger-scale changes of this kind may result in localized extinction, invasions, and a shift in their bio-geographic pattern. This can lead to a significant change in species richness, which is thought to be the primary factor disrupting the marine ecosystem and biodiversity. The function of ecosystems, community structure, and biodiversity can all be impacted by the climate in aquatic environments.

4.2.4. Economical Crisis of the Fish Farmers Due to Climate Change

The lives of fish farmers are directly impacted by changes in the aquatic environment. The disruption of fish production results in financial losses for farmers. In addition to global warming, another issue affecting farmers' lives is cyclones. Every year, a cyclone coupled with flooding and a lot of rain causes a significant issue for farmers, particularly in India's coastal states. The fact that there are now more powerful tropical cyclones in the Indian Ocean raises serious concerns. The cyclone is intensifying due to variables like high humidity, warm sea temperatures, and atmospheric instability. The Indian Ocean's temperature has also risen as a result of global warming, which has encouraged destructive cyclones. A cyclone known as Fani made landfall in West Bengal, Odisha, and Andhra Pradesh in May 2019. It destroyed five lakh houses in 14 districts and damaged boats, jetties, and fishermen's shelters along the coast. The losses were projected to be worth 12,000 crores in Odisha alone. In the seafood industry, there was a 60–70% decrease in shrimp production. The most recent storm, Cyclone Amphan, made landfall in Bangladesh and eastern India in May 2020, specifically in West Bengal. This was the first super storm to hit the Bay of Bengal since the 1999 storm that devastated Odisha and killed over 9,000 people. West Bengal's coastal regions, including East Midnapur, North and South 24 Parganas, Kolkata, Hoogly, and Howrah, were impacted by amphan. The West Bengal Chief Minister stated that the state had suffered damages totaling 1 lakh crore rupees (15.38 billion USD) and that there were over 86 fatalities. In particular, the Sundarban areas were severely damaged, with millions of homes suffering damage and villages flooding as a result of breached embankments. Both the locals and the fishermen need years to heal from these experiences. They lack a place to live, a boat for fishing, and the money to repay the loans, which eventually has an impact on their mental health and occasionally triggers suicidal thoughts.

4.2.5. Adaptation and Mitigation Measures to Reduce the Effects of Climate Change

Advance preparation and integration of anticipated future climate alterations into current decision-making processes is referred to as adaptation. This encompasses the utilization of environmentally friendly materials, increased tree planting, and preserving nature to the utmost extent. Conversely, the proactive measures taken to minimize the likelihood of climate change happening, as well as lessening its impact in case it does occur, is called mitigation. Significantly decreasing our carbon footprint and related activities should be a top priority. The reduction of environmental pollution should be implemented promptly, before it reaches a point where remedial action becomes ineffective. Immediate implementation of the following strategies is suggested.

Adaptation of forest conservation strategies: Forests are crucial to preserving the ecosystem's balance. In order to maintain the local population's standard of living, we should forbid the use of forestland for non-forest purposes and work to conserve and stop the destruction of forest land through afforestation and reforestation.

Climate study integration into the curriculum at the school level: With the support of government initiatives, we can raise awareness among the youth by integrating climate-related studies into the conventional curriculum. This will contribute to the development of consciousness in the younger generation from the outset, greatly expanding this culture locally, state-wide, and nationally.

Slowing down population growth: As more people live longer, more people are becoming burdened, particularly in developing nations like India. It is now a significant barrier to the advancement of social and economic development. Therefore, slowing down the population growth curve in developing nations is necessary to lessen the population pressure on the region in order to combat climate change.

Integration of climate issues with economic planning: To promote the integration of climate issues with economic planning and management, local, state, and federal policies and programs pertaining to climate protection should be included.

4.2.6. Recommendation for Better Management of Fisheries Against Climate Changes

- The ecosystem approach should be comprehensive, sound, integrated, compact and revised to make complete management of sand oceans of coasts, fisheries and aquaculture.
- Environmental friendly aquaculture and fishing practices to be undertaken.
- Fuel-efficient aquaculture and fishing practices are to be undertaken.
- Integration of climate-proof aquaculture with other sectors.
- Overfishing and excess fishing capacity should be eliminated through the implementation of reduced subsidy systems.

- Risk assessments should be proper and accurate at the local level.
- Exploration of the carbon sequestration process by aquatic ecosystems.

4.2.7. Steps for Sustaining the Fish Production and Economy Against Climate Change

Practical solutions are required to close a significant knowledge gap pertaining to climate change, aquaculture management, and fisheries. Proper adaptation and mitigation measures should be put into place, as well as institutional and human capacity building, to assess the risk that climate change poses to coastal communities. Thus, carefully controlled aquaculture and fisheries may give rise to a robust and fruitful ecosystem. To promote the development of a robust and productive ecosystem, cross-sectoral responsibility for the responsible use of coastal and catchment areas is necessary. Furthermore, since young people are the foundation of our society, youth participation in all policies and decision-making processes of aquaculture and fisheries at the continental and national levels should be institutionalized effectively.

4.2.8. Positive Effects of Climate Change on the Aquatic Environment

- **Slowed Down the Winter Death Rate of Aquatic Organisms:** Water temperature is one of the most crucial factors in determining the survival of aquatic animals. Many years ago, especially before the drastic climate change, winters were too cold to maintain the minimum metabolic rate of the aquatic organisms and the consequent death rate has increased rapidly. As a result of climate change, the average temperature of the water body increased so rapidly that winter has now become bearable considerably. So, the number of deaths due to winter temperature–falling has decreased.
- **Reduction of the Fuel Cost of the Aquatic Environment:** As a result of climate change, heat energy becomes available and affordable at a cheaper rate. So, the demand for fuel in the aquatic environment has decreased and the consequent cost of fuel has also become cut down.
- **Growth in Aquaculture Production:** Some thermophilic organisms living in the aquatic environment demand high temperatures to maintain their metabolic rate at an optimum level. The excess heat that is introduced as a result of climate change meets the demand of those aquatic organisms. So, in that way, climate change benefited the overall aquacultural yield.

Ecosystems on land and in water are both seriously threatened by climate change. Today's population explosion is fueled by a random increase in the use of fossil fuels, industrialization, deforestation, and profit-driven capitalism. These factors can compound each other to cause climate change. Sea level rise, temperature increases in the air and water, and related water intrusion due to climate change and global warming all have a significant influence on the aquaculture industry. The fish farmers' ability to make a living is directly impacted by this alteration in the aquatic environment or a decline in fish production. Therefore, if the required steps are taken to reduce environmental pollution as soon as possible, this situation can be corrected. To combat economic instability and preserve harmony with the environment, researchers, economists, policymakers, and farmers should work together. One thing to keep in mind is that, to defend ourselves against impending threats, we must preserve nature.

4.3. OCEAN ACIDIFICATION AND ITS EFFECTS ON MARINE LIFE

Human activities result in the atmospheric release of carbon dioxide (CO_2), which causes atmospheric warming and subsequent climate change. The oceans absorb between one-third and half of the CO_2 that humans release into the atmosphere. This contributes to slowing down climate change and atmospheric warming, but it also directly affects seawater's chemical composition, a phenomenon known as "ocean acidification" (Figure 4.1).

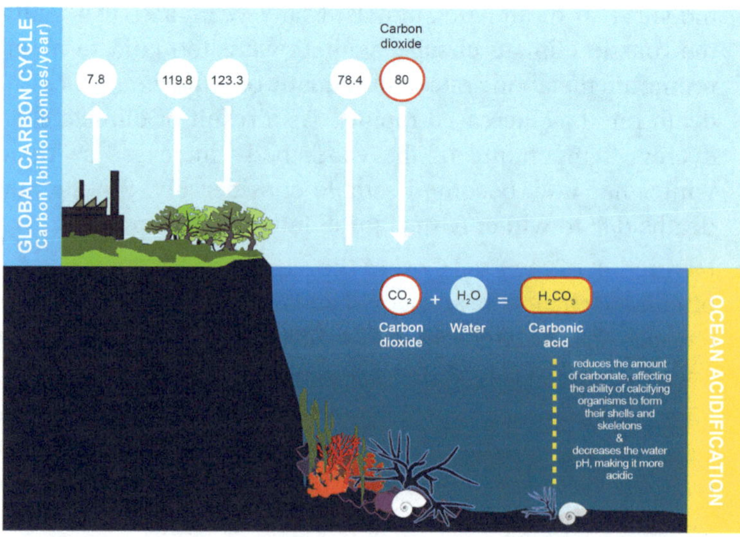

Figure 4.1. *Some of the carbon dioxide emitted to the atmosphere by human activities is absorbed by the oceans. When carbon dioxide combines with water in the*

ocean it forms carbonic acid, which makes the ocean more acidic and may reduce the ability of calcifying organisms to form their shells and skeletons.

Source: https://coastadapt.com.au/sites/default/files/coastadapt_image/T213_Figure-1.gif. Adapted from J. Cook, skepticalscience.com.

The oceans have taken up to half of the CO_2 released into the atmosphere since about 1850. As a result, from 8.2 to 8.1, the average pH of ocean surface waters has decreased by roughly 0.1 units. This translates to a 26% rise in ocean acidity, which is a change occurring at a rate that is approximately 10 times faster than it has in the previous 55 million years (Figure 4.2).

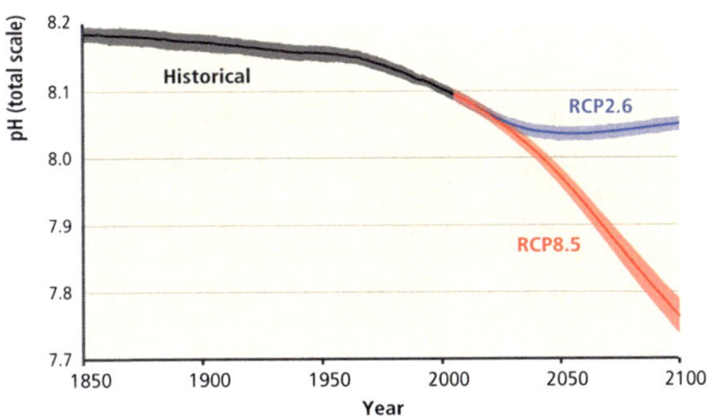

Figure 4.2. *Global mean ocean surface pH from 1850 to 2100, from climate models. The modelled historical trend shows an overall decrease of about 0.1 pH units (black). Projections up to 2100 are shown for high emission scenarios (RCP8.5, red) and low emission scenarios (RCP2.6, blue).*

Source: https://coastadapt.com.au/sites/default/files/coastadapt_image/T213_Figure-2.gif. Gattuso et al., 2014, Fig. OA1b.

Future increases in atmospheric CO_2 will have a direct impact on the extent of ocean acidification. Seawater's acidity could rise by 0.4 units by the end of the century if greenhouse gas emissions keep going at their current rate (the RCP8.5 trajectory).

There won't be uniform ocean acidification around the world. It is anticipated that polar seas and upwelling areas, which are frequently found on the western coasts of continents, will acidify more quickly than temperate or tropical areas. The ecosystem will have a major impact on the pH. Certain kinds of shells can be corroded by the acidic water in some areas of the Arctic, and in California, corrosive events have occasionally happened. In a few decades, the majority of surface waters will be continuously corrosive (Figure 4.3).

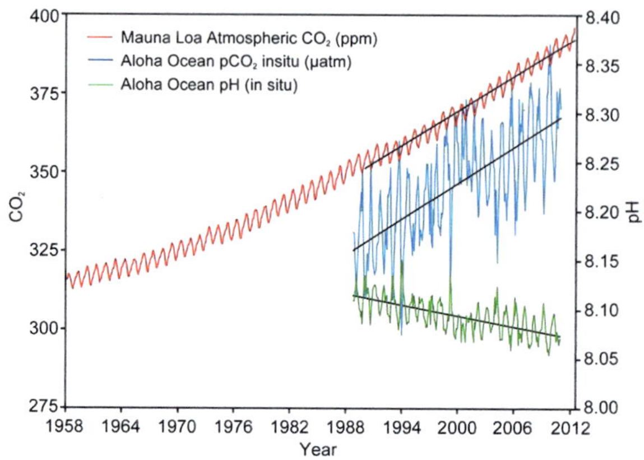

Figure 4.3. *Atmospheric CO_2 concentrations and ocean pH values. Atmospheric CO_2, shown in blue (seasonal variations) and red (long-term smoothed trend), is measured at Mauna Loa, Hawai'i. Ocean pH values (green and orange) are from the ocean to the north of Hawai'i (Station Aloha). As CO_2 accumulates in the ocean, the water becomes more acidic (the pH declines).*

Source: https://coastadapt.com.au/sites/default/files/coastadapt_image/T2I3_Figure%203.jpg. Modified from Feely et al., 2009.

4.3.1. The Effects of Ocean Acidification on Marine Organisms and Ecosystems

The amount of carbonate, a crucial component of seawater, decreases due to ocean acidification. This hinders the formation of shells and skeletons in marine organisms, including coral and some types of plankton, and may cause shells that already exist to disintegrate. Seawater's pH today varies greatly, and an organism can withstand shifts in pH levels over the course of its existence. Since lifetime exposure to lower pH levels poses a risk, the sustained nature of the change is the issue with ocean acidification. The degree of adaptation that calcifying organisms can achieve will depend on how quickly the environment becomes more acidic.

Different species are affected by ocean acidification in different ways. Higher CO_2 concentrations in the ocean may help some algae and seagrass by accelerating their growth and photosynthetic rates. On the other hand, other marine animals like mollusks, corals, and certain types of plankton will suffer from an increasingly acidic environment (Figure 4.4). These skeletons and shells may lose some of their strength or density. This could slow the rate of recovery for coral reefs and make them more susceptible to storm damage.

Figure 4.4. *A mollusk shell dissolves under acidic conditions. The shell almost completely dissolves after 45 days when placed in seawater with pH and carbonate levels projected by models for the year 2100.*

Source: By NOAA Environmental Visualization Laboratory (EVL) – Impacts of ocean acidification – NOAA, Public Domain, https://commons.wikimedia.org/w/index.php?curid=49165506.

Ocean acidification may also cause changes in the growth, development, abundance, and survival of marine organisms. The majority of species appear to be more susceptible during their formative years. For example, young fish might have problems finding a good place to live.

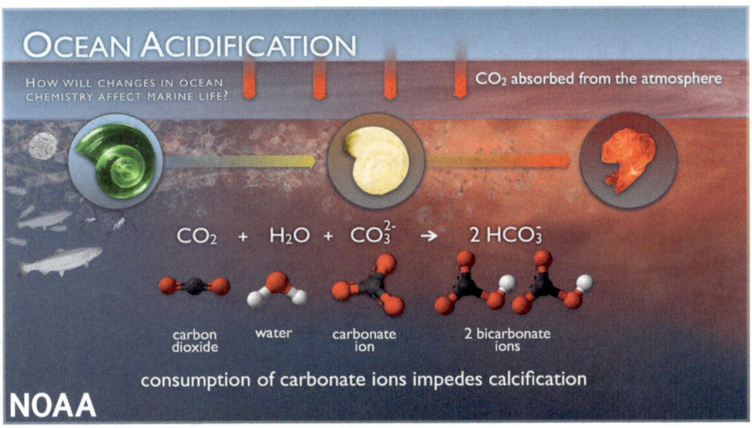

Figure 4.5. *Summary of effects of ocean acidification.*

Source: By NOAA – https://www.noaa.gov/education/resource-collections/ocean-coasts/ocean-acidification, Public Domain, https://commons.wikimedia.org/w/index.php?curid=126147944.

Research indicates that this century will see significant changes in ocean ecosystems due to ocean acidification, regardless of the varying reactions, whether positive or negative, within and between marine groups. The combination of these changes with other newly discovered climate-related risks, like the decline in ocean oxygen levels, or "ocean deoxygenation," which is already having an impact on marine life in some areas, could exacerbate these changes (Long et al., 2016) (Figure 4.5).

4.3.2. The Effects on Human Societies

Human societies, which rely on the products and services these ecosystems provide, will be impacted by changes in marine ecosystems. Significant revenue drops, job and livelihood losses, as well as other indirect economic costs, could be the effects on society.

Socioeconomic impacts associated with the decline of the following ecosystem services are expected:

- **Food:** Ocean acidification has the potential to affect food security. Commercially and ecologically important marine species will be impacted, although they may respond in different ways. Mollusks such as oysters and mussels are among the most sensitive groups. By 2100, the global annual costs of mollusk loss from ocean acidification could be over US$100 billion for a business-as-usual (RCP8.5) CO_2 emissions pathway.
- **Coastal Protection:** Marine ecosystems such as coral reefs protect shorelines from the destructive action of storm surges and cyclones, sheltering the only habitable land for several island nations. This protective function of reefs prevents loss of life, property damage, and erosion, and has been valued at US$9 billion per year.
- **Tourism**: This industry could be severely affected by the impacts of ocean acidification on marine ecosystems (e.g., coral reefs). In Australia, the Great Barrier Reef Marine Park attracts about 1.9 million visits each year and generates more than A$5.4 billion for the Australian economy.
- **Carbon Storage and Climate Regulation:** The capacity of the ocean to absorb CO_2 decreases as ocean acidification increases. More acidic oceans are less effective in moderating climate change.

4.4. SEA-LEVEL RISE CAUSED BY CLIMATE CHANGE

One of the most important consequences of climate change is the rise in sea level. The world's attention has been drawn to the high predicted rates of sea level rise in the future. Countries in low-lying areas and small islands in particular worry that coastal erosion and flooding will reduce their land area and, in the worst case scenario, force a significant portion of their population to migrate abroad. Therefore, as the effects of climate change become more apparent, this issue has garnered increased attention on a global scale.

Physical phenomena like tides and waves cause the sea level to fluctuate over time and space. The height of the sea surface averaged over a time period, say a month or a year, long enough to eliminate tide and wave fluctuations is known as mean sea level at a given location. 1) The global mean sea level is distributed spatially as well. Global mean sea level (GMSL) is the average sea level across all of the world's oceans. Because phenomena that predominate at regional and local scales alter the global mean change, local mean sea level changes typically diverge from GMSL changes. We use the term "sea-level change (rise)" to refer to the general change (increase) in mean sea level, including both local and global variations. When it's important to distinguish between the two, the rise in global (local) mean sea level is referred to by a specific term, such as "global (local) sea-level rise."

Sea level fluctuations can have a variety of causes, not just ones connected to climate change. It is commonly known that over the course of the last several 100,000 years during the Holocene, there has been a significant fluctuation in the mean sea level caused by the alternating glacial and interglacial periods. There are also much shorter-term variations in sea level, such as tide, waves, and tsunamis. This fluctuation in mean sea level reached approximately 120 m. Sea level change brought on by climate change stands out among such a broad range of variations. Scientific research is expected to address numerous questions regarding its potential significant effects. These include: How much will future climate change-related GMSL rise? How serious are the effects of the anticipated sea-level rise, and can human society adequately address them? Finally, how much did climate change contribute to past GMSL rise?

First, a summary of the current state of the observed sea level rise, its causes, and its anticipated future rise is provided. Next, this essay will look at the effects of sea level rise and other climate change factors from both a global and a Japanese perspective. We'll talk about anticipated solutions to the negative effects at the end.

The Intergovernmental Panel on Climate Change (IPCC) has conducted a systematic assessment of the most recent scientific understanding. Since the IPCC's founding in 1988, four significant assessment reports have been released. The First to Fourth Assessment Reports project future sea-level rises by 2100 to be 31–110 cm (Business as usual scenario), 13–94 cm, 9–88 cm, and 18–59 cm, respectively. The IPCC employed various climate models and scenarios for greenhouse gas (GHG) emissions in each assessment. This is the main cause of the variations in the anticipated increase in mean sea level. Specifically, the values provided in the 2007 Fourth Assessment Report (AR4) are comparatively low. This is because there was insufficient reliable understanding of the dynamics of the Greenland Ice Sheet (GIS) and West Antarctica Ice Sheet (WAIS), including their collapse and outflow, which prevented the assessment from accounting for their contributions. Because of this, some scientists hypothesized that these figures might represent the low end of the sea level rise that is yet to come.

There has been advancement in many areas pertaining to the problem of sea level rise since the release of the IPCC AR4. More precise and trustworthy sea-level change data for the historical variations in the global mean sea level (GMSL) from the 19th to the 21st centuries became available with the advancement of analysis techniques applied to satellite altimeter data combined with tide gauge data (Cazenave and Llovel, 2010; Church and White, 2011; Ray and Douglas, 2011). Global warming contributes to an increase in the heat content of the ocean, which in turn causes the seawater to expand thermally, a major factor in the mean sea level rise. Additionally, many methods were created to adjust historical data from expendable bathythermographs (XBT) to generate a more accurate estimate of the variations in ocean heat content. Though precise quantitative evaluations of the ice sheets in Greenland and Antarctica remain challenging, our knowledge of the primary mechanisms driving the mass increase or decrease of these ice sheets has advanced. New sea-level rise projections have been made extensively for the 21st century and longer time horizons beyond this century. These projections are based on the development of process-based models, such as oceanic and atmospheric general circulation models (AOGCMs). Semi-empirical models, which lack a physical scientific foundation, are developed for the projection in tandem with process-based models by analyzing historical data on global mean temperature and mean sea level. Many impact assessments have been conducted in the field to comprehend the possible effects of sea level rise. Many climate change impact projects, including ones pertaining to the coastal zone, have been put into action in Japan.

4.4.1. Sea-Level Rise Since the End of the 19th Century

First, we'll look at sea level variations that have been seen over the last 100 years. Two methods have been used to make the observations: radar altimeters mounted in satellites and tide gauge observation.

Since extensive and consistent observations are required for entire oceans, it is difficult to determine the height of the global mean sea level directly. The Permanent Service for Mean Sea Level (PSMSL), overseen by

Proudman Oceanographic Laboratory, Bidston Observatory in the UK, has compiled the data gathered by tide gauges. Presently, PSMSL receives data from roughly 1900 tide gauges spread across nearly 200 countries. Tide gauges are disproportionately found in the Northern Hemisphere, despite their vast number. Select tide gauges, situated on stable ground, are used for the analysis of a long-term change in the global mean sea level because uncorrected tide gauge data also reflect land movement at the observing location. Furthermore, during the current interglacial period, the earth's crust has been continuing to rebound—a visco-elastic response to the melting of large glaciers. Glacier Isostatic Adjustment (GIA) is the term for this. Consequently, in order to estimate the mean sea level, the GIA land movement must be corrected. Lemke and others. (2007) found that the GIA correction varies from a few tenths of a millimeter per year in the far field to approximately 1 mm/year (or more) close to former ice sheets; the GIA-related error in tide-gauge based GMSL change is calculated to be 0.15 mm/year.

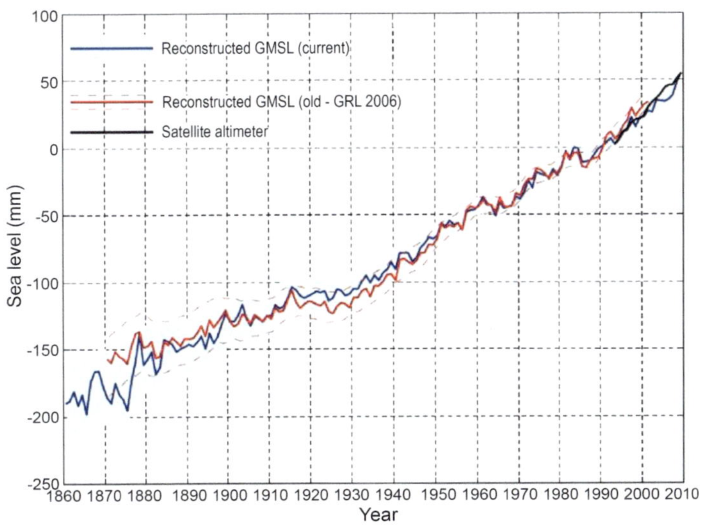

Figure 4.6. *Global mean sea level from 1860 to 2009. Blue line: estimated from coastal and island sea-level data with one standard deviation error (shading). Red line: estimated by Church and White (2006) for 1870–2001 (solid line) with one standard deviation error (dashed lines). Black line: satellite altimeter data since 1993.*

Source: Mimura N. (2013). Sea-level rise caused by climate change and its implications for society. Proc Jpn Acad Ser B Phys Biol Sci.;89(7):281-301. doi: 10.2183/pjab.89.281. PMID: 23883609; PMCID: PMC3758961.

Topex/Poseidon, a satellite altimeter launched in 1992, was the first instrument used for satellite observations. This observation is still being made with the help of later satellites like Jason-1 and Jason-2. The satellite altimeter provides the separation between the sea surface and a fixed reference surface, usually an ellipsoid. Since the early 1990s, the primary method for accurately and consistently measuring sea level has been satellite altimetry. This is because of its benefits over tide gauges, which include regular worldwide coverage with frequent revisit cycles and the absence of land movement influence. The Topex/Poseidon altimeter had relatively large errors at first, but data proccssing and altimetry system adjustments have significantly increased the estimate's precision.

Similar studies, like those by Cazenave and Llovel (2010) and Church and White (2011), summarize the modern GMSL change. The result obtained by Church and White (2011) is introduced here because they show close results. The long-term trend of GMSL from 1860 to 2010 is summarized in Figure 4.6, which was created by superimposing the two sources of observed data. Figure 4.6 shows that, up until 1930, the rate of increase in GMSL was not very high, but that it has since accelerated. According to Church and White (2011), there was a 21 cm increase in GMSL between 1880 and 2009, and between 1993 and 2009, the rate of increase nearly doubled to 3.2 ± 0.4 mm/year, compared to 1.7 ± 0.2 mm/year between 1900 and 2009.

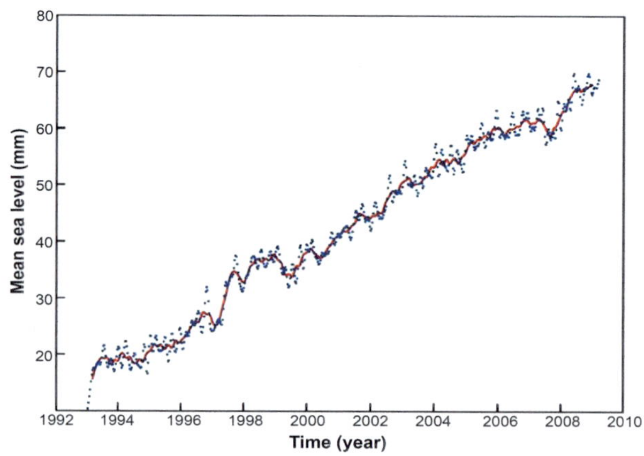

Figure 4.7. *Global mean sea level from satellite altimetry for 1993 to 2008. Blue dots are raw 10-day data. The red line corresponds to a 90-day smoothing of the raw data. The −0.3 mm/year GIA correction has been applied.*

Source: Mimura N. Sea-level rise caused by climate change and its implications for society. Proc Jpn Acad Ser B Phys Biol Sci. 2013;89(7):281-301. doi: 10.2183/pjab.89.281. PMID: 23883609; PMCID: PMC3758961.

Figure 4.7, which relates to the last section of Figure 4.6, displays the change in GMSL from 1993 to 2008 as determined by satellite altimeters. This graph shows a sharp rise in GMSL between 1997 and 1999, followed by a decline in 2008. These erratic variations are thought to be connected to a strong El Niño in 1997–1998 and a La Niña in 2007. The influence of the Pacific Decadal Oscillation (PDO) and the North Atlantic Oscillation (NAO) was also noted by Cazenave and Llovel (2010). These are clearly more pronounced, longer-term variations that last between 10 and 20 years. It is clear that both such natural variations and anthropogenic climate change contribute to the GMSL fluctuation. Furthermore, it should be mentioned that the rate of increase in GMSL since the altimeter record began in 1993 has been above the sea level projections in the Third and Fourth Assessment Reports of the IPCC.

4.4.2. Factors Causing Changes in Sea Level, and Their Contributions

4.4.2.1. Factors Causing Changes in Sea Level

What influences mean sea level fluctuations over an extended period of time is the next query. We will look at the various factors affecting mean sea level to provide an answer. The two main variables that affect the size and shape of ocean basins are variations in the overall volume of sea water and movements of the earth and ocean bottom. Dynamic elements including winds, atmospheric pressures, ocean currents, and waves also have an impact. Thus, there are many factors, with various temporal and spatial scales as shown below.

- Factors related to changes in the volume of sea water contained in ocean basins
    - Thermal expansion of sea water
    - Growth/decay of land-based glaciers and ice caps
    - Growth/decay of ice sheets, such as the Greenland Ice Sheet (GIS) and West Antarctica Ice Sheet (WAIS)
    - Terrestrial water storage, *i.e.*, dam reservoirs, lakes, etc., and depletion of groundwater
- Factors related to changes in the size and shape of the ocean basins
    - Isostatic adjustment of land mass, especially Glacier Isostatic Adjustment (GIA)
    - Tectonic movement including ground subsidence/uplift associated with earthquakes
    - Ground subsidence/uplift due to compaction of the ground, pumping up of ground water, etc.

  - o Sediment inflow from land
- Other factors causing local/temporal changes in sea level
  - o Changes in ocean currents
  - o Changes in atmospheric pressure
  - o Tide, tsunami, storm surges and waves
  - o Natural inter-annual variations, such as the Pacific Decadal Oscillation (PDO)

Numerous of the previously mentioned factors are individual factors that are directly related to changes in sea level. Typically, these elements contribute to variations in sea level as a result of various geophysical events or phenomena. The glacial-interglacial cycle is the most prominent of these phenomena that alters sea level. Over the previous 500,000 years, this has happened six times. Similar temporal changes are revealed by estimated mean sea level and air temperature in Antarctica. During the last interglacial period, which lasted between 130 and 120,000 years B.C., the global mean sea level was several meters higher than it is today. This was 140,000 years ago. B. P. About 120,000 years ago marked the beginning of the last glacial epoch. B. P, and during the glacial maximum 20,000 years, the mean sea level was roughly 120 m below the current level. B.P. 15000 years B. P. the current interglacial period began, and as the ice sheets melted, sea levels began to rise as well. Furthermore, between 7,000 and 6,000 years B, during the warm period. B. P., sea level was two to three meters higher than it is now. The visco-elastic response of the plate on which the Japanese islands were located to the increased GMSL is the cause of this higher sea level, which is known as the Jomon Transgression in Japan. In summary, these data show a large variation of GMSL on a geological time scale (Figure 4.8).

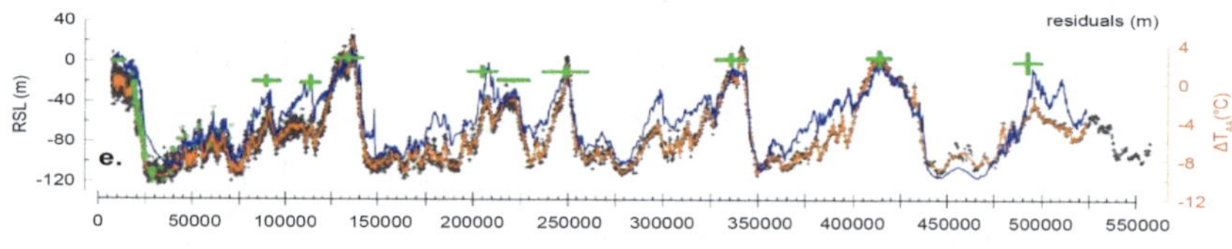

Figure 4.8. *Sea level (blue, green: scale on the left) and Antarctic air temperature (orange, gray: scale on the right) over the last 550,000 years, from paleo-records (Left origin is the present-day). The green symbols are coral and speleothem-based sea-level markers.*

Source: Mimura N. Sea-level rise caused by climate change and its implications for society. Proc. Jpn. Acad. Ser. B Phys. Biol. Sci., 2013;89(7):281-301. doi: 10.2183/pjab.89.281. PMID: 23883609; PMCID: PMC3758961.

This focuses on the change in mean sea level that occurred between the end of the 20th century and 100 years ago, which is a much shorter temporal scale than the alternating glacial and interglacial periods. The most recent scientific findings on sea-level change over a 100-year timescale will be reviewed in the sections that follow, with an emphasis on significant direct causes.

4.4.2.2. Factors Related to Changes in the Volume of Sea Water Contained in Ocean Basins

The melting of land-based ice and the thermal expansion of sea water are the main causes of the recent sea level change. The latter involves Antarctica's and Greenland's ice sheets melting and releasing their contents. Heat is transferred from the atmosphere to the ocean when an increase in the concentration of greenhouse gases in the atmosphere raises air temperature. The surface mixed layer, where strong wind mixing of surface waters results in nearly uniform water temperature and density, absorbs the majority of the ocean's heat. The heat is then transferred from the mixed layer to the deep layer via the thermocline. Approximately 90% of the heat that the earth has absorbed over the past 40 years is thought to have been stored in the ocean due to its 1000 times greater heat capacity than that of the atmosphere. Compared to the atmosphere, this heat storage is 20 times greater. Thus, one of the main indicators of global warming is the rise in the ocean's heat content. It should be remembered that the ocean's slow process of heat convection and diffusion means that it will take millennia for the water to reach a thermal equilibrium. This indicates that while the air temperature stabilizes much more quickly, the ocean's ability to absorb heat will last for more than 1000 years.

Sea water expands thermally as a result of higher water temperatures brought on by increased heat storage. Thus, a key contributing factor to the average sea level rise brought on by global warming is the ocean's capacity to store heat. We must ascertain the sea water temperature's spatial distribution in order to calculate the ocean's thermal expansion. The vertical temperature distribution has been observed using mechanical bathythermographs (MBT) and expendable bathythermographs (XBT). While data from MBT observations is available from World War II to approximately 1990, XBT observations began after the mid-1960s. In the modern era, the temperature and salinity of the ocean are regularly monitored using ARGO floats. There has been a suggestion that the XBT and MBT data had a significant positive temperature bias. For instance, when an XBT descends into the ocean, the vertical positions are determined using the XBT's falling velocity formula, which results in errors when measuring the temperature at depth. Extensive research has been conducted to produce new estimates for the historical variations in the ocean heat content (e.g., by correcting errors and performing statistical analyses on the available XBT and MBT data) (Ishii and Kimoto, 2009; Levitus et al., 2012).

Figure 4.9 shows a revised estimate for the ocean heat content (OHC) for two depth layers; 0–2000 m (red) and 700–2000 m (black). The OHC of the 0–2000 m layer shows an overall increasing trend with a decrease from 1965 to 1970 and stagnation from 1980 to 1990. It is indicated that the global linear trend of OHC 0–2000 m is 0.43×10^{22} J/year from 1955 to 2010, which corresponds to a total increase in heat content of 24.0×10^{22} J, and a mean temperature increase of 0.09□. From Figure 4.9, the 700–2000 m layer is responsible for about one-third of the ocean warming of the 0–2000 m layer, indicating that major ocean warming has so far taken place in the relatively shallow ocean. Levitus et al. (2012) also conclude that the heat storages for 0–2000 m and 0–700 m corresponds to a mean sea-level rise of 0.54 mm/year and 0.41 mm/year, respectively, from 1955 to 2010.

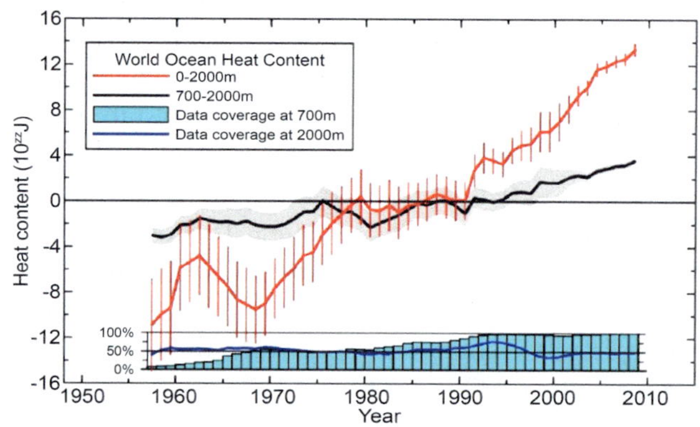

Figure 4.9. *Time series for the World Ocean of ocean heat content for the 0–2,000 m (red) and 700–2,000 m (black) layers based on pendatal (five year) running mean analyses. Reference period is 1955–2006. Each pentadal estimate is plotted at the midpoint of the 5-year period. The vertical bars represent ±2 × S.E. about the pentadal estimate for the 0–2000 m estimates and the grey-shaded area represent ±2 × S.E. about the pentadal estimate for the 0–700 m estimates. The blue bar chart at the bottom represents the percentage of one-degree square areas (globally) that have at least four pentadal one-degree square anomaly values used in their computation at 700 m depth. Blue line is the same as for the bar chart but for 2000 m depth.*

Source: Mimura N. Sea-level rise caused by climate change and its implications for society. Proc Jpn Acad Ser B Phys Biol Sci. 2013;89(7):281-301. doi: 10.2183/pjab.89.281. PMID: 23883609; PMCID: PMC3758961.

Melting land-based ice is another significant contributor to sea level rise. This is caused by two things: the melting of ice caps and mountain glaciers, as well as the melting and retreat of ice sheets in Greenland and Antarctica. There are many glaciers in the world; the World Glacier Monitoring Service has recorded over 130,000 of them in its World Glacier Inventory. It is believed that the water content stored in glaciers is equal to 0.5 GMSL rise. Due to the heightened sensitivity of high mountain glaciers and ice caps to temperature increases, there has already been an acceleration of melting and runoff, resulting in the retreat of glaciers and the creation of glacier lakes in areas like the Himalayas. However, the amount of ice in the West Antarctic Ice Sheet (WAIS) and Greenland Ice Sheet (GIS) is equal to a sea level rise of 3–5 meters and approximately 7 meters, respectively. As a result, these ice sheets' actions will have a significant impact on future sea level rise. Numerous investigations have concentrated on this issue since the 2007 release of IPCC AR4. The outflow of ice to the outer coastal regions and surface melting/accumulation determine the balance of the ice sheets. Snow accumulation gives ice sheets their mass, which melting causes them to lose. Exodus from the shoreline represents an additional mechanism of loss. Surface melting and total mass loss in Greenland are detected by gravity field observations made with the Gravity Recovery and Climate Experiment (GRACE) satellite. Svendsen along with others. (2013) documented an acceleration of the GRACE-observed GIS mass loss.

The fact that melt water has a lubricating effect is one that garners more attention. Meltwater enters crevasses in the summer to reach the lower layers between the glacier and base rock. This could have lubricating effects. The ice sheet's outflow could quicken if this phenomenon is widespread. The long-term total breakdown of the GIS is another possible risk.

We'll talk about the West Antarctic Ice Sheet (WAIS) next. Even though there is another ice sheet in East Antarctica, we will concentrate on WAIS since it is thought to be more susceptible to outflow. In the past, it was thought that as global warming raised the atmosphere's water vapor content and precipitation, the WAIS grew as a result of more snowfall, which helped to lower the global mean sea level. However, it has recently come to light that the dynamics of the ice shelf that stretches from the glacier edge to the Antarctic Ocean are closely linked to the outflow of the ice sheet. Sea water heated the ice shelf from below as the sea level rose recently, which accelerated melting. This could eventually lead to a faster outflow of the ice sheet by weakening the barrier preventing it. Long-term changes in the WAIS dynamics could hasten the GMSL rise, even though there are still a lot of unknowns and more observations and research are required to get a solid understanding. Thus, areas of emphasis for estimating future sea-level rise are ice shelf behavior and outflow (e.g., Stocker et al., 2010).

Shifts in land-based water storage are another factor. This covers a variety of storage methods, including soil moisture, ground water, wet lands, and reservoirs hidden behind dams. The rate of sea level rise will decrease with an increase in terrestrial water storage. Dam reservoirs are thought to slow the rise in sea level because hundreds of thousands of dams have been built worldwide. Even though it is challenging to estimate human activities precisely, such as irrigation, groundwater extraction, and dam reservoirs, Gornitz (2001) calculated that the water stored behind dams was equal to –0.33 to –0.27 mm/year of sea-level fall (a negative rate indicates sea-level fall). Pokhrel et al. (2012) proposed that, between 1961 and 2003, the use of ground water, artificial reservoir water impoundment, changes in terrestrial water storage caused by climate change, and loss of water from closed basins were the main causes of the 0.77 mm/year global sea-level rise or roughly 42% of the total sea-level rise that occurred during that time. However, other studies demonstrate that increased freshwater use from groundwater extraction and large inland lake extraction, like the Aral Sea, has exacerbated sea-level rise. Changes in the natural hydrological cycle and human demands for and usage of water will result from climate change. As a result, assessing and accounting for the impact of hydrological processes in sea level change projections continues to be difficult.

4.4.2.3. Contribution of Major Factors

Many factors, including the thermal expansion of seawater, the melting and outflow of land-based ice, the storage of water on land, etc., contribute to sea level change. Extensive research has been conducted because it is crucial to evaluate these factors' contributions in order to project future sea level changes. The term "sea-level budget" refers to the comparison between the estimated and observed sea-level rise. In the field of climate science, closing the gap between these values has long been a challenge. A selection of the sea-level budget summaries from Church et al. (2011) is displayed in Table 4.1. as along with data from Llovel and Cazenave (2010).

Table 4.1. *Contributions of Components to the Budget of Global Mean Sea-Level Rise (Unit: mm/year)*

Period	**1972–2008 (Church et al., 2011)**	**1993–2008 (Church et al., 2011)**	**1993–2007 (Cazenave & Llovel, 2010)**	**2003–2007 (Cazenave & Llovel, 2010)**
Thermal expansion	0.80 ± 0.15	0.88 ± 0.33	1.0 ± 0.3	0.25 ± 0.8
Glaciers & ice caps	0.67 ± 0.03	0.99 ± 0.04	1.1 ± 0.25	1.4 ± 0.25

Greenland	0.12 ± 0.17	0.31 ± 0.17	0.4 ± 0.15	0.5 ± 0.15
Antarctica	0.30 ± 0.20	0.43 ± 0.20	0.3 ± 0.15	0.5 ± 0.15
Terrestrial storage	−0.11 ± 0.19	−0.08 ± 0.19	—	−0.2 ± 0.1
Sum of estimated components	1.78 ± 0.36	2.54 ± 0.46	2.85 ± 0.35	2.45 ± 0.85
Observation	2.10 ± 0.16	3.22 ± 0.41	3.3 ± 0.4	2.5 ± 0.4
Difference (Observed–estimated)	0.32 ± 0.39	0.69 ± 0.62	0.45	−0.05

Note: The uncertainty range of Cazenave & Llovel (2010) computed estimates is one standard deviation.

Sourc: Mimura N. Sea-level rise caused by climate change and its implications for society. Proc Jpn Acad Ser B Phys Biol Sci. 2013;89(7):281-301. doi: 10.2183/pjab.89.281. PMID: 23883609; PMCID: PMC3758961.

Estimates for various time periods from the late 20th century to 2008 are shown in Table 4.1. In the years 1972–2008, thermal expansion accounted for about 40% of the observed rise; the combined effect of thermal expansion and glacier and ice cap melting accounts for about 75% of the rise. The observed rate of GMSL rise is nearly the same in both estimates, i.e., 3.22 mm and 3.3 mm per year, in that order. The two estimates of the rates of GMSL rise (2.54 mm/year and 2.85 mm/year) are quite similar, despite the fact that they do not match the observations. Moreover, they provide comparable estimates for the fractional contributions of each component: roughly 10% for the Antarctic and Greenland ice sheets, and 30% for thermal expansion, glaciers, and ice caps. Comparisons between the two eras indicate that when ice contributions rise, the percentage of thermal expansion falls from 40% to 30%. It is also evident that, especially after 1993, the Greenland and Antarctic ice sheets played a significant role in closing the sea-level budget. Only data from Cazenave and Llovel (2010) are available for the recent period of 2003–2007, and they demonstrate a decrease in the observed rate of sea-level rise during this time.

As demonstrated above, significant progress has been made in reconstructing the historical sea-level budget, reducing the discrepancy between the estimated and observed sea-level rise. Reanalyzing data from ocean observations and developing observations and models of related phenomena allowed for this advancement. But the budget for sea level rise is also subject to change. Future research will face challenges in producing more precise and consistent estimates of the contributions.

4.4.2.4. Regional Distribution and Relative Sea-Level Rise

Thus far, we have looked at the average global mean sea level across all oceans. In actuality, though, there is a spatial distribution to sea level fluctuations. The mean sea level change's regional distribution is depicted quite clearly in Figure 4.10. Local variations in the density structure, ocean currents, low pressure system frequency, etc., are blamed for such a distribution. Take note of that figure, the distribution for other periods is probably going to differ from Figure 4.10, which represents the average distribution for the years 1993 to 2001.

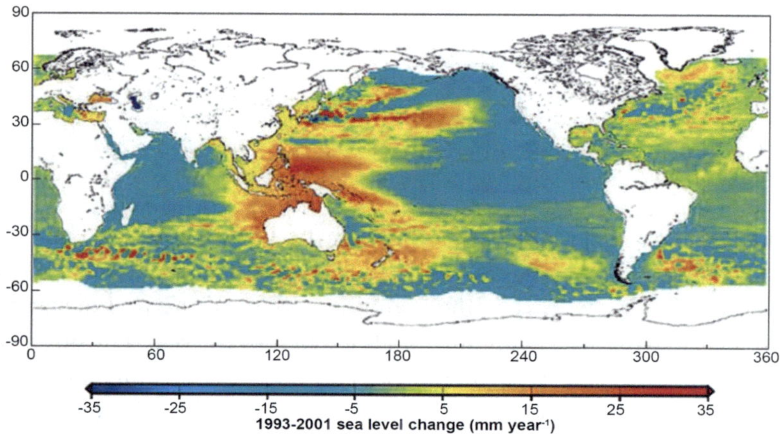

Figure 4.10. *Spatial patterns in sea level trends over 1993–2001 observed by satellite altimetry.*

Source: Mimura N. Sea-level rise caused by climate change and its implications for society. Proc Jpn Acad Ser B Phys Biol Sci. 2013;89(7):281-301. doi: 10.2183/pjab.89.281. PMID: 23883609; PMCID: PMC3758961.

Moreover, the real variations in the mean sea level in a given area are a result of both crustal and land movement locally, as well as variations in ocean volume. Relative sea-level rise (RSLR) is the term used to describe situations in which certain factors raise the sea level. Since the external force affecting coastal zones is the rise in sea level relative to land, estimating the RSLR is essential to assessing the effects of sea level rise. Glacier Isostatic Adjustment is a common crustal movement (GIA). This is the earth's crust's long-term rebound from the areas that were covered in massive ice sheets during the glacial era. There is evidence of this movement in Siberia, the northern region of North America, northern Europe, and the area surrounding Antarctica. Since GIA is still in place today, sea levels fall relative to land rather than rising.

For instance, from 1880 to 2010, Stockholm, Sweden, saw a sea level fall of −3.82 mm/year, and from 1883 to 2009, Bergen, Norway, saw a sea level fall of −0.54 ± 0.21 mm/year. Assuming a viscoelastic body (e.g., a heavy ice sheet), the response of the crust to the disappearance of the sheet is calculated to estimate the distribution of GIA (Peltier, 2009). Conversely, land subsidence is a common local occurrence. Because large river deltas are made of loosely deposited sediment, they experience subsidence as a result of consolidation. Land subsidence is also a result of human activity, such as the extraction of groundwater for irrigation and industrial use. For instance, the Ganges Delta in Bangladesh and the Chaophraya Delta in Thailand experienced significant subsidence (17 mm/year between 1960 and 1994). This made storm surges and flooding more likely. Aside from these, other factors that lead to local changes include plate tectonics and earthquake-induced uplift and subsidence. The Ojika Peninsula in Miyagi Prefecture underwent a 1.2 m subsidence as a result of the Great East Japan Earthquake in 2011.

Finally, we will examine the variations in sea levels surrounding Japan. The changes in Japan over the last 100 years are depicted in Figure 4.11. Because of the significant tectonic movement, tidal stations situated in areas with less active land movement were selected to prepare Figure 4.11. The record shows a fluctuation spanning almost 20 years, with the relative maximum occurring around 1950. The mean sea level began to rise once more after 1990. The recent increase is happening at a faster rate than the global average of 3.2 mm/year, at 4.8 mm/year. The long-term record displayed in Figure 4.11 may have long-term fluctuations for which there are plausible explanations, such as variations in the temperature of the water and the Kuroshio current's paths around Japan. Despite several studies, the cause of the fluctuation has not yet been identified.

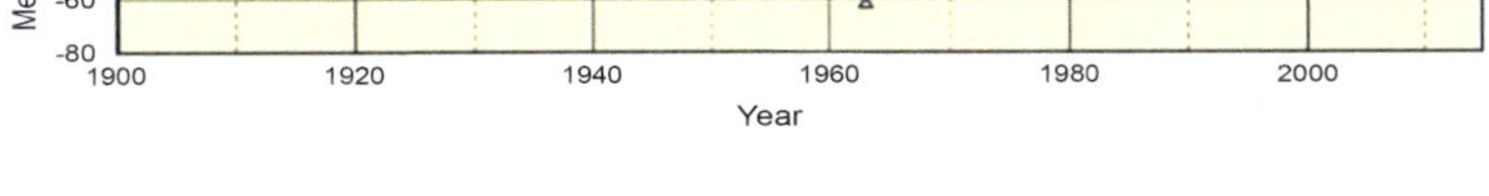

Figure 4.11. *Sea-level change around Japan (1906–2011).*

Source. Mimura N. Sea-level rise caused by climate change and its implications for society. Proc Jpn Acad Ser B Phys Biol Sci. 2013;89(7):281-301. doi: 10.2183/pjab.89.281. PMID: 23883609; PMCID: PMC3758961.

REFERENCES

1. Attrill, M., & Power, M. (2002). Climatic influence on a marine fish assemblage. *Nature, 417*, 275–278.
2. Babcock Hollowed, A., Hare, S. R., & Wooster, W. S. (2001). Pacific basin climate variability and patterns of northeast Pacific marine fish production. *Progress in Oceanography, 49*, 257–282.
3. Baede, A. P. M. (Ed.). (2007). Annex I glossary. In S. Solomon, D. Qin, M. Manning, Z. Chen, M. Marquis, K. B. Averyt, M. Tignor, & H. L. Miller (Eds.), *Climate Change 2007: The Physical Science Basis* (pp. 941–954). Cambridge University Press.
4. Bhatia, K. T., Vecchi, G. A., Knutson, T. R., Murakami, H., Kossin, J., Dixon, K. W., et al. (2019). Recent increases in tropical cyclone intensification rates. *Nature Communications, 10*(1), 635.
5. Billé, R., & Coauthors. (2013). Taking action against ocean acidification: A review of management and policy options. *Environmental Management, 52*(4), 761–779.
6. Brown, C. J., Saunders, M. I., Possingham, H. P., & Richardson, A. J. (2013). Managing for interactions between local and global stressors of ecosystems. *PLoS ONE, 8*(6).
7. Bryant, D., Burke, L., McManus, J., & Spalding, M. (1998). *Reefs at Risk: A Map-Based Indicator of Threats to the World's Coral Reefs*. World Resources Institute.
8. Buddemeier, R. W., Kleypas, J. A., & Aronson, R. B. (2004). *Potential Contributions of Climate Change to Stresses on Coral Reef Ecosystems*. Pew Center on Global Climate Change.
9. Cazenave, A., & Llovel, W. (2010). Contemporary sea level rise. *Annual Review of Marine Science, 2*(1), 145–173.
10. Cheung, W. W. L., Lam, V. W. Y., & Pauly, D. (2008). *Modelling Present and Climate-Shifted Distribution of Marine Fishes and Invertebrates*. Fisheries Centre Research Report, University of British Columbia.
11. Cheung, W. W., Lam, V. W., Sarmiento, J. L., Kearney, K., Watson, R., & Pauly, D. (2009). Projecting global marine biodiversity impacts under climate change scenarios. *Fish and Fisheries, 10*(3), 235–251.
12. Cheung, W. W., Lam, V. W., Sarmiento, J. L., Kearney, K., Watson, R. E. G., Zeller, D., & Pauly, D. (2010). Large-scale redistribution of maximum fisheries catch potential in the global ocean under climate change. *Global Change Biology, 16*(1), 24–35.

13. Chhotoray, S. (2019). Cyclone Fani took us 20 years behind, snatched our livelihood: Distressed Chilika fishermen await aid. *News 18*. Retrieved from: https://www.news18.com/news/india/cyclone-fani-has-snatched-our-livelihood-taken-us-20-years-behind-say-chilika-fishermen-2150891.html (accessed on 16 October 2024).
14. Church, J. A., & White, N. J. (2006). A 20th century acceleration in global sea-level rise. *Geophysical Research Letters, 33*, L10602.
15. Church, J. A., & White, N. J. (2011). Sea-level rise from the late 19th to the early 21st century. *Surveys in Geophysics, 32*(4, 5), 585–602.
16. Church, J. A., Gregory, J. M., Huybrechts, P., Kuhn, M., Lambeck, K., Nhuan, M. T., Qin, D., & Woodworth, P. L. (2001). Changes in sea level. In *Climate Change 2001: The Scientific Basis* (pp. 639–693). Cambridge University Press.
17. Church, J. A., Gregory, J. M., Huybrechts, P., Kuhn, M., Lambeck, K., Nhuan, M. T., Qin, D., & Woodworth, P. L. (2001). Changes in sea level. In J. T. Houghton, Y. Ding, D. J. Griggs, M. Noguer, P. J. van der Linden, X. Dai, K. Maskell, & C. A. Johnson (Eds.), *Climate Change 2001: The Scientific Basis* (pp. 639–693). Cambridge University Press.
18. Church, J. A., White, N. J., Coleman, R., Lambeck, K., & Mitrovica, J. X. (2004). Estimates of the regional distribution of sea-level rise over the 1950 to 2000 period. *Journal of Climate, 17*(13), 2609–2625.
19. Church, J. A., White, N. J., Konikow, L. F., Domingues, C. M., Cogley, J. G., Rignot, E., Gregory, J. M., van den Broeke, M. R., Monaghan, A. J., & Velicogna, I. (2011). Revisiting the earth's sea-level and energy budgets from 1961 to 2008. *Geophysical Research Letters, 38*, L18601.
20. Cochrane, K., Young, C. de, & Bahri, T. (2009). Climate change implications for fisheries and aquaculture. *FAO Fisheries and Aquaculture Technical Paper, 530*, 212.
21. Cooper, N., Barber, P. C., Bray, M. C., & Carter, D. J. (2002). Shoreline management plans: A national review and an engineering perspective. *Proceedings of the Institution of Civil Engineers: Water Maritime Engineering, 154*, 221–228.
22. DEFRA. (2001). *Shoreline Management Plans: A Guide for Coastal Defence Authorities*. Department for Environment, Food and Rural Affairs, London.
23. Dulvy, N. K., Rogers, S. I., Jennings, S., Vanessa, S., Dye, S. R., & Skjoldal, H. R. (2008). Climate change and deepening of the North Sea fish assemblage: A biotic indicator of warming seas. *Journal of Applied Ecology, 45*, 1029–1039.

24. Easterling, W. E., Aggarwal, P. K., Batima, P., Brander, K. M., Erda, L., Howden, S. M., et al. (2007). Food, fibre, and forest products. In *Climate Change 2007: Impacts, Adaptation and Vulnerability* (pp. 273–313). Cambridge University Press.
25. Epstein, P. R. (2000). Is global warming harmful to health? *Scientific American, 283*(2), 50–57.
26. FAO. (2008). Climate change implications for fisheries and aquaculture. In *The State of Fisheries and Aquaculture 2008* (pp. 87–91). FAO.
27. Feely, R. A., Fabry, V. J., & Guinotte, J. M. (2007). Ocean acidification of the North Pacific Ocean. *PICES Press, 2*. Retrieved from: http://www.pices.int/publications/pices_press/volume16/v16_n1/pp_22-26_Acidification_f.pdf (accessed on 16 October 2024).
28. Feely, R. A., Sabine, C. L., & Fabry, V. J. (2006). Carbon dioxide and our ocean legacy. Retrieved from: http://www.pmel.noaa.gov/pubs/PDF/feel2899/feel2899.pdf (accessed on 16 October 2024).
29. Field, C. B., Mortsch, L. D., Brklasich, M., Forbes, D. L., Kovacs, P., Patz, J. A., et al. (2007). North America. In M. L. Parry, O. F. Canziani, J. P. Palutikof, P. J. Hanson, & C. E. Linden (Eds.), *Climate Change 2007: Impacts, Adaptation and Vulnerability* (pp. 617–652). Cambridge University Press.
30. Fitchett, J. (2019). Why the Indian Ocean is spawning strong and deadly tropical cyclones. *Down To Earth*. Retrieved from: https://www.downtoearth.org.in/news/climate-change/why-the-indian-ocean-is-spawning-strong-and-deadly-tropical-cyclones-64489 (accessed on 16 October 2024).
31. Gattuso, J. P., Brewer, P. G., Hoegh-Guldberg, O., Kleypas, J. A., Pörtner, H. O., & Schmidt, D. N. (2014). Cross-chapter box on ocean acidification. In C. B. Field, V. R. Barros, D. J. Dokken, K. J. Mach, M. D. Mastrandrea, T. E. Bilir, M. Chatterjee, K. L. Ebi, Y. O. Estrada, R. C. Genova, B. Girma, E. S. Kissel, A. N. Levy, S. MacCracken, P. R. Mastrandrea, & L. L. White (Eds.), *Climate Change 2014: Impacts, Adaptation, and Vulnerability. Part A: Global and Sectoral Aspects* (pp. 131–133). World Meteorological Organization. Retrieved from: https://www.ipcc.ch/pdf/assessment-report/ar5/wg2/WGIIAR5-IntegrationBrochure_FINAL.pdf.
32. Gattuso, J. P., Frankignoulle, M., Bourge, I., Romaine, S., & Buddemeier, R. W. (1998). Effect of calcium carbonate saturation of seawater on coral calcification. *Global Planetary Change, 18*, 37–47.
33. Genner, M. J., Sims, D. W., Wearmouth, V. J., Southall, E. J., Southward, A. J., Henderson, P. A., et al. (2004). Regional climatic warming drives

long-term community changes of British marine fish. *Proceedings of the Royal Society of London, Series B: Biological Sciences, 271*, 655–661.

34. Geospatial Information Authority of Japan. (2011). Crustal movement associated with the earthquake on March 11. Retrieved from: http://www.gsi.go.jp/chibankansi/chikakukansi40005.html (accessed on 16 October 2024).
35. Gornitz, V. (2001). Impoundment, groundwater mining, and other hydrological transformations: Impacts on global sea level rise. In B. C. Douglas, M. S. Kearney, & S. P. Leatherman (Eds.), *Sea Level Rise: History and Consequences* (pp. 97–119). Academic Press.
36. Grandcourt, E. M., & Cesar, H. S. (2003). The bio-economic impact of mass coral mortality on the coastal reef fisheries of the Seychelles. *Fisheries Research, 60*(2), 539–550.
37. Group I of the Intergovernmental Panel on Climate Change. (2001). In R. T. Watson (Ed.), *Climate Change 2001: The Scientific Basis*. Cambridge University Press.
38. Hashimoto, N., Kawai, H., & Matsuura, K. (2005). Analysis of typhoon characteristics in the future under global warming with the use of RCM20 data and stochastic typhoon model. *Annual Journal of Coastal Engineering, JSCE, 52*, 1221–1225.
39. Hibino, T. (2001). Sea level fluctuation in the north-western Pacific and characteristics of the sea-level change around Japan. *Annual Journal of Coastal Engineering, JSCE, 48*, 446–450.
40. Hiddink, J. G., & Hofstede, R. T. (2008). Climate-induced increases in species richness of marine fishes. *Global Change Biology, 14*, 453–460.
41. Hooper, D. U., Chapin, F. S. III, Ewel, J. J., et al. (2005). Effects of biodiversity on ecosystem functioning: A consensus of current knowledge. *Ecological Monographs, 75*, 3–35.
42. Hughes, T. P., Baird, A. H., Bellwood, D. R., Card, M., Connolly, S. R., Folke, C., et al. (2003). Climate change, human impacts, and the resilience of coral reefs. *Science, 301*(5635), 929–933.
43. IGBP, IOC, SCOR. (2013). *Ocean Acidification Summary for Policymakers: Third Symposium on the Ocean in a High-CO_2 World*. International Geosphere-Biosphere Programme. Retrieved from: http://www.igbp.net/download/18.30566fc6142425d6c91140a/1385975160621/OA_spm2-FULL-lorez.pdf (accessed on 16 October 2024).
44. IPCC CZMS. (1990). Strategies for adaptation to sea-level rise. In *Climate Change: The IPCC Response Strategies* (pp. 128–159). Island Press.

45. IPCC. (1990). *Climate Change: The IPCC Scientific Assessment*. Report prepared by Working Group I (Houghton, J. T., Jenkins, G. J., & Ephraums, J. J., Eds.). Cambridge University Press.
46. IPCC. (1995). *Climate Change 1995: The Science of Climate Change*. Contribution of Working Group I to the Second Assessment Report of the Intergovernmental Panel on Climate Change (Houghton, J. T., Meira Filho, L. G., Callender, B. A., Harris, N., Kattenberg, A., & Maskell, K., Eds.). Cambridge University Press.
47. IPCC. (2000). *Special Report on Emission Scenarios*. A special report of Working Group III of the Intergovernmental Panel on Climate Change (Nakicenovic, N., & Swart, R., Eds.). Cambridge University Press.
48. IPCC. (2001). *Climate Change 2001: The Scientific Basis*. Contribution of Working Group I to the Third Assessment Report of the Intergovernmental Panel on Climate Change (Houghton, J. T., Ding, Y., Griggs, D. J., Noguer, M., van der Linden, P. J., Dai, X., Maskell, K., & Johnson, C. A., Eds.). Cambridge University Press.
49. IPCC. (2001). Summary for policymakers. *A Report of Working Group I of the IPCC*.
50. IPCC. (2007). Climate change 2007: Synthesis report. In R. K. Pauchauri & A. Reisinger (Eds.), *Contribution of Working Groups I, II, and III to the Fourth Assessment Report of the Intergovernmental Panel on Climate Change*. Geneva, Switzerland: IPCC.
51. IPCC. (2007a). *Climate Change 2007: The Physical Science Basis*. Contribution of Working Group I to the Fourth Assessment Report of the IPCC (Solomon, S., Qin, D., Manning, M., Chen, Z., Marquis, M., Averyt, K. B., Tignor, M., & Miller, H. L., Eds.). Cambridge University Press.
52. IPCC. (2007b). *Summary for policymakers*. In *Climate Change 2007: The Physical Science Basis*. Contribution of Working Group I to the Fourth Assessment Report of the Intergovernmental Panel on Climate Change (Solomon, S., Qin, D., Manning, M., Chen, Z., Marquis, M., Averyt, K. B., Tignor, M., & Miller, H. L., Eds.). Cambridge University Press.
53. IPCC. (2013). Summary for policymakers. In T. F. Stocker, D. Qin, G. K. Plattner, M. Tignor, S. K. Allen, J. Boschung, A. Nauels, Y. Xia, V. Bex, & P. M. Midgley (Eds.), *Climate Change 2013: The Physical Science Basis*. Cambridge University Press. Retrieved from: https://www.ipcc.ch/pdf/assessment-report/ar5/wg1/WG1AR5_SPM_FINAL.pdf (accessed on 16 October 2024).
54. Ishii, M., & Kimoto, M. (2009). Reevaluation of historical ocean heat

content variations with time-varying XBT and MBT depth bias corrections. *Journal of Oceanography, 65*, 287–299.

55. Isobe, M. (2013). Impact of global warming on coastal structures in shallow water. *Ocean Engineering*, in press.
56. Jackson, J. B. C., Kirby, M. X., Berger, W. H., Bjorndal, K. A., Botsford, L. W., Bourque, B. J., Bradbury, R. H., Cooke, R., Erlandson, J., Estes, J. A., Hughes, T. P., Kidwell, S., Lange, C. B., Lenichan, H. S., Pandolfi, J. M., Peterson, C. H., Steneck, R. S., & Tegner, M. J. (2001). Historical overfishing and the recent collapse of coastal ecosystems. *Science, 293*, 629–638.
57. Japan Meteorological Agency, Global Environment and Marine Department. (2012). Long-term trend of sea level around Japan. Retrieved from: http://www.data.kishou.go.jp/kaiyou/shindan/a_1/sl_trend/sl_trend.html.
58. Jones, G. P., McCormick, M. I., Srinivasan, M., & Eagle, J. V. (2004). Coral decline threatens fish biodiversity in marine reserves. *Proceedings of the National Academy of Sciences, 101*, 8251–8253.
59. Kitoh, A., Ose, T., Kurihara, K., Kusunoki, S., Sugi, M., & KAKUSHIN Team-3 Modeling Group. (2009). Projection of changes in future weather extremes using super-high-resolution global and regional atmospheric models in the KAKUSHIN Program: Results of preliminary experiments. *Hydrological Research Letters, 3*, 49–53.
60. Klein, R. J. T., & Tol, R. S. J. (1997). Adaptation to climate change: Options and technologies—An overview paper. *Technical Paper FCCC/TP/1997/3*. UNFCC Secretariat.
61. Kleypas, J. A., Buddemeier, R. W., Archer, D., Gattuso, J. P., Langdon, C., & Opdyke, B. N. (1999). Geochemical consequences of increased atmospheric CO_2 on coral reefs. *Science, 284,* 118–120.
62. Klyashtorin, L. (1998). Long-term climate change and main commercial fish production in the Atlantic and Pacific. *Fisheries Research, 37,* 115–125.
63. Komine, H., & Duc, D. M. (2011). Geotechnical and geoenvironmental engineering measures to adapt and mitigate global warming impacts. *VNU Journal of Science, Earth Sciences, 27*(1S), 104–113.
64. Krishnakumar, P. K. (2019). Shrimp production in Odisha hit by Cyclone Fani. *The Economic Times.* Retrieved from: https://economictimes.indiatimes.com/news/economy/agriculture/shrimp-production-in-odisha-hit-by-cyclone-fani/articleshow/69370625.cms?from=mdr (accessed on 16 October 2024).

65. Kroeker, K. J., Kordas, R. L., Crim, R., Hendriks, I. E., Ramajo, L., Singh, G. S., Duarte, C. M., & Gattuso, J. P. (2013). Impacts of ocean acidification on marine organisms: Quantifying sensitivities and interaction with warming. *Global Change Biology, 19*(6), 1884–1896.

66. Lemke, P., Ren, J., Alley, R. B., Allison, I., Carrasco, J., Flato, G., Fujii, Y., Kaser, G., Mote, P., Thomas, R. H., & Zhang, T. (2007). Observations: Changes in snow, ice and frozen ground. In S. Solomon, D. Qin, M. Manning, Z. Chen, M. Marquis, K. B. Averyt, M. Tignor, & H. L. Miller (Eds.), *Climate Change 2007: The Physical Science Basis. Contribution of Working Group I to the Fourth Assessment Report of the IPCC* (pp. 337–383). Cambridge University Press.

67. Levitus, S., Antonov, J. I., Boyer, T. P., Baranova, K., Garcia, H. E., Locarnini, R. A., Mishonov, A. V., Reagan, J. R., Seidov, D., Yarosh, E. S., & Zweng, M. M. (2012). World ocean heat content and thermosteric sea level change (0–2000 m), 1955–2010. *Geophysical Research Letters, 39,* L10603.

68. Lombard, A., Garric, G., Penduff, T., & Molines, J. M. (2009). Regional patterns of observed sea level change: Insights from a 1/4° global ocean/sea-ice hindcast. *Ocean Dynamics, 59,* 433–449.

69. Long, M. C., Deutsch, C., & Ito, T. (2016). Finding forced trends in oceanic oxygen. *Global Biogeochemical Cycles, 30*(2), 381–397.

70. Macpherson, E. (2002). Large-scale species-richness gradients in the Atlantic Ocean. *Proceedings of the Royal Society of London Series B, 269,* 1715–1720.

71. Maruyama, Y., & Mimura, N. (2010). Global assessment of climate change impacts on coastal zones with combined effects of population and economic growth. *Selected Papers of Environmental Systems Research, JSCE, 38,* 255–263.

72. McLean, R. F., Tsyban, A., Burkett, V., Codignotto, J. O., Forbes, D. L., Mimura, N., Beamish, R. J., & Ittekkot, V. (2001). Coastal zones and marine ecosystems. In J. J. McCarthy, O. F. Canziani, N. A. Leary, D. J. Dokken, & K. S. White (Eds.), *Climate Change 2001: Impacts, Adaptation, and Vulnerability. Contribution of Working Group II to the Third Assessment Report of the IPCC* (pp. 343–379). Cambridge University Press.

73. Meehl, G. A., Stocker, T. F., Collins, W. D., Friedlingstein, P., Gaye, A. T., Gregory, J. M., Kitoh, A., Knutti, R., Murphy, J. M., Noda, A., Raper, S. C. B., Watterson, I. G., Weaver, A. J., & Zhao, Z. C. (2007). Global climate projections. In S. Solomon, D. Qin, M. Manning, Z. Chen, M. Marquis, K. B. Averyt, M. Tignor, & H. L. Miller (Eds.), *Climate Change*

2007: The Physical Science Basis. Contribution of Working Group I to the Fourth Assessment Report of the IPCC (pp. 747–845). Cambridge University Press.

74. Mimura, N., & Nobuoka, H. (1996). Verification of the Bruun rule for the estimation of shoreline retreat caused by sea-level rise. *Coastal Dynamics 95,* ASCE, 607–616.

75. Mimura, N., Inoue, K., Kiyohashi, M., Izumiya, T., & Nobuoka, H. (1994). Assessment of sea-level rise impact on sandy beaches (2): Verification of predictive model and national assessment. *Proceedings of Coastal Engineering, JSCE, 41,* 1161–1165.

76. Minar, M. H., Hossain, M. B., & Shamsuddin, M. D. (2013). Climate change and coastal zone of Bangladesh: Vulnerability, resilience, and adaptability. *Middle-East Journal of Scientific Research, 13*(1), 114–120.

77. Moss, R., Babiker, M., Brinkman, S., Calvo, E., Carter, T., Edmonds, J., Elgizouli, I., Emori, S., Erda, L., Hibbard, K., Jones, R., Kainuma, M., Kelleher, J., Lamarque, J. F., Manning, M., Matthews, B., Meehl, J., Meyer, L., Mitchell, J., Nakicenovic, N., O'Neill, B., Pichs, R., Riahi, K., Rose, S., Runci, P., Stouffer, R., van Vuuren, D., Weyant, J., Wilbanks, T., Ypersele, J. P., & Zurek, M. (2008). Towards new scenarios for analysis of emissions, climate change, impacts, and response strategies. *IPCC,* Geneva.

78. Mueter, F. J., & Litzow, M. A. (2008). Sea ice retreat alters the biogeography of the Bering Sea continental shelf. *Ecological Applications, 18*(2), 309–320.

79. Nandi, J., & Thakur, J. (2020). Amphan transforming into super cyclone, first after deadly 1999 super cyclone in Bay of Bengal. *Hindustan Times.* Retrieved from: https://www.hindustantimes.com/india-news/amphan-transforming-into-super-cyclone-first-after-deadly-1999-super-cyclone-in-bay-of-bengal/story-12vMmByNipIQwibsJwHJIO.html (accessed on 16 October 2024).

80. Narita, D., Rehdanz, K., & Tol, R. S. (2012). Economic costs of ocean acidification: A look into the impacts on global shellfish production. *Climatic Change, 113*(3, 4), 1049–1063.

81. Nicholls, R. J., & Cazenave, A. (2010). Sea-level rise and its impact on coastal zones. *Science*, *328*, 1517–1520.

82. Nicholls, R. J., Hoozemans, F. M. J., & Marchand, M. (1999). Increasing flood risk and wetland losses due to global sea level rise: Regional and global analyses. *Global Environmental Change*, *9*, S69–S87.

83. Nicholls, R. J., Wong, P. P., Burkett, V. R., Codignotto, J. O., Hay, J. E.,

McLean, R. F., Ragoonaden, S., & Woodroffe, C. D. (2007). Coastal systems and low-lying areas. In M. L. Parry, O. F. Canziani, J. P. Palutikof, P. J. van der Linden, & C. E. Hanson (Eds.), *Climate Change 2007: Impacts, Adaptation and Vulnerability. Contribution of Working Group II to the Fourth Assessment Report of the IPCC* (pp. 315–356). Cambridge University Press.

84. Nichols, R., Hoozemans, F. M. J., & Marchand, M. (1999). Increasing food risk and wetland losses due to global sea level rise: Regional and global analyses. *Global Environmental Change*, *9*, S69–S87.

85. NOAA. (2013). Sea Levels Online. Retrieved from: http://tidesandcurrents.noaa.gov/sltrends/sltrends.html (accessed on 16 October 2024).

86. Noguchi, K., Suwa, Y., Gomi, H., & Matsufuji, E. (2010). For the setting of the benchmark value of previous sea level rise along Japan coasts for climate change adaptation. *Journal of JSCE, Ser. B2 (Coastal Engineering)*, *66*, 1256–1260. (in Japanese)

87. Noonan, S. H., & Fabricius, K. E. (2015). Ocean acidification affects productivity but not the severity of thermal bleaching in some tropical corals. *ICES Journal of Marine Science: Journal du Conseil*, *73*(3), 715–726.

88. Oouchi, K., Yoshimura, J., Yoshimura, H., Mizuta, R., Kusunoki, S., & Noda, A. (2006). Tropical cyclone climatology in a global warming climate as simulated in a 20 km-mesh global atmospheric model: Frequency and wind intensity analyses. *Journal of Meteorological Society of Japan*, *84*, 259–276.

89. Orr, J. C., Fabry, V. J., Aumont, O., Bopp, L., Doney, S. C., Feely, R. A., & Gattuso, J. P. (2005). Anthropogenic ocean acidification over the twenty-first century and its impact on calcifying organisms. *Nature*, *437*(7059), 681–686.

90. Parmesan, C., & Yohe, G. (2003). A globally coherent fingerprint of climate change impacts across natural systems. *Nature*, *421*, 37–42.

91. Parry, M. L., Canziani, O. F., Palutikof, J. P., van der Linden, P. J., & Hanson, C. E. (2007). *Climate Change 2007: Impacts, Adaptation and Vulnerability: Working Group II Contribution to the Fourth Assessment Report of the IPCC*. Cambridge University Press.

92. Peltier, W. R. (2009). Closure of the budget of global sea level rise over the GRACE era: The importance and magnitudes of the required corrections for global isostatic adjustment. *Quaternary Science Reviews*, *28*, 1658–1674.

93. Perry, A. L., Low, P. J., Ellis, J. R., & Reynolds, J. D. (2005). Climate

change and distribution shifts in marine fishes. *Science*, *308*, 1912–1915.

94. Peterson, A. T., Ortega-Huerta, M. A., Bartley, J., Sánchez-Cordero, V., Soberón, J., Buddemeier, R. H., & Fa, J. E. (2002). Future projections for Mexican faunas under global climate change scenarios. *Nature*, *416*, 626–629.
95. Plagányi, É. (2019). Climate change impacts on fisheries. *Science*, *363*(6430), 930–931.
96. Poff, N. L., Brinson, M. M., & Day, J. W. (2002). Aquatic ecosystems and global climate change: Potential impacts on inland freshwater and coastal wetland ecosystems in the United States. Retrieved from: www.pewclimate.org (accessed on 16 October 2024).
97. Pokhrel, Y. N., Hanasaki, N., Yeh, P. J. F., Yamada, T. J., Kanae, S., & Oki, T. (2012). Model estimates of sea-level change due to anthropogenic impacts on terrestrial water storage. *Nature Geoscience*, *5*, 389–392.
98. Project Team for Comprehensive Projection of Climate Change Impacts. (2008). *Global Warming Impacts on Japan–Latest Scientific Findings* (pp. 1–95). Retrieved from: http://www.nies.go.jp/s4_impact/pdf/S-4_report_2008eng.pdf (accessed on 16 October 2024).
99. Project Team for Comprehensive Projection of Climate Change Impacts. (2009). *Global Warming Impacts on Japan–Long-Term Climate Stabilization Level and Impact Risk Assessment* (pp. 1–40). Retrieved from: http://www.nies.go.jp/s4_impact/pdf/S-4_report_2009eng.pdf (accessed on 16 October 2024).
100. Rahmstorf, S. (2007). A semi-empirical approach to projecting future sea level rise. *Science*, *315*, 368–370.
101. Ralbam, V. (2010). Impact of climate change on Indian marine fisheries and option for adaptation. In *Lobster Research in India* (pp. 65–71). Kochi, India: CMFRI.
102. Ray, R. D., & Douglas, B. C. (2011). Experiments in reconstructing twentieth-century sea levels. *Progress in Oceanography*, *91*, 496–515.
103. Robinson, A., Calov, R., & Ganopolski, A. (2012). Multistability and critical thresholds of the Greenland ice sheet. *Nature Climate Change*, *2*, 201–205.
104. Roessig, J. M., Woodley, C. M., Cech, J. J., & Hansen, L. J. (2004). Effects of global climate change on marine and estuarine fishes and fisheries. *Reviews in Fish Biology and Fisheries*, *14*(2), 251–275.
105. Rohling, E. J., Grant, K., Bolshaw, M., Roberts, A. P., Siddall, M., Hemleben, C., & Kucera, M. (2009). Antarctic temperature and global sea

level closely coupled over the past five glacial cycles. *Nature Geoscience*, *2*(7), 500–504.

106. Root, T. L., Price, J. T., Hall, K. R., Schneider, S. H., Roxenzweiq, C., & Pounds, J. A. (2003). Fingerprints of global warming on wild animals and plants. *Nature*, *421*, 57–60.
107. Rosenzweig, C., Karoly, D., Vicarelli, M., et al. (2008). Attributing physical and biological impacts to anthropogenic climate change. *Nature*, *453*, 353–357.
108. Sala, O. E., Chapin, F. S. III, Armesto, J. J., et al. (2000). Global biodiversity scenarios for the year 2100. *Science*, *287*, 1779–1784.
109. SCOR. (1991). The response of beaches to sea-level changes: A review of predictive models. *Journal of Coastal Research*, *7*, 895–921.
110. Secretariat of the Convention of Biological Diversity. (2014). *An Updated Synthesis of the Impacts of Ocean Acidification on Marine Biodiversity* (Technical Series No. 75). Montreal: Hennige, S., Roberts, J. M., & Williamson, P. (Eds.).
111. Stachowicz, J. J., Fried, H., Osman, R. W., & Whitlatch, R. B. (2002). Biodiversity, invasion resistance, and marine ecosystem function: Reconciling pattern and process. *Ecology*, *83*, 2575–2590.
112. Stocker, T., Dahe, Q., Plattner, G. K., Tignor, M., Allen, S., & Midgley, P. (2010). *IPCC Workshop Report: Workshop on Sea Level Rise and Ice Sheet Instabilities*. The IPCC Working Group I Technical Support Unit, University of Bern, Switzerland.
113. Sugawa, T., Udo, K., Mimura, N., & Mano, A. (2011). Projection of shoreline retreat due to sea-level rise along Japanese coasts. *Journal of JSCE, Ser. B2 (Coastal Engineering)*, *67*, 1196–1200. (in Japanese)
114. Suzuki, T. (2009). Estimation of inundation damage caused by global warming in three major bays and western parts of Japan. *Global Environment, Association of International Research Initiatives for Environmental Studies*, *14*(2), 231–236. (in Japanese)
115. Svendsen, P. L., Andersen, O. B., & Nielsen, A. A. (2013). Acceleration of the Greenland ice sheet mass loss as observed by GRACE: Confidence and sensitivity. *Earth and Planetary Letters*, *364*, 24–29.
116. Thomas, C. D., Williams, S. E., Cameron, A., et al. (2004). Biodiversity conservation: Uncertainty in predictions of extinction risk. *Nature*, *430*, 1, 2.
117. Turley, C., & Gattuso, J. P. (2012). Future biological and ecosystem impacts of ocean acidification and their socioeconomic-policy implications.

Current Opinion in Environmental Sustainability, *4*(3), 278–286.

118. Walther, G. R., Post, E., Convey, P., Menzel, A., Parmesan, C., Beebee, T. J., et al. (2002). Ecological responses to recent climate change. *Nature*, *416*(6879), 389.

119. Watson, R. T., Zinyowera, M. C., Moss, R. H., & Dokken, D. J. (2001). *IPCC Special Report: The Regional Impacts of Climate Change: An Assessment of Vulnerability*. United Nations: IPCC.

120. Wittmann, A. C., & Pörtner, H. O. (2013). Sensitivities of extant animal taxa to ocean acidification. *Nature Climate Change*, *3*(11), 995–1001.

121. Wong, P. P., Losada, I. J., Gattuso, J. P., Hinkel, J., Khattabi, A., McInnes, K. L., Saito, Y., & Sallenger, A. (2014). Coastal systems and low-lying areas. In C. B. Field, V. R. Barros, D. J. Dokken, K. J. Mach, M. D. Mastrandrea, T. E. Bilir, M. Chatterjee, K. L. Ebi, Y. O. Estrada, R. C. Genova, B. Girma, E. S. Kissel, A. N. Levy, S. MacCracken, P. R. Mastrandrea, & L. L. White (Eds.), *Climate Change 2014: Impacts, Adaptation, and Vulnerability. Part A: Global and Sectoral Aspects. Contribution of Working Group II to the Fifth Assessment Report of the Intergovernmental Panel on Climate Change* (pp. 361–409). Cambridge University Press.

122. Woodworth, P. L., & Player, R. (2003). The permanent service for mean sea level: An update to the 21st century. *Journal of Coastal Research*, *19*, 287–295.

123. World Bank. (2000). *Cities, Seas and Storms: Managing Change in Pacific Island Economies. Vol. IV: Adapting to Climate Change*.

124. World Glacier Monitoring Service. (2013). *World Glacier Inventory*. Retrieved from: http://www.gtn-g.org/data_overview.html (accessed on 16 October 2024).

125. Worm, B., Barbier, E. B., Beaumont, N., et al. (2006). Impacts of biodiversity loss on ocean ecosystem services. *Science*, *314*, 787–790.

126. Yasuda, T., Ando, K., Mori, N., & Mase, H. (2009). Prediction and stochastic modeling of future typhoon characteristics based on AGCM projections under a global warming scenario. *Journal of JSCE, Ser. B2 (Coastal Engineering)*, *65*, 1281–1285.

127. Yasuda, Y., Nakajo, S., Kim, S. Y., Mori, N., Mase, H., & Horsburgh, K. (2011). Evaluation of storm surge risk directly based on climate change projection. *Journal of JSCE, Ser. B2 (Coastal Engineering)*, *67*, 1171–1175.

128. Yasuhara, K., Komine, H., Murakami, S., Chen, G., & Mitani, Y. (2010). Effects of climate change on geo-disasters in coastal zones. *Journal of Global Environmental Engineering, JSCE*, *15*, 15–23.

Chapter **5**

Aquatic Biodiversity Monitoring and Assessment

Contents

INTRODUCTION

Approximately 0.01% of the globe's total water volume is found in surface freshwater habitats, occupying a minuscule 0.08% of the Earth's surface. Estimates suggest that between 50,000 to 100,000 different species might call these subterranean freshwater reservoirs home. The rich biodiversity of freshwater sources provides a broad spectrum of essential services. Regrettably, these ecosystems are among the most at risk of extinction due to human activities and infrastructure development, which are exacerbated by the effects of climate change. Aquatic biodiversity encompasses the diverse range of life and ecosystems in the world's freshwater, tidal, and marine regions, as well as the complex interactions between them. This includes all unique species, their habitats, and the intricate relationships that exist between them. Aquatic biodiversity is comprised of phytoplankton, zooplankton, aquatic plants, insects, fish, birds, and mammals.

Many synthetic and natural chemicals have been created and used over the past few decades without consideration for the potential effects that their release, along with any contaminants or degradation products, may have on the environment. The overall load of pollutants has undoubtedly contributed to the observed changes in the structure and function of aquatic ecosystems, even though the majority of these chemicals did not have an immediate impact on the aquatic environment. For instance, the massive development of pesticides following World War II and the consequent detrimental effects on consumers (birds and other organisms) of contaminated prey, along with the emergence of diseases like Itai-ltai and Mina Mata caused by toxic metals, led to a shift in the focus of environmental pollutants after the publication of Rachel Carson's book Silent Spring in 1962. Ever since, there has been a growing interest in the early identification of chemical pollutants' undetectable environmental effects.

The purpose of aquatic biological monitoring is to enable the protection of aquatic ecosystems, to provide accurate and trustworthy information on the potential effects of chemicals that may be present in the water as a result of human activity, and, in particular, to provide scientific guidance for legislation and enforcement.

The two new global issues (ozone depletion, greenhouse effect, global climate change, etc.) and the recognition of the intricacy of ecosystems in environmental policy indicate the need for more knowledge about biological systems. Beginning in the 1960s, the majority of monitoring initiatives focused on chemical and physical parameters. It has come to our attention that while the majority of monitoring programs are capable of giving a sensitive indication of the concentration of substances chosen for study, vital biological information has been absent from this data. Negative effects are implied by pollution, and these are typically evaluated in light of a biological system. Bioindicators, also known as biomonitors, are biological objects that are used to monitor the state of the environment and changes brought about by environmental pollution. These objects span a wide range of biological systems and functions, from population dynamics to subcellular processes.

The focus is on biological variable monitoring because these systems can be inexpensive, require less complex equipment, and—above all—reflect the integrated expression of pollution load. However, in order to identify the causal factors, parallel chemical monitoring has to be carried out in all cases.

5.1. AQUATIC BIOMONITORING

The science of determining the ecological state of rivers, lakes, streams, and wetlands by observing the organisms—fish, invertebrates, insects, plants, and algae—that call these places home is known as aquatic biomonitoring. Even though aquatic biomonitoring is the most popular type, this method can be used to study any ecosystem.

5.1.1. Definition and Functions of Biomonitoring

Generally speaking, monitoring is the process of repeatedly observing one or more environmental elements for a specific purpose, in accordance with predetermined schedules in terms of both space and time, and using comparable techniques for environmental sensing and data collection. By using (aquatic) organisms as natural monitors, biomonitoring offers factual and supplementary information about historical and current environmental behavior trends. In general, the monitoring systems can be regulatory, similar to control, or descriptive in terms of emission (concentration or effects). According to this, these systems have some common functions:

- Signal function is used for the detection of adverse changes in the environment, for early warning, and in searching for the causes of changes (localization of sources and emissions).
- Prediction function deals with the prediction of autonomous changes in the environment and contribution to the prediction of the effect of human activities to improve the environmental (measures) or any other factor producing some load on the environment.
- Control function is frequently used for the observation of measurements or the achievement of goals concerning environmental quality.

- Scientific research for environmental management as activity-effect studies, or construction of compound balances and a response to the specific questions regarding the control of government concerning policy subjects, all correspond to the instrument function.

Biological responses are used in biological monitoring, also known as biomonitoring, to evaluate environmental changes, most of which are brought about by human activity. Programs for biomonitoring can be quantitative, semiquantitative, or qualitative. Biological monitoring can be classified as semicontinuous or continuous depending on the type of biosensors used to detect changes in water quality, the response variables provided by the organisms (chemical, electrical, computer analysis, etc.), and the nature of the chemical compounds being monitored (fluctuating qualitatively, quantitatively, or both). Programs for monitoring water quality across the board are increasingly utilizing biomonitoring, a useful assessment tool. Biomonitoring comes in a variety of forms, depending on the objectives of the project. Surveillance before and after a project is finished or before and after a toxic substance enters the water is one kind of biomonitoring. The other kind of biomonitoring makes sure that rules and regulations are followed or that the quality of the water is preserved.

Indicators, indicator species, or indicator communities are used in biomonitoring. Fish, algae, and/or benthic macroinvertebrates are typically employed. A few aquatic plants have also been employed as pollutant indicator species, such as those that have been enriched with nutrients. There are benefits and drawbacks to every approach. Among biomonitors, macroinvertebrates are the most commonly utilized. Certain organisms have been shown to exhibit biochemical, genetic, morphological, and physiological changes that are indicative of specific environmental stressors and can be employed as such. Environmental conditions are reflected in the existence or absence of an indicator, or an indicator species or community. The disappearance of a species is not as significant as it would appear because factors other than pollution can cause it to disappear (e.g., competition, or geographical obstacles that kept it from ever existing at the location). More evidence of pollution than the absence of a single species can be found when several species of different orders with comparable tolerance levels that were previously present at the same site are absent. Knowing which species should be present at the site or in the system is obviously essential.

Human society uses monitoring to try and alert the environment to changes brought about by human activity. Its goal is to keep the ecosystem's biota under quality control. Early implementation of this prospective signaling is crucial to averting unfavorable changes in the ecosystem. This suggests that monitoring should only provide the appearance of protection; when harmful effects are discovered, prompt corrective action should be taken. Deep core sediment and ice samples are used to record environmental conditions that existed before the Industrial Revolution, while retrospective monitoring is used to describe historical environmental conditions (e.g., museum species). There are already a lot of biological and chemical monitoring techniques described; however, many more will be developed in the future. They all agree on the need to foresee potential chemical risks to people and the environment in advance.

5.1.2. Purpose of Aquatic Biomonitoring

An essential technique for evaluating aquatic organisms and their environments is aquatic biomonitoring. It can show the ecosystem's general health and status, identify environmental trends and the effects of various stressors, and assess the impact of different human activities on the aquatic environments' general health. The environment is significantly impacted by water pollution and general stresses to aquatic life. The primary causes of pollution in rivers, lakes, and oceans are human-caused incidents and actions, including surface runoff, littering, oil spills, sewage, ocean mining, and nuclear waste.

Keeping an eye on aquatic life can help you monitor and comprehend the ecosystems of nearby land. Swift alterations in the surroundings, such as pollution, have the potential to transform ecosystems and community assemblages, thereby posing a threat to aquatic or near-water species. The size and health of aquatic populations affect terrestrial species because many aquatic species are food sources for these species.

5.1.3. Relation Between Biomonitoring and Toxicology

Ecotoxicology is a unique combination of toxicology and ecology that aims to study and describe the effects of toxic substances on the ecosystem. Biomonitoring of environmental pollution involves the use of ecotoxicological approaches and methods.

Ecology studies the interactions and relationships between individuals and species that make up a particular biological community, known as an ecosystem, as well as the relationships that organisms have with the physical environment. The microbes, plants, and animals that live in a river, lake (reservoir), or sea are known as aquatic ecosystems. They are distinguished by their dynamic equilibrium and well-defined interrelationships. The physical and chemical makeup of the water as well as the water itself define the quality of the environment that the living inhabit. The water's chemical and physical properties exhibit notable fluctuations and changes that are constantly influenced by the surrounding environment. Significant changes in temperature, precipitation, water supply, and evaporation can affect these elements. To a certain extent, aquatic organisms adjust to these shifts by changing aspects of their lifestyle that are directly related to their physical and chemical surroundings, such as feeding, moving, reproducing, and metabolism.

Toxicological studies the negative effects that different chemicals have on living things; these effects are referred to as toxic effects. When a chemical is toxic, it interferes with metabolism at the cellular and molecular levels, disrupting life processes. Toxic substances are typically found in very small amounts within the ecosystem and are regarded as pollutants. Anthropogenic pollutants are toxic substances that end up concentrated, produced, or released into the environment as a result of human activity. Some of them are byproducts of biological systems (bacteria, plants, and animals), as well as inorganic or organic products and byproducts of the chemical industry and other technological processes. Some are natural elements of the geosphere (toxic metals), which over time appear in the environment in concentrations surpassing the geochemical background.

Pollutants in the environment can have a significant impact on freshwater and marine ecosystems, disrupting the relationships between various species as well as the organisms that live there. Although the effects can be seen at the individual level in certain sensitive species, they can also be applied to the ecosystem as a whole. As a result, research on the relationship between toxicology and ecology, specifically the detection of environmental pollutants and their biological effects, covers the sub-individual, individual, population, and ecosystem levels in equal measure.

5.1.4. Origin and Types of Toxic Chemicals in the Environment

Not merely for didactic purposes, the sources of anthropogenic environmental pollution—which are the results of human activity—were previously separated into four categories: industrial, agricultural, community, and airborne. Along with the primary activity types that cause the pollution, this type of grouping also refers to the specificity of the release and potential counteraction strategies to remove unfavorable effects. For instance, a large number of toxic substances are produced as a result of industrial activity. Mining for ore, metals and other elements found in the geosphere, as well as their separation, concentration, and procession, are all done through industrial technologies. Many, many toxic substances become soluble or enter the air as a result of processing metal, producing different chemicals, using and treating wastes and byproducts, and so on. A significant portion of environmental pollution is caused by dissolved materials, which are discharged into natural waters as parts of treated or raw waste waters in varying concentrations. The use of closed, waste-free technologies, the purification of contaminated industrial waters at the site of production, and the implementation of countermeasures to lessen the release of pollutants into the air provides the best defense against these kinds of chemical pollution.

In actuality, pollution that arises as a consequence of agricultural technologies also has industrial roots. Due to their toxic biological effects on pests, herbicides, fungicides, and pesticides are used in plant protection; however, when they end up in rivers and lakes, they are deemed toxic pollutants due to the detrimental effects they have on the aquatic life and ecosystem. Since toxic chemicals used in plant protection or animal husbandry are applied widely over a wide area in the field, it is nearly impossible to collect and remove residues or toxic metabolites that appear as pollutants. In this instance, limiting the amount of chemicals used and using medications that quickly deteriorate after providing therapeutic benefits can help reduce pollution.

Toxic materials that leak out of settlements and into the environment are regarded as having a communal origin. This group includes common household chemicals like detergents, bactericides, cosmetics, medications, stain removers, and everyday drugs. It also includes waste products from heating and transportation (coal and oil combustion), metal instrument corrosion, and construction waste. A significant quantity of hazardous materials is released into the atmosphere by wind and washing out of waste dumps and spoils, eventually ending up in surface waters. Canalization, sewage water purification, waste collection, and appropriate waste treatment are some ways to reduce settlement pollution in the environment. Limiting the use of persistent toxicants can also help reduce pollution coming from municipalities. Runoff from roads, settlements, and other sources significantly contaminates rivers and lakes.

The main causes of air pollution are industrial and societal. However, because pollutants released into the atmosphere travel great distances and have fallout that occurs far from their source in the form of dry sedimentation or precipitation, it is a unique and independent category. It is possible for both organic and inorganic chemicals to be among the harmful substances found in the air. The amount and quality of pollutants can be measured where they are actually present, but the only location where air pollution can be reduced is the source of the pollution. Because of these unique conditions, it necessitates not only local or governmental regulation and intervention but frequently intergovernmental as well. Toxic chemicals occurring as pollutants can be grouped into two main categories:

- Substances of natural origin.
- Substances produced exclusively by human activity.

The first group includes some naturally occurring nonmetallic elements as well as heavy metals and their compounds. Mercury, cadmium, lead, chromium, aluminum, cobalt, nickel, and a few other elements that are crucial to life, like copper, zinc, and sulfur, are the most significant toxic metals. Within this group, radioactive elements constitute a unique category. Low-background pollution is a naturally occurring phenomenon that has been caused by metals being released and rocks weathering over Earth's history. Since the dawn of human civilization, these elements have been extracted in vast quantities through mining and metallurgy, and with their use comes the potential for environmental contamination. Every time metals are worked with and used for industrial, agricultural, or communal purposes, wastes and byproducts are created that could, in greater or lesser amounts, end up in the environment. They may become soluble in ionic form or through chemical or microbial processes; when these substances interact, they may be harmful to living things.

The chemical industry creates thousands of compounds that are used in homes, businesses, pharmacies, and agriculture; a large portion of these compounds are hazardous to the environment. There may be negative effects from the compounds themselves or from their metabolic products. Because the toxicity of inorganic and organic compounds containing metals is dependent on the metal ion, they exhibit a strong correlation with the former group. The most significant toxic compounds are the other substances, which are organic pollutants devoid of metal ions. These include dioxins, nitrosamines, cyanides, polycyclic aromatic hydrocarbons, polychlorinated biphenyls, chlorinated hydrocarbons, and organic esters.

5.1.5. Forms and Manner of Occurrence of Pollutants in the Aquatic Ecosystem

Depending on where they come from, the majority of toxic pollutants end up in surface waters in dissolved form at varying concentrations. Others enter lakes and rivers, where they combine with solid waste that is attached to particles and settles as sediment. The solubility of substances in water is frequently pH dependent; therefore, when the pH of the water changes, toxic substances that are dissolved in it may precipitate or become free of their bonds and enter the solution. Microbial activity and complex formation may lead to similar processes. This could lead to changes in the concentration of dissolved toxic substances in the water, which would alter their accessibility for accumulation in living things and their efficacy for biological interference.

Water movements, sediment mixing, and diffusion cause hazardous materials to typically be dispersed evenly over a wide area, frequently in a vertical direction as well. However, the concentration is higher near point sources that are diagnostically significant for locating the pollution's source. Toxic substance stratification may exist in more compacted sediments, offering insight into the pollution's historical trajectory. Additionally, living things may contribute to the dispersion and mixing of pollutants by physically upsetting the water and benthos as well as through chemical turnover caused by their metabolism.

There are several ways in which dissolved toxic substances can be absorbed by organisms. Through absorption or the involvement of active transport mechanisms, they can pass through the membranes and enter the interior space. Specific membrane permeability is necessary for absorption, but an energy-intensive carrier mechanism is required for transport. Pollutant accumulation may involve both of these processes.

Fishes and macroinvertebrates also absorb toxic substances that have been adsorbed to particles. Benthic organisms that feed in the sediment and planktonic animals that filter the suspended elements are the first to do so. Toxic substances may be released, solubilized by the stomach and intestinal processes, and then pass through the endothelium to enter the extracellular fluid inside the cell. Toxic substances may be directly absorbed by rooted aquatic plants from the sediment. Pollutant accumulation occurs because the release of toxic substances proceeds more slowly than their uptake. Through the food chain and the ingestion of contaminated organisms, a great deal of pollution is absorbed and stored in animals (biomagnification). Toxic substances can accumulate in living organisms and increase their toxic effects, but they can also be detoxified by binding, deposition in an inactive form, or excretion, which releases them back into the water or sediment.

5.1.6. Effects of Toxic Pollutants on Aquatic Ecosystems

5.1.6.1. Pollution Effects on River Ecosystems

Rivers are open systems because of their strong directionality and interactions caused by drainage basins. The effects of stream water pollution on communities are ultimately determined by the unique characteristics of riverine ecosystems. The presence of living things in flowing water is primarily determined by temporal dimension and spatial gradients.

- The longitudinal gradients-increasing depth and width of the riverbed, decreasing stream velocity and sediment grain size, and increasing nutrient load from headwaters to estuaries are a result of the aggregated effects of hydrological, morphological, geological, geographical, and climatic factors of the catchment areas. These abiotic factors of the longitudinal gradients fundamentally determine the ecological zonation of communities manifested in the structure and function of the river continuum.

- The aquatic zone of the riverine ecosystem is interlinked laterally with the riparian (littoral) and terrestrial (floodplains, wetlands) zones. A great variety of different habitats result from the transverse gradients of stream velocity, erosion, and sedimentation.
- The third spatial dimension appears in the vertical gradients of deepwater, in the sediment and interstitial waters of the river. The latter region is the indispensable life space of the special groundwater biota and the location of essential chemical processes or bank filtration of drinking water.
- Temporal variation-the fourth dimension-may be natural (floods, seasonal changes in water regimen) or human-induced. In the latter case, the time scale may be very short, and the effect of human impact (e.g., lack of natural events because of damming or fragmentation of a river) may be very drastic.

The habitats and species composition of the communities along the rivers and watercourses vary greatly as a result of all these gradients. It is possible to view the aquatic community as an integrated response to all biological relationships and environmental factors that are currently in place. In addition to hydro-morphological dynamism, other factors that have a significant impact on the riverine ecosystem's temporary state include land use within the catchment area, eutrophication, toxic pollution, canalization, and other river regulation techniques. During the ecological water management monitoring activities, it is necessary to simultaneously address natural and man-made stress factors to assess the degree of ecological integrity of river ecosystems.

On the one hand, we need to improve our ecological understanding of the composition and primary roles of river ecosystems. Correct selection of ecologically representative and sensitive enough parameters for biological and ecological water quality assessment requires an understanding of the nature of underlying processes. On the other hand, defining the natural variety in stream waters requires the development of a biotypological classification for the rivers. In the context of water management, the conceptual and scientifically sound categorization of running waters involving reference sites and aquatic communities might be the only foundation for practical biomonitoring.

While the continuum concept views the watercourse as a continuum of communities, the classical concept divides the river into distinct zones. Sequential communities preserve the constant flow of matter and energy along the river despite structural and functional differences between them. It is debated whether a continuum concept can be applied to large rivers, where discontinuum is the general consensus. If one believes that classification is the executable, practical method of evaluating water quality zone by zone, while a continuum approach is a conceptual foundation for comprehending the function of river ecosystems, then the two ideas may be complementary.

As many factors as possible must be included in a realistic biotypological classification: it must take into account terrestrial, riparian, and interstitial zones (ecotops) in addition to the running water body; it must be valid at various scales (local-regional, national-international, transboundary, etc.); and last but not least, it must adopt a holistic approach with respect to the entire watershed area.

The diffuse and point-like pollution mixes with the geochemical background in large rivers. At times, the distinct impacts are difficult to distinguish from one another (e.g., the downstream portion of the river under study, the combined load of point sources upstream manifests as a diffuse effect). The plume effect, which is caused by variations in morphology, current, and convection in response to the current spatial and temporal gradients in the river, makes the point sources, or so-called hot spots, even more difficult to locate.

The relative approaches that appear to be trustworthy among the complex and variable conditions of the river ecosystem are based on comparison with controls. The best illustration of this kind of evaluation is the translocation of aquatic organisms to monitor pollutants, which is one of the most popular active biomonitoring techniques. While it is impossible to completely eradicate the uncertainties that come with studying local organisms, translocation techniques can greatly lessen them. Big rivers are not only larger than smaller streams, but they also have different water quality characteristics, which can vary even within the river's length and width. As a result, when validating the methods in large rivers, careful research and confirmation are required for the smaller stream methods.

5.1.6.2. Pollution Effects on Lake Ecosystems

Although the general features of lake ecosystems are similar to those of rivers, they possess some unique qualities that are important for tracking toxic pollution. Like rivers, lakes serve as recipients of an open system for the catchment area; however, because of their limited or nonexistent outflow, lakes are closed systems where materials carried by inflow waters or precipitation stay for an extended period of time. Even though the amount of pollution in inflow waters varies, dissolved or particulate chemicals disperse relatively quickly, and marked concentration gradients cannot persist even if there are variations in the mixing of substances caused by currents, wind effects, and diffusion depending on the size and depth of the lake. However, local variations may be compared to the entire lake, particularly in large lakes, at the mouth of rivers or streams, or at the point sources (settlements, harbors, waste water treatment plants). This is less indicative of the chemical makeup of the water itself, but it can be found in the sediment as a result of the gradual dispersion of materials on the bottom's surface.

Like rivers, the amount of precipitation has a significant impact on the level and effects of pollution in shallow or small lakes. However, in large lakes, precipitation is not a significant factor except at the location of the immediate source of the pollution. Moreover, evaporation plays little role in raising the concentration of pollutants in large, deep lakes. As a result, changes in water level in the Great Lakes cannot be expected to cause abrupt, noticeable shifts in the concentration of harmful substances. On the other hand, the concentration of harmful substances in the water can increase dramatically as a result of the mixing and dispersing of sediment.

Unlike rivers, lakes lose a significant portion of their water through evaporation rather than outflow or water takeout. As a result, any toxic materials that enter the ecosystem with inflow water will remain there for all time. Only water outflow, sediment excavation, and the removal of biological products (fish, reed) containing accumulated toxicants can result in a reduction of toxic chemicals. Since accumulation, binding, and deposition in the sediment and biomass are the primary routes of inactivation, the concentration of pollutants in the water has only slightly increased. Toxic chemicals circulate throughout the ecosystem as a result of the majority of living things in the lake dying and decomposing and the accumulated toxic substances returning time and time again to the water, sediment, and living things.

5.1.6.3. Accumulation and Synergism Resulting in Augmentation of the Effect

Unmistakably, the accumulation of toxic substances in surface waters has an ecological impact: living things are put at risk as their concentration rises. However, toxic substances in dissolved form only cause harm to biological systems when they come into direct contact with regulatory mechanisms. Toxic chemicals that form insoluble complexes, bind, or adsorb to particles sink to the bottom and increase sediment pollution without having an immediate negative ecological impact. Such an ideal state is not endless because of the microbial, enzymatic, and chemical processes occurring in the sediment; even complexes and bound forms mobilize and become more or less soluble. In addition, the entire ecosystem, particularly benthic organisms, contributes to the mobilization of inactive toxicants through their metabolism. However, aquatic life also contributes to the deposition, accumulation, binding, inactivation, and sedimentation of dissolved toxicants in water.

The main mechanisms involved in toxicity are the absorption and build-up of toxic substances within living things. While chemicals associated with particles are accessible through feeding and the alimentary system, where they may become soluble and suitable for absorption, all dissolved substances have the potential to enter the organism to some extent through the membranes separating it from the surrounding water. This is also how compounds that have been previously incorporated into prey organisms are absorbed. Consequently, the concentration of the toxicants rises in the blood and intercellular space, where they may have detrimental effects on physiological and biochemical processes through their binding to membranes and enzymes. In addition, contaminants that bind to particular sites may become inactive and undergo detoxification, internal deposition, or excretion as a means of removal. Because the concentration of chemicals in the environment and food determines their uptake and accumulation in the organism, the concentration of chemicals in the organism will indicate the degree of environmental pollution.

Toxic substance accumulation affects physiological effects in addition to indicating the degree of environmental pollution. A certain degree of pollution may be permanently maintained within the organism by the deposited toxic substances; however, the accumulated toxicants are not permanently bound and may be mobilized by internal processes, which can have a toxic effect. This indicates that contaminants stored within the organisms serve as a constant source of danger.

The waters are not devoid of different toxic substances; rather, as is typical, a greater number of substances coexist, even though their concentrations may vary greatly. However, the toxic concentrations of different substances differ somewhat. In addition, hazardous chemicals may have additive or antagonistic effects on one another, particularly if their targets are the same. Subthreshold concentrations may become effective due to the synergism of substances; this can only be revealed in biological studies and not by chemical measurements of the concentrations.

5.1.6.4. Mechanisms and Consequences of the Action of Pollutants

By interacting with internal regulatory mechanisms, toxic substances have an adverse effect on the physiological processes of living things. The biological effect is contingent upon the pollutant's ability to interact at the molecular level with an organism's regulatory system, thereby altering, disrupting, or inhibiting physiological mechanisms. Adverse effects arise when the toxic chemical agent binds to membranes, enzymes, or internal or external receptors. This alteration in function results in a change in the regulation of living processes. When a toxic substance acts on a single system, like photosynthetic activity, cell metabolism, movement regulation, food uptake and digestion, respiration, excretion, hormonal and neuronal regulation, et al., the effect can be very specific. However, in other situations, a toxic substance may act simultaneously on multiple targets. The organism may die as a result of the toxic substances' lethal biological effects in severe situations or at high concentrations. In each case, a customized strategy and thorough laboratory research are needed to clarify the precise mechanisms underlying the effects.

Examining and elucidating ecological impacts differ slightly. Seldom do the detrimental effects of pollutants on the ecosystem cause every organism to deteriorate or decay to the same extent. Extreme pollution can occasionally cause damage to many species at once, but because different species differ in terms of reactivity, tolerance, and defense mechanisms, there is never a complete extinction in these situations. Even within the same species, there may be variations in sensitivity. The physiological significance of the pollutant target varies amongst organisms, which contributes to the variation in response across species. Additionally, certain species have compensatory or protective mechanisms in place to offset the negative effects of environmental degradation on other organisms.

The procedure is even more complicated in cases of persistent or sublethal pollution. The accumulation and removal or neutralization of the pollutants may be important in these situations, as well as the different ways that different species and developmental stages respond to oxygen deprivation in relation to temperature, food availability, and other factors. In certain animals, resistance can arise and have long-lasting effects due to individual physiological adaptation or selection acting in a way that causes the adaptation to be inherited genetically by the entire population. A species' ability to withstand pollution may also have an impact on the organism that preys on it. When an organism exhibits resistance, it means that it either releases the pollutant quickly or does not accumulate it, preventing the predator from ingesting a significant amount of the toxic substance. On the other hand, predators will be put in greater danger by eating food containing concentrated toxicants if the mechanism of resistance is some kind of inactivation process within the organism, such as binding and/or deposition in the tissues. Feeding on animals that respond to heavy metal pollution by producing metallothionein and binding to metals causes this to occur. Understanding these particular characteristics and variances is crucial for comprehending how environmental pollution affects the aquatic ecosystem. Biological monitoring and laboratory testing can provide light on these properties.

The diverse ways in which different species react to pollutants cause deviations in the ecosystem's rate of abundance; certain species may decline while others may grow and take center stage. Predators' ability to feed themselves is negatively impacted when sensitive species populations decline or go extinct, which is seen as a loss in the food chain. Overall, this process has a negative impact on biodiversity, leading to a decline in it. Anthropogenic pollution is one of the many factors that can cause changes in an ecosystem, whether those changes are in the biodiversity or species proportion. In some cases, this has been demonstrated, while in others, it is only suspected. Basic knowledge about the mechanism of action and estimation of potential effects on the ecosystem can be obtained from biological monitoring and thorough elucidation of the effects of toxic substances.

5.1.6.5. Biological Versus Chemical Monitoring in Water Quality Assessment (Advantages and Limitations)

Sound policy development and implementation are nearly impossible without an understanding of how changes in water quality affect aquatic life and resources. The key issue in determining water quality is whether to use chemical or biological techniques, so it's important to highlight a few aspects of this inquiry. It is impractical to conduct a compound-by-compound chemical analysis of surface water or effluent to detect every possible chemical. Based on available toxicity data, chemical analysis yields figures that need to be translated into potential biological effects. Chemical data may be lacking for some constituents, particularly trace metabolites and reaction products, and the chemical approach is unable to account for any potential additive, synergistic, or antagonistic effects. It is always important to remember that toxicity is a manifestation of the medium's chemical and physical properties as well as the compound's interest qualities.

Since it was discovered that chemical analysis of complex wastes was insufficient to predict their toxicity, tests known as toxicity bioassays were developed to assess the potency of water pollutants as toxicants—a biologically significant characteristic. However, using these bioassays calls for a great deal of extrapolation, and we have to figure out which changes are harmful (steps in the disintegration of organisms) and which are adaptive (necessary to the continued success of organisms changed by pollution).

Compared to chemical analysis, biological investigations of a contaminated lake or stream have many advantages. As natural observers, plants and animals keep a record of the current circumstances. One aspect of the issues with chemical analysis that were previously mentioned may be resolved by biological testing. They allow for the control of one limiting parameter, namely effluent toxicity, and integrate the effects of all effluent components. Compared to a comprehensive, in-depth chemical analysis, they might be more resource-focused and suggest a likely biological response in the environment. They could also identify antagonistic, synergistic, or additive interactions.

They do have some limitations, though, as it is frequently impossible to adequately express precision and reproducibility. The process of biomonitoring can take a lot of time. The results do not reveal the cause of the toxic effect in the absence of more information. At the moment, there is a lack of development in the process of translating laboratory results into the real world.

Biological index changes are a sign of changing environmental conditions. The concept of an index implies that whether or not the utilized waters are endangered, the biological indices themselves don't offer any concrete proof. Therefore, not all modifications to the environment can be classified as pollution, and not all modifications to biological indices can be interpreted as proof that the environment has undergone pollution-related changes. However, changes in biological indices brought on by a decline in water quality may very well be proof of pollution when the water has been designated for specific uses and a recognized aquatic community survives in environments that allow those uses.

According to the hierarchical concept, data collected at one time scale might not be useful at a higher scale, or it might be partially useful. It is not possible to estimate the temporal and spatial variation of hydraulics of a large river system, for example, from measurements made on the flow and volume of water in a brook within a given time frame. Thus, the size of the process at hand also affects the kind of data required for monitoring. Extrapolation techniques are not easily applicable to solve temporal and spatial problems because the characteristics of each hierarchical level exceed the integrated sum of the characteristics of the lowest level that is closest in the hierarchy.

Finding the biological monitoring systems' scientific justification, dependability, general industry, regulatory agency, and academic community acceptance of the methodology, and cost are the future objectives of this investigation. If the methodology is proven to be sound, we need to investigate why industry and regulatory bodies do not adopt it more widely.

Any strategy for protecting water quality must combine chemical and biological monitoring in an integrated manner. Chemical and physical monitoring are not superseded by biological monitoring. Each of them offers convergent informational strands that support one another without contradicting one another. The chemical monitors would determine the type of problem, while the biomonitors would show whether one was present.

5.2. BIOINDICATORS IN AQUATIC ENVIRONMENT

Organisms or biological processes known as bioindicators can be used to identify and/or measure the effects of pollutants on the environment by indicating changed environmental conditions. An estimate of the natural state or the extent of contamination that exists in that specific ecosystem would be indirectly provided by the evaluation of the bioindicators' status. An aquatic ecosystem's bioindicators, which can be used to gauge the contaminants present in the water, include fish, algae, macrophytes, zooplankton, bivalve mollusks, seabirds, and bivalve mollusks. A biological system, such as plankton, can quickly adapt to changes in its surroundings and provide insight into the overall health of its surroundings. Variations in animal populations may point to detrimental alterations brought on by pollution of the environment. Rapid increases in pollution are encouraged by the growth of industry and agriculture. Bioindicators can be a model or an index of measures that describe the health of an ecosystem.

A microbe, plant, animal, or fungus can all be considered bioindicators. Water pollution can be identified in many ways using microbial indicators, such as bioluminescent bacteria. Plants are sensitive instruments for stress recognition and prediction in the environment. The creatures and the groups they belong to are able to keep an eye out for any changes that might point to an issue with their ecosystem. Changes may affect an individual's chemical, physiological, behavioral, or ecological health. One way to think about ecological health is in terms of ecosystems, which preserve structural and functional qualities. It can be broadened to encompass a variety of facets of human health and welfare. In an ecosystem, every organism has the capacity to report on the state of its surroundings. The environmental health of an ecosystem can be monitored using specific physiological and behavioral changes in bioindicators. Before choosing any of these particular alterations as bioindicators, they must be thoroughly investigated as they vary from organism to organism. When an organism or association of organisms known as bioindicators is exposed to environmental samples, like soil or water, scientists typically track any changes that result from the exposure. Bioindicators can reveal how long an issue may last as well as the cumulative effects of various pollutants on the ecosystem. For instance, a) a large marine organism's abundance or a darkening of coral pigmentation may indicate that a reef has been exposed to low-quality water for several weeks or months, and b) reduced photosynthesis in plants or coral may indicate stress from herbicide exposure.

5.2.1. Response of Organism-Types

- **Ecological Changes:** involving changes in population density, key species and species diversity
- **Behavioral Changes:** changes in feeding activities, bacterial mobility or web spinning
- **Physiological Changes:** accumulation of heavy metal, CO_2 production, BOD and microbial activity

5.2.2. Criteria for Selecting Bioindicators

- The indicator should have a casual relationship to the ecological significant endpoint
- The indicator should have specific dose responsiveness to specific stressors, i.e., should be sensitive and specific
- It should be a wide temporal and spatial distribution
- Available throughout the year, sedentary, capability, hardy nature, small and easily cultivable, indigenous or representative
- It should have results which are transparent and reproducible
- Easy to collect and should be cost-effective

5.2.3. Factors Governing Bioindicator Distribution in Aquatic Environment

- Oxygen availability
- Temperature and salinity fluctuations
- Sediment and substrate type
- Presence of pollutants such as pesticides, acidic materials, toxic gases and heavy metals.
- Food availability

5.2.4. Common Bioindicators in Aquatic Systems

- Microbes
- Phytoplankton and algae
- Macro-invertebrates (Ex: Bivalve and Gastropods)

- Zooplankton (Ex: Rotifer, Cladocerans and Copepods)
- Macrophytes and
- Other animals (Ex: Fishes, Mammals and birds)

5.2.5. Types of Bioindicators and Their Uses

- **Microbial Indicators**: The health of an aquatic or terrestrial ecosystem can be determined by looking at microorganisms. Due to their quick growth, microorganisms can respond to physicochemical and biological changes as well as even minute concentrations of pollutants. Microorganisms can be sampled more easily than other macro-organisms because they are more common. As a result of exposure to benzene and cadmium contaminants, microorganisms produce new proteins called stress proteins, which can be used as early warning indicators (Parmar et al., 2016). An early warning system based on these stress proteins may be employed. For instance, water can be tested using microbial indicators, and water can be tested for environmental pollutants using bioluminescent bacteria. Toxins cause the bacteria's cellular metabolism to be inhibited or disturbed in the water. This influences how much light the bacteria release. This exam is much faster than the traditional ones, taking anywhere between five and 30 minutes to finish. Still, it merely points to the existence of a toxin that is responsible for the organism's alteration.
- **Phytoplankton and Algae**: An ecosystem's ability to support a particular type of plant or other vegetative life can reveal vital information about the state of the surrounding ecosystem. Phytoplankton is a helpful indicator of water quality and has been successfully used to observe water contamination (Wu, 1984). The best indicators of water quality, especially in lakes, are plankton because of their extreme sensitivity to changes in the environment. Monitoring the lake's water quality during periods of high phosphorus and nitrogen centralization is one of the reasons plankton are taken into consideration in lakes. Diatoms are a potent indicator of environmental change and have become a preferred indicator in monitoring studies (Dixit et al., 1992). Lichens are frequently observed on tree trunks and rocks. In response to alterations in the climate, air quality, and forest structures, they adapt their behavior. The absence of lichens in a forest could be a sign of environmental stressors like high sulfur dioxide, sulfur-based pollutants, and nitrogen concentrations (Figure 5.1).

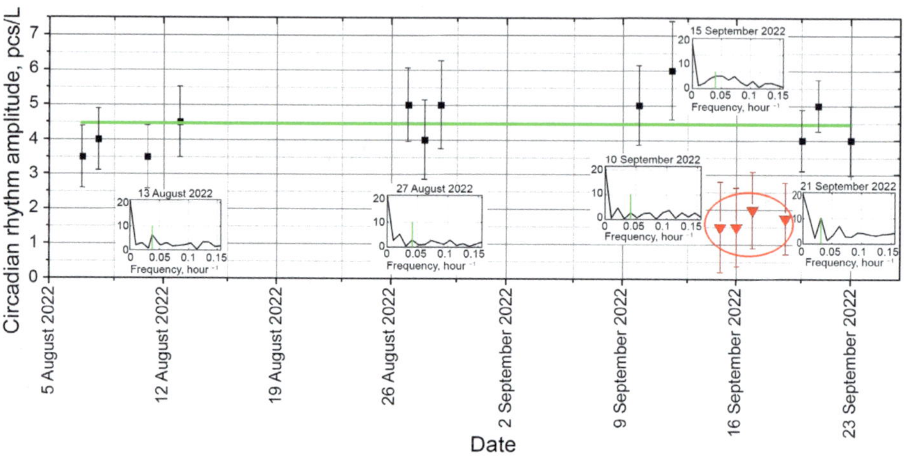

Figure 5.1. *Plankton indicating pollution.*

Source: Dyomin V, Davydova A, Kirillov N, Kondratova O, Morgalev Y, Morgalev S, Morgaleva T, & Polovtsev I. Monitoring bioindication of plankton through the analysis of the fourier spectra of the underwater digital holographic sensor data. Sensors. 2024; 24(7):2370. https://doi.org/10.3390/s24072370.

- **Animal Indicators**: An animal population's rise or fall could be a sign of pollution harming the ecosystem. For instance, animal species that depend on these food sources will become fewer in number if pollution results in the depletion of significant food sources. Other methods of animal indication include tracking the amount of toxins in animal tissues or the rate at which malformations appear in animal populations, in addition to keeping an eye on the size and quantity of specific species.

Since macro-invertebrates are large enough to be seen with the unaided eye, they are preferred as bioindicators. For instance, bivalve fish (also known as "Mussel Watch") and crustaceans are the best for researching metal pollution. Since 1986, the common blue mussel Mytilusedulis has been used in the United States to track metal pollution as part of the Mussel Watch Program. Due to their widespread distribution, ease of transfer to other areas for "active" biomonitoring, high abundances, and rapid accumulation of numerous xenobiotics (such as DDT, PAH, and PCBs) over the course of three years of long-term monitoring, these bivalves are a good indicator of coastal pollution.

- **Zooplankton**: Development and movement of zooplankton are influenced by both biotic and abiotic parameters, including temperature, salinity, pollutants, stratification, and stratification.

dietary restrictions, predatory behavior, and competition. As such, it is recognized as a superior bioindicator for assessing the level of contamination in any body of water, both coastal and oceanic (Parmar et al., 2016).

- **Macrophytes**: Macroalgae or seaweeds (e.g., *Fucus* spp.) gather metals, herbicides, and a wide range of other substances from seawater. In inland wetlands and water bodies, *Typha* sp. serves as an indicator for the pollutants nickel (Ni) and cadmium (Cd). Juncus sp. has been found to be a reliable indicator of urban water runoff's zinc (Zn) pollution (Sumampouw and Risjani, 2014). Heavy metals such as cadmium, copper, iron, lead, zinc, and manganese in rivers can be measured using floating *Eichhornia crassipes* for mercury, non-rooted submerged *Ceratophyllum demersum*, and rooted submerged *Potomogeton crispies* (Jain et al., 2010). You can also access heavy metals in ponds by using liverworts and moss. Cadmium sensitivity has also been demonstrated by *Wolffia globosa*, which may be utilized as a cadmium contamination indicator in lakes, ponds, and swamps.

5.2.6. Plankton as Indicators of Fish Production in Ocean Environment

Sardinella longiceps, the oil sardine, are indicated by *Fragilaria oceanica*, Herring fisheries are indicated by Coccolithophores, and Lesser sardine is indicated by *Hemidiscus hardmanianus*.

5.2.7. Recent Studies in Aquatic Bioindicators

The eutrophication status of the Cochin Estuarine and coastal waters has been observed to be indicated by the loricate ciliates, specifically *Tintinnopsis beroidea* and *T. uruguayensis* (Sivasankar et al., 2018). As an indicator species for tracking genotoxic agents in estuarine environments, backwater clams have been identified (D'Costa et al., 2018). The levels of trace elements (As, Cu, Mn, Cd, Zn, and Ni) in sediments can be determined using seagrass (*Halophila stipulacea*) (Bonanno and Raccuia, 2018). In the marine environment, limpets (*Patella* sp.) can be used to monitor metals, hydrocarbons, and organochlorine (Reguera et al., 2018).

5.2.8. Advantages

- Biological impacts can be determined
- Suitable to monitor synergetic and antagonistic impacts of various pollutants on a creature

- Early-stage diagnosis as well as harmful effects of toxins on plants, as well as human beings, can be monitored
- Easy to count, due to their prevalence
- Comparatively cheaper than other techniques which involve costly procedures and equipment
- The bioindicator is helpful, objective, straightforward, and reproducible

5.2.9. Significance

- Practical way of assessing the health of the environment
- Indicates general toxicity of the environment and conservation of natural resources
- Monitor the progress of environmental cleanup and test substances, like drinking water for the presence of contaminants
- Planktonic monitors unite biological, physical, chemical factors, and are utilized as an important part for evaluating health status of water bodies
- Bioindicators can be utilized at various scales, from the cell to the environmental level, for assessing the changes taking place in a specific biological community

Bioindicators hold great promise for predicting natural disasters, mitigating pollution, exploring and conserving natural resources, and promoting sustainable development while minimizing harm to the biosphere. It can be used to forecast environmental changes promptly and the effects of human activity, especially pollution. They are also a way to keep an eye on or identify any detrimental effects that industry is having on the aquatic environment.

5.3. BIOINDICATORS IN MONITORING ENVIRONMENTAL POLLUTION

It has taken around a century for the field of water pollution biology to develop. Up until the 1950s, this field's growth was uneven and sluggish. This decade turned out to be a watershed. Resources and public interest came together to sufficiently fund the necessary biological research. The most progress in our understanding of the biology of contaminated waters has been made in the last 20–30 years.

There is no end to the ways that pollution indicators can be monitored, so the topic cannot be covered in detail. To come up with the best strategy to reflect an early monitoring technique and significant changes in the chemical and/or physical conditions of the environment, we need to use our imagination. For biological monitoring, nearly any species, community, or parameter can be employed. The conditions at each site will determine the appropriate selection. Apart from providing scientific evidence for the ecological significance of the selected species and/or parameters, the selected sensor should yield dependable results and be relatively easy to handle. If monitoring techniques are to be used extensively, they shouldn't be too hard for technicians to use on a daily basis.

The concept of biological indicators of pollution originated with the early works (pollution zones, saprobic systems). Pollution indicators, which are indicator organisms indicative of environmental quality, were required like that of saprobic systems. The dynamics and function of communities and ecosystems can be used to identify changes that occur as a result of exposure to toxic pollutants. At the population, organism, and suborganism levels—a lower level of biological organization—effects will also be discernible. Visible effects include the extinction of vulnerable species, the rise of more resilient species, modifications to the metabolism of oxygen, the cycling of nutrients, the movement of energy within ecosystems, and the functioning of feedback mechanisms.

From the ecosystem level to the subspecies level, the data gathered from the biomonitors may be arranged based on the degree of biological complexity. The following synopsis provides pollution indications for various biological organization levels.

5.3.1. Pollution Indication on the Ecosystem Level

Physical, chemical, and biological factors interact to create any given aquatic habitat. Toxic compounds and other man-made chemical factors can have a significant impact on aquatic ecosystems, primarily through direct action on vulnerable species. However, through prey-predator-competitor interactions, the direct effect can have a multitude of indirect effects. Ecosystems' structure and/or functioning can reveal both direct and indirect effects. This means that one can use these parameters as a means of detecting effects, (e.g., some diversity index as a pollution indicator).

The various indicator organisms must be taken into account as biological associations or communities rather than as individual organisms when assessing the validity of aquatic organisms as indicators of pollution levels and water quality. Ecosystems are always changing because they are susceptible to evolutionary developments. Ecosystems' short-term stability, or homeostasis, is the outcome of cybernetic interactions between their numerous constituent parts. Several different kinds of disruptions (e.g., by contaminants). Because of this homeostasis, it is very challenging to distinguish between changes in ecosystem structure caused by pollutants present or by homeostasis. Accordingly, it is only possible to draw conclusions about the effects of pollutants when they surpass normal variation.

The distribution of ecosystems, biomass, Chi a, plant pigments, RNA, DNA, and ATP are all potential biomonitor parameters that can be used to assess biodiversity (variations in the number of taxonomic groups and organisms) and the structural responses of ecosystems. As functional parameters of the ecosystems, decomposition, primary production, community metabolism, energy and matter flow, increased energetic costs, food web alteration, and so forth, should signal pollutant disruption of the nutrients cycle.

Leeuwangh critically evaluates the benefits and drawbacks of bioindication techniques at the ecosystem level, noting that our current understanding of how communities and ecosystems respond to chemical perturbations is still quite limited. As a result, it is hard to make a recommendation with absolute certainty based on parameters that accurately forecast the negative effects of chemical stress. While the aforementioned parameters have occasionally been employed to describe ecosystem responses, there is no guarantee that they will always be useful for identifying hazards early on.

5.3.2. Pollution Indication on the Population Level

It is also possible to analyze various population-level indicator parameters, such as mortality, size distribution, population extinction, changes in composition, dominance switches, variations in patterns of diversity and similarity, declines in biomass and abundance, changes in spatial structure, and successional influences, to track changes in the population after pollution.

5.3.3. Pollution Indication on the Species Level

Sentinel species that are susceptible to environmental stressors are believed to have an early role in identifying them. Regretfully, no species is more vulnerable to any given chemical than another. The best species to monitor are those that are representative of all major taxa, exposure pathways, and physiological traits. In situ biotest mortality most accurately characterizes environmental toxicity. Toxicants can bioaccumulate through food or water absorption. The rate of absorption, metabolism, and excretion determines the final degree of accumulation. Biochemical reactions include hormone levels, enzyme activity, metallothionein induction, macromolecules, and enzyme induction as sublethal effects follow exposure to a toxicant. Bioindication and monitoring often make use of morphological effects (damage to cells and tissues, deformities), physiological processes, and behavior responses (avoidance-preference reactions, swimming behavior, feeding behavior, respiration activity, valve movements).

5.3.4. Biomonitoring Techniques at Different Levels of Biological Communities

In many places where other samplers are ineffective, artificial substrate samplers offer an alternative method of sampling. They get a fauna that is less variable between samples and equally diverse. This is a useful technique for assessing and monitoring water pollution biologically. For instance, in streams, periphyton is frequently more helpful than plankton in assessing the effects of pollution because it more precisely represents the circumstances at a particular sampling site. The development of aquatic biology as a quantitative field closely corresponded with the development of artificial substrate samplers. Using periphyton (algae, fungi, and bacteria), macroinvertebrates (insect larvae, snails, free-living flatworms, amphipods, isopods, oligochaetes, worms, bryozoans, hydroids, and sponges), and rocky intertidal and subtidal communities, artificial substrate samplers provide a representative sample of native communities, primarily in rivers and streams. The most common artificial substrates are multiplate substrates, microscope slides, and barbecue baskets filled with rocks.

Since they are anticipated to be overly sensitive to signs of sluggish recovery processes, indicators or scoring systems are helpful in assessing recovery from toxic stress. Environmental elements unrelated to pollution can have an impact on the fauna and, consequently, the index values or score. Monitoring rivers with artificial substrates at less frequent intervals than with other methods can help determine recovery from toxic stress because lithoreophilic insects are thought to have the best (re)colonization possibilities.

Because they are comparatively immobile, benthic macroinvertebrates are frequently chosen as indicators of the water quality in aquatic ecosystems. The kind and quantity of species, as well as the number of specimens per species, are typically taken into consideration as a biological measure of pollution stress (e.g., biodegradable organic wastes). Toxic pollution's effects and the aquatic community's reaction to this kind of stress could be very dissimilar. The findings suggest that while a species' tolerance is pollutant-specific, there are very few differences in how susceptible a species is to toxic conditions as a result of pollution by multiple toxicants. As a result, one should have serious doubts about the validity of biological systems that rely on the distribution of macrobenthos to categorize surface waters contaminated with different chemicals.

As sentinel and sedentary organisms, bivalves are known to be able to concentrate certain sets of pollutants from ambient waters, including heavy metals, halogenated hydrocarbons, petroleum hydrocarbons, and transuranic elements. The following arguments sum up the benefits of using mussels to measure environmental pollutants: mussels are sedentary, filter-feeding organisms that are widely distributed (e.g., *Mytilus edulis* in marine environments, *Dreissena polymorpha* and Unionidae in freshwaters). Their low metabolic activity and good accumulation capability (they could have a concentration factor of 10,000–100,000) make them useful for indicating ambient xenobiotic levels. Large populations of mussels make them easy to sample, they can be moved around with ease, and since they are commercial products, mussels and oysters are directly related to public health.

The problem of temporal variability (seasonal concentration alterations, changes in biochemical activity, filtration rates, etc.) and the inter-comparability of data from different species (*Mytilus edulis* and *M. californianus*) were also highlighted by a surveillance program of US coastal waters as early uncertainties of this method. In order to address this issue, it makes more sense to observe a particular location over the course of a few years in order to identify hotspots, highly industrialized areas, and background levels. Numerous monitoring programs in the marine environment along the Dutch coastal waters have made use of the translocation technique, which has many benefits. Large samples of continuously submerged mussels of a chosen class are moved from a location that is comparatively unpolluted to the target area and suspended from buoys inside cages using a translocation technique. By using statistically similar groups of animals derived from common stock and knowing the chosen period of exposure, the resolution power between geographical locations is increased. This allows exposure locations to be selected independently of the monitored species' natural occurrence.

The Microtox test was created to gauge aqueous sample toxicity that is acute and nonspecific. A strain of the bioluminescent bacterium Photobacterium phosphoreum serves as the test organism in this instance. This marine bacteria produces luciferase, a compound that releases light energy as a byproduct of the Krebs cycle. Less luciferin is produced and luminescence is reduced whenever a toxic substance enters the bacteria and affects metabolism or membrane integrity. It is believed that the decrease in light production correlates with the sample's level of toxicity. This Microtox test's primary benefits are its quick response time and tiny sample volume requirement. The Microtox system's results showed a decent degree of correlation with the outcomes of commonly used standard tests, like the acute Daphnia and fish tests. The toxicity of organic compounds dissolved in the Rhine River and the toxicity of industrial wastewaters in Germany were both successfully monitored using this technique. The nitrification rates were introduced by Botermans and Admiraal as a gauge for the ecological recovery of the Rhine. Because nitrifying bacteria are sensitive to chemical pollution, concentrations of both organic and inorganic chemicals in the ppm range may have an adverse effect on their capacity to oxidize ammonium ions to nitrate. In highly industrialized and urbanized areas, nitrification predominates over other nitrogen processes, so chemically inhibiting nitrification in a river may cause issues related to ammonia buildup. If not, the trend of the river's ecological recovery could be observed by observing the rising nitrification rates that coincide with the falling levels of harmful substances during that time.

5.4. STATIC AND DYNAMIC APPROACHES IN BIOLOGICAL MONITORING

Both static and dynamic methods can be used in biological monitoring of the pollution and toxic impact threatening freshwater ecosystems. A static approach takes into account and measures the amount of pollution (pollutant presence and concentration) in organisms, assuming that this is a good indicator of the condition of the environment they live in. It is possible to trace the variation in pollution with repeated measurements. Native populations of organisms living in a biotop are gathered for measurement in the most basic type of static monitoring, known as passive monitoring, without the need for any additional manipulation. Plants or animals are specifically relocated from an unpolluted biotop to an area to be studied as part of static, but active, monitoring. The accumulation of toxic substances, which reflects the degree of pollution in the experimental area, is then measured. When it comes to biological monitoring, a dynamic approach takes into account the negative effects of pollutants on human health as well as the regulatory mechanisms that affect individuals, species communities, and the ecosystem's overall functioning.

5.4.1. Static Methods in Monitoring the State of the Environment

The concentrations of toxic substances accumulated in organisms are taken into account as a value characteristic of the surrounding environment when evaluating environmental pollution. Based on experimental results demonstrating a positive correlation between the concentration of toxic substances in the immediate environment and in the organisms residing therein, this assumption is made. It has been shown for both persistent organic compounds and toxic metals. However, this assumption is only true at concentrations that are below lethal. A pollutant's concentration that is near or above the lethal level will seriously impair physiological functions, such as the accumulation of substances, and the concentration that can be measured in living things will not accurately represent the pollution in the biotope.

Static monitoring, which takes into account environmental pollution, works best for determining the concentration of persistent substances and should only be used sparingly to regulate readily degradable chemicals. Toxic metal ions and a wide range of organic pollutants, including polychlorinated biphenyls, organometals, and chlorinated and polycyclic aromatic hydrocarbons, are examples of persistent pollutants that accumulate in organisms and have a toxic effect. Measuring the concentration of these compounds in plant and animal tissues is known as static monitoring.

It is crucial to understand that various organisms may absorb the same contaminant to varying degrees from their shared environment. For instance, research has revealed that the cadmium, copper, lead, and mercury concentrations in the bodies of the creatures that inhabit Lake Balaton in Hungary vary, with chironomids being more likely to accumulate mercury, crustaceans to accumulate copper, and mollusks to accumulate cadmium. Microbes, plants, and animals can all absorb and store pollutants in very different ways. Even different body parts and organs can accumulate toxic substances in different ways. The muscles of the fish (pike-perch and bream) in Lake Balaton had the highest concentration of mercury of any organ, while the gills and liver had the highest concentrations of lead and cadmium. Additionally, different parts of aquatic plants have varying capacities for storing and accumulating pollutants. All of this relates to the idea that an organism's concentration of a pollutant indicates not only the amount of pollution present in the surrounding environment but also the organism's specificity and capacity for accumulation.

It is possible to determine whether an organism is a strong or poor biological indicator. Organisms with a high accumulator capacity and the capacity to retain substances taken up from the environment for extended periods of time are regarded as good bioindicators.

Standardized measurement methods, appropriate result evaluation, and the selection of appropriate bioindicators are the cornerstones of the application and practical use of biological techniques in environmental monitoring. When selecting suitable static biomonitors, one should take in consideration the following:

- The species used for monitoring should be widespread and available at any or most of the time in the studied aquatic ecosystem.
- It should be available for collection in the necessary amounts.
- It should have a high capacity for accumulation and good storing ability of chemical substances.

- It should be tolerant of low-level pollution.
- In case a special part or organ is used as a biomonitor, its isolation should be simple and the quantity sufficient for chemical determination.

Any organism that satisfies these requirements can be employed for both terrestrial and aquatic biomonitoring as a static bioindicator. The most often utilized freshwater organisms for monitoring the pollution of the aquatic environment are fish, plankton, and crustaceans living in open water, as well as higher plants like algae, seaweed, and reed. Additionally, the larvae of chironomides and mussels belong to the benthic animal kingdom. When it comes to plants, in addition to species identification, it's important to take into account each plant's developmental stage during the vegetation period and the portion of the plant being measured. Algae samples taken from the water and plankton made up of crustaceans typically contain a variety of species rather than just one. Even when the level of environmental pollution is the same, variations in the species composition can lead to variations in the pollutants' detected concentrations. A quantitative analysis of the sample's species can help to minimize error. When using biological objects to monitor environmental pollution, additional individual specificities or circumstances need to be taken into account. One of these is the relationship—which can be either positive or negative depending on the substance—between the animal's size and age and the concentration of pollutants in it. Larger organisms (higher plants, mollusks, fish) should only have a specific, defined portion of their bodies—not the entire body—used for measurement. This is partially because it is impractical and difficult to prepare a homogenate from the entire animal, and partially because it is advantageous to take advantage of the fact that different organs accumulate substances to varying degrees, allowing one to choose the organs with the highest capacity for accumulation.

Every time samples are prepared and a measurement is made, the same protocol should be followed. It is important to carefully consider the identity of the methodical circumstances when comparing the results with previous data or values obtained in other laboratories, as variations in this regard may lead to different values and incorrect conclusions. Such errors can be prevented by using real samples as reference materials; however, it is nearly hard to quantitatively verify results that were previously obtained using different techniques later on.

Typically, there's a significant disparity in pollution levels recorded for individuals of the same species, collected from the same habitat, despite meticulous sample collection and measurement processes. To achieve reliable results, a substantial number of concurrent measurements (ranging from 5 to 10 samples) are necessary. Given that the sample size should still be manageable, it appears sensible to select a handful of the most reliable bioindicators for routine use, as they can provide insights into the pollution level and its timely fluctuations within the examined ecosystem.

5.4.2. Case Studies for Static Monitoring: Detection of Metal Pollution in the River Danube with Passive and Active Monitoring Techniques

The 1980s and 1990s saw the use of aquatic organisms' capacity for bioaccumulation during bioaccumulation monitoring to identify levels of heavy metals in the ecosystem of the Danube River in Hungary.

The filamentous green alga *Cladophora glomerata* was employed in the early 1980s to regulate the concentrations of heavy metals in the Danube River. It was well known that this attached, periphytic species was a good indicator of heavy metals (Whitton et al., 1989). However, we had to investigate this ability and the circumstances under which it is valid in a sizable river that has been heavily industrially polluted and eutrophicated. We discovered that the alga's heavy metal concentration is influenced by its developmental stage and vertical orientation within the littoral zone by exposing it to an artificial floating substrate (AFS). These temporal and spatial variations were highly dependent upon and correlated with Cladophora's dry weight. The ambient concentration of heavy metals turned out to be the most significant external factor that directly affects the heavy metal level in the algae, aside from the weight dilution effect, which is the most significant internal factor. Other physicochemical variables (such as pH, ORP, conductivity, light, temperature, and flood) typically have an indirect impact on the uptake and accumulation process through changes in dry weight. Scientists could track the pollution events after resolving all of these basic queries and comparing the real concentration of heavy metals in Cladophora to the baseline level typical of the alga on the examined river section. In addition to acting as a trend and signal in long-term monitoring, this could also act as a control function. Researchers were able to conclude that, at the start of the 1980s, the Danube's studied reach was industrially, moderately polluted after analyzing our data and comparing them to the literature.

The attached biomonitor, the Cladophora, demonstrated to be a reliable indicator of the dissolved form of heavy metals (e.g., Ag, Cd, and Co), which are readily absorbed and come from nearby pollution sources. This is an example of the source orientation. The Cladophora web's *Trichoptera* inhabitants are occasionally one- or two-times greater concentrators of specific heavy metals (e.g., Ag, Cd, Cu, and Pb) than the alga itself, despite the fact that attached algae have shown to be an extremely sensitive and source-oriented indicator of the dissolved phase of heavy metals. As collector organisms, this group of macroinvertebrates favors suspended particulate matter (SPM), which serves as a pre-concentrator of heavy metals. As a result, their metal concentrations are typically higher than those of other macroinvertebrate feeding groups. If not, animals' mobility might call into question the reliability of their ability to identify and define the sources of pollution.

An active biomonitoring device using zebra mussels was progressively developed in our field experiments on the Danube, which began at the beginning of the 1990s, to monitor heavy metal pollution. To lessen the issues brought on by variability and, in the end, provide a workable method that can be incorporated into the water-quality monitoring systems of major rivers, the applicability of the selected approach has been examined from a variety of angles over the years. The Dreissena basket was created with the help of some of the earlier knowledge that other specialists who had studied smaller streams had acquired. The ultimate objective was to create a device that could be effectively utilized in big rivers like the Danube. 100–100 zebra mussel specimens were placed inside nets with an 8 mm mesh size. The outer containers for these mussels-in-net units were perforated cylindrical plastic tubes measuring 30 cm in length, 10 cm in diameter, and 6 mm in mesh size. Similar to shields, these latter shield mussels from mechanical harm while still allowing water and food to flow through them appropriately. Lastly, ropes and weights were used to anchor these two Dreissena-basket units 30 cm below the water's surface.

Using carabiners makes it simple to fasten all of the containers, ropes, and anchors together. The Dreissena basket's flexible, module-like structure makes its components simple to transport and assemble in the field, and it also guarantees the mussels are handled gently. Since the inner units, mussels-in-nets, are easy to transport in containers filled with water, it is simple to replace the animals in the field during successive periods. Both in deep and low Danube waters, this inexpensive and simple-to-make instrument proved to be appropriate in many ways.

A certain amount of the variability typical of native populations could be reduced by assessing and applying the findings of lethal and sublethal condition indices determined during monitoring operations. We could distinguish between seasonal variations and condition changes by computing the discrepancies between the measured and computed body weight using the correlation equations between shell length and body weight. We recorded both the shorter and longer events of pollution by recovering the animals in toto and replacing them with fresh stocks of mussels in accordance with the 5–7-weeklong consecutive periods. By using Dreissena baskets, they were able to identify the Danube's hot spots as well as its longer polluted sections. The majority of the heavy metals under study were found to be polluted in the 1990s in the sections of Medve-Nagymaros (1,806–1,695 rkm), Paks-Mohacs (1,532–1,447 rkm), and the entering border section (1,842 rkm). These areas are close to Hungary's and the Slovak Republic's most developed areas. During the monitored period, Co, Ni, and Pb could be regarded as regular pollutants due to their high concentration levels in mussels. The periodic high deviations also suggested that these metals appeared sporadically. Without a doubt, the comparatively high Fe concentration in zebra mussels was caused by the Danube's typical high Fe content of suspended matter rather than pollution.

The steadily improved Dreissena basket's flexible design makes it a very manageable tool that can be utilized for active biomonitoring in the constantly shifting hydrological conditions of a big river. This apparatus has the potential to mitigate the majority of methodological issues and lessen the variability that comes with passive techniques. As a final suggestion, it might be suggested to incorporate the widespread use of Dreissena-basket—a dependable, affordable, and manageable tool for active biomonitoring—into the drainage area-scale water quality monitoring systems of major rivers.

In comparison to concentration values, heavy metal body-burden values in mussels are equally significant because they attest to the actual process of cumulative accumulation along the river's longitudinal section. While some authors view the concept of body burden as artificial, our experiences—which are similar to those of Mersch and Pihan—indicate that it could be quite helpful. By assessing the dry weight, concentration, and body burden values all at once, it is possible to identify the pollution brought on by cumulative accumulation over time and space, as well as to isolate the high concentration values that result from dry weight reduction rather than pollution.

5.4.3. Dynamic Methods in Monitoring the State of the Environment

One can assess the stability or variation in the degree of environmental pollution based on external sources as well as those arising from timely variations in the pollution process bound to seasonal, yearly, or multiple years changes by comparing the concentration of toxic substances in indicator organisms obtained by repeated measurements.

In addition to using static methods, such as measuring the pollutants' concentration in organisms on a regular basis, dynamic, functional tests can also be used to track temporal changes. By tracking changes in both the population's overall and individual members' quality of life, dynamic approaches assess and mirror the existence and intensity of pollution. Functional, dynamic tests can be used to measure physiological, biochemical, and morphological indices as well as mobility, feeding, gas exchange, growth, reproduction, reactivity, and other functions.

Because of their physiological reaction to pollutants, mussels are not only excellent accumulators of toxic substances and thus ideal organisms for static biomonitoring, but they can also be employed in the dynamic monitoring of water pollution. One measurable and highly reliable indicator of water quality is the variation in the movement of the valve and the length of the active and rest periods. Additionally, water emissions from the outflow siphon can be used to monitor the quality of the water by indicating the presence of pollutants in the water. Examples of functional biomonitors include the freshwater rotifer's swimming pattern, the fiddler crab's color changes, and changes in a bivalve's oxygen consumption and biochemical composition caused by toxicants. In addition to growth and larval mortality, changes in morphology and enzyme activity can also indicate exposure to pollutants.

Dynamic biomonitors can be employed at the ecosystem level to measure population size, species diversity, species ratios, and species community structure. Any of these traits can be studied in natural field settings or under simulated, artificial conditions. Pollutants and changes in the level of pollution of the aquatic environment can affect any of these traits. It is necessary to determine the control status and possess in-depth knowledge of the species characteristics in order to use the population for monitoring. It's not an easy task, particularly since both the ecosystem and the individual are always changing and not static. Numerous variables can lead to modifications, and particular studies are needed to identify the anthropogenic pollutants' selective influence. The population's reaction to environmental contaminants can be explained by two main theories: either by molecular changes leading to genotoxic effects or by differences in the sensitivity of different species to toxicants causing selective deterioration in the population structure. For instance, it has been proposed that the alterations in the community structure of redbreast sunfish populations are due to chromosomal damage brought on by mercury, PCBs, and other substances found in industrial effluents. The death of a fathead minnow population in outdoor microcosm experiments showed a strong correlation with brain acetylcholinesterase activity, which is inhibited by organophosphorus insecticides used in agricultural pest management (Parmar, Rawtani, and Agrawal, 2016).

5.5. BIOLOGICAL EARLY WARNING SYSTEMS

Constant monitoring allows for the quick identification of pollutant responses at stress levels as low as those determined by long-term testing to be physiologically safe. Biological early warning systems (BEWSs) are methods that enable the prediction of the toxicity of chemical mixtures that fluctuate both qualitatively and quantitatively. Within BEWSs, organisms exposed to the water under observation on a frequent or continuous flow basis provide an automated means of detecting short-term changes in toxicity.

Naturally, BEWSs require parameters that yield fast answers—not necessarily highly precise or highly specific. However, trends in ecosystem development necessitate extreme caution when choosing appropriate parameters that indicate the trends one wants to pursue. BEWSs are effective methods for monitoring the toxicity of industrial effluents and receiving waters both inside plants and at the streamside. They can also shield aquatic life in receiving waters from unanticipated exposure to unfavorable conditions. In addition to more conventional chemical analysis-based monitoring techniques, BEWSs can be helpful. The number of parameters in these analyses is constrained, and the measured concentrations don't always yield the needed data (i.e., the potential harm that pollution may cause to aquatic life).

Automated biomonitors use living things as sensors—from bacteria and invertebrates to vertebrates—to self-regulate or self-act in the assessment of water quality.

The following criteria can be summarized for BEWSs as fundamental operational requirements:

- Operation of the system should be continuous and automatic, so the variable biological response can be quantifiable through computer analysis.
- The system must detect a wide range of developing toxic conditions rapidly and reliably.
- The number of false alarms due to nontoxic variations in water quality should be minimal.
- Appropriate methods for the analysis of data must be developed (i.e., statistical determination of the normal range of response variables and of differences between normal variation and individual and/or diurnal variations).
- The monitoring system should be relatively easy to operate and the results easy to interpret.
- Aquatic organisms as biosensors should be inexpensive, and easily cultured (e.g., commercially available).
- The apparatus should require as little maintenance and skilled human labor as possible (Parmar, Rawtani, and Agrawal, 2016).

5.5.1. Types of Most Frequently Used BEWSs

5.5.1.1. Fish Monitor

Fish respiration is a perfect biomonitoring system for evaluating the quality of water since it can continuously and automatically identify a low-cost range of acute and chronic toxicological hazards with high reliability and speed. The majority of toxicants appear to be detectable by the monitor at concentrations at least 10 times higher than the no observed effect level (NOEL). It is acknowledged that the system is most likely useful for preventing unintentional spills that result in acutely toxic conditions when the detection limits are contrasted with the NOEL for the same compounds.

Monitoring respiratory frequency is generally more sensitive than other available techniques (positive rheotaxis, bacteria's oxygen consumption, algae's oxygen production, and Daphnia's immobilization inhibition). The use of fish respiration is debatable for natural waters, but feasible for in-plant monitors (industries, sewage treatment plants, and strategic watershed locations). Monitoring effluent quality in conjunction with other biological techniques and suitable multi-detectional chemicophysical monitoring systems appears to be the most successful application, provided that the response and detection (false detection, lag time, etc.) are precisely defined.

5.5.1.2. Daphnia Monitor

The system's foundation is the identification of alterations in water flea activity following a pollution effect. The number of infrared light beam blocking events per unit of time brought on by Daphnids in the test water column serves as the basis for detection. Before the Dynamic Daphnia Test period, a few real-world issues (such as sediment removal, food concentration, and water temperature) had to be resolved.

5.5.1.3. Mussel Monitor

Both freshwater and marine bivalves have the ability to detect high pollution levels and can close their valves to prevent stress from pollutants. The repeatability and reversibility of the valve closure imply that this response may be used in detecting rising toxicity conditions of the water, even though such an effective avoidance behavior may conflict with the outcomes of accumulation monitoring programs.

Due to the widespread distribution of *Dreissena polymorpha*, this species' valve response is utilized as a BEWS. In general, the response to pollutants can be categorized into two distinct behavioral patterns: a strong increase in activity (incomplete closure) and an increase in the frequency of complete valve closure at regular intervals. In *Unio pictorum*, as opposed to *Dreissena polymorpha*, the periods of activity are brought on by contraction of the foot muscles, the mussel makes its way through the sediment's upper layer. This organism can serve as a reliable indicator of chemical pollution, as demonstrated by its ability to isolate itself from the harmful effects of toxic concentrations.

In contrast to the trout system, which is thought to be the most sensitive system currently in use, the mussel system shows a slight decrease in sensitivity to most compounds. Sloof conducted a thorough investigation into the average detection limits of BEWSs based on the respiratory frequency of fish (*Salmo gairdneri*) and the valve closure response of mussels (*Dreissena polymorpha*).

Further advantages of the mussel system are as follows:

- A more direct method of measuring a response and less susceptible to noise problems, and false alarms.
- Changes in temperature and photoperiod do not interfere with the behavior of the mussels.
- Easier to handle (e.g., no feeding is required).
- Mussels are self-maintaining.
- More feasible for widespread industrial application; can be installed and operated at a much lower cost (ca. 20% of fish monitoring).

In addition to being used in site-specific applications at drilling platforms in the sea to check for potential chemical leaks, mussels can also be used in the monitoring of industrial or sewage effluents, drinking water stations' intakes, and contaminated bodies of water or contamination gradients.

5.5.1.4. Multi-Sensor Monitors

It was suggested that multiple systems be implemented simultaneously as part of the Rhine Action Program (RAP). A system utilizing water fleas and another utilizing a photosynthetic organism (cyanobacteria or algae) ought to be put into place in addition to a fish monitor. The systems were set up at the Rhine River's Lobith monitoring station.

The respiration activity of the microorganisms is measured (semi) continuously by the BASF toxicometer, a type of model treatment plant. If the respiration activity significantly decreases, an alarm is triggered. The toxicometer was designed to show acute and substantial intoxication right away, but no gradual or long-term degradation of the purification process. Toxic wastewater can be directed to protect the activated sludge basins if the toxicometer is placed far enough away from the sewage treatment plant (about 3.5 km).

5.5.2. Evaluation of Advantages and Shortages of BEWSs

- Those species that usually are easy to culture and/or maintain under artificial conditions are often not the most sensitive to toxicants. No species is equally sensitive to all possible pollutants. One possible solution to the problem of species sensitivity in a single-species system is to incorporate a sensor that is known to be sensitive to the toxicant likely to occur in the water body being monitored. This, of course, presumes that the toxicants of concern are largely known beforehand (e.g., industrial wastewater). Certain species, such as Daphnia pulex, could be an overprotective sensor for a particular water body depending on biota indigenous to the system and the physical habitat. Generally, it can be said that problems of species sensitivity of BEWS may need to be observed on a case-by-case basis.
- With frequent rotation and replacement of sensor organisms, the problem of sensor acclimation can be solved, too.
- Maximum distance needs to allow sufficient time for alarm (hydraulics and turnover time).
- The monitored response must be one that changes as a pre-lethal symptom of poisoning, and it must be conducive to automation allowing for reliable detection of duration from normal conditions.
- The more easily observed responses are generally the less sensitive indicators. Therefore, an important conceptual issue in the development of any BEWS is the exact purpose for which the system is designed, along with an appropriate compromise of the trade-offs in cost versus system sensitivity.
- In a particular system (e.g., utilizing frequency, amplitude, cough pattern, and general activity of fish), the more behavioral and biological information utilized from the sensor, the more sensitive and rapid the response of the system as a whole.
- Considering the validation of the BEWSs, there is a big question for the "in-plant" system and how this signal of deleterious biological effect relates to the response of natural systems. All deleterious biological effects require corrective action even though the sensor may be somewhat more sensitive than the biota in the natural receiving systems.

It is emphasized that acceptable water quality is still not guaranteed even in the event that a BEWS does not sound an alarm. It is still possible for toxic compounds to exist, albeit most likely in trace amounts or at levels that become harmful only after extended exposure. The detection limits of these systems are typically far from the levels of environmental toxicants. Furthermore, the monitor systems' compound-specific sensitivity is the subject of scant information. Furthermore, disasters are frequently identified by fish kills upstream of the monitoring station before biomonitors identify hazardous conditions. Experience to date suggests that one should exercise caution when accepting the usefulness of BEWSs in alerting people to hazardous conditions in naturally occurring surface waters. Nonetheless, these biomonitors can be effectively used as in-plant monitors for the surveillance of effluent quality at large industries and sewage treatment plants. They can thus be very helpful in preventing catastrophic spills from harming the aquatic ecosystem (Parmar, Rawtani, and Agrawal, 2016).

5.6. REMOTE SENSING AND GIS IN WATER QUALITY ASSESSMENT

Water is a precious natural resource that is necessary for both human survival and the wellbeing of ecosystems. Water serves many purposes, including providing water for mining operations, livestock, irrigated agriculture, and domestic and industrial use. However, the quality of the water has drastically declined due to anthropogenic activities and the growth of industrial development. The quality of river water can degrade as a result of home wastewater, industrial effluents, and agricultural waste. Three categories can be used to describe the characteristics of water: physical, biological, and chemical. Water quality is currently assessed by in situ measurements and the collection of water samples for later laboratory analysis. Although these measurements are precise for a specific location and time, they do not provide a spatial or temporal picture of the water quality across a large area.

As a result, monitoring programs that make use of GIS and remote sensing (RS) are required to identify potential contamination threats and to provide efficient response at all levels. In comparison to in situ methods, which limit measurement to specific sampling sites, remote sensing and GIS techniques are more efficient, affordable, and useful for tracking water quality parameters in freshwater bodies (lakes, rivers, groundwater, and reservoirs) and coastal levels. Satellite remote sensing's synoptic view and GIS's analytical power combine to provide a technologically sound approach to researching different aspects of water resources. Because satellite remote sensing can cover a large area at a relatively low cost and is repetitive in nature, it is an obvious source of data for mapping and map revision. The following sections provide a brief overview of these two techniques before going over how RS and GIS are used in water quality studies.

5.6.1. Remote Sensing Techniques

The general definition of remote sensing is the ability to observe an object without being in close proximity to it. By observing the earth's surface from various platforms, including satellites and aircraft, remote sensing systems enable the collection and analysis of data about the environment and resources over vast distances. Satellite imagery offer many advantages over conventional survey techniques:

- Areal synoptic coverage (gives areal information as against point information through conventional techniques)
- Repetitive global coverage (for monitoring change)
- Real-time processing
- Sensing of surrogates rather than the desired specific observation
- Multispectral coverage
- More automation, i.e., less human error

The fundamental idea behind remote sensing is that different objects reflect, emit, and absorb radiation at different electromagnetic wavelength ranges in different intensities based on their physical characteristics. It is feasible to distinguish between various ground object types (e.g., water, dry soil, wet soil, vegetation, rocks, etc.) and map their distribution on the ground by using data from one or more wavelength ranges. Aerial photography is one of the key remote sensing techniques used. Since aerial photography makes up a very small portion of the electromagnetic spectrum, it has essentially stayed in the visible region. Nearly the entire electromagnetic spectrum can now be used thanks to advances in remote sensing techniques.

All signals that are gathered by remote sensing devices originate from electromagnetic radiation, or EM. The energy source varies based on the properties of the sensor. The majority of systems rely on solar energy to provide all the electromagnetic energy required to image land surfaces. We refer to these systems as passive sensors. Active sensors are other types of sensors that produce their own energy, transmit it in a specific direction, and then record the amount of energy that is reflected back by objects in the signal path.

Alterations in the energy levels of electrons, the acceleration of electrical charges, the decay of radioactive materials, and the thermal motion of atoms and molecules can all produce electromagnetic energy. The sun's nuclear reactions generate electromagnetic radiation (EM) that spans the entire spectrum and travels through space without significantly altering its nature until it reaches the atmosphere.

5.6.1.1. Image Interpretation

The primary goal of image interpretation is to glean details about the features that an image displays. It is described as the process of looking at pictures in order to recognize objects and determine their meaning. Information extraction relies on the experience, observational skills, creativity, and patience of the image analyst. It also depends on how well he comprehends the fundamentals of an image. Any graphic image can also be digitally represented so that, in the event that the image is not available in pictorial form, the patterns of brightness in the image form an array of numeric values that can be easily added, subtracted, multiplied, divided, and generally subjected to statistical manipulations that are challenging or impossible. Remotely sensed data are subjected to a wide range of operations through digital analysis, which produces information or enhanced data. It's important to keep in mind that digital analysts still interact with people occasionally. In contrast to manual interpretations, it requires substantial analyst input during decision-making, resulting in information delivery at a faster rate and with higher-quality output.

For man-machine interaction, image processing systems need a variety of input/output devices, a central processing unit (CPU), data storage devices, and system consoles. Sufficient image display capabilities are essential for image processing.

Remote sensing applications necessitate image processing software, which can be broadly categorized into data input, pre-processing, image display, image enhancement and filtering, classification, and image output routines.

An idealized sequence for digital analysis can be broken up into four specific groups:

a. Pre-processing
b. Enhancement,
c. Georeferencing and Mosaic
d. Analysis and Classification

5.6.2. GIS Techniques

GIS is a computer-based technology used to manage digitalized geographic data. Its purpose is to gather, store, process, and analyze data that is dispersed geographically. It includes attribute data as well as geometry data, including coordinates and topographical details. With a focus on maintaining and exploiting the intrinsic qualities of spatial data, GIS allows us to present results in both graphic and report form.

5.6.3. Main Functions of GIS

Although it is theoretically possible to perform all GIS operations manually, only very small research areas will be able to accomplish many tasks due to their high time requirements. The tedious jobs can be easily completed by utilizing computers, their graphics capabilities, and GIS software. One major flaw in the early computerized map handling concepts was their inability to handle spatial features in addition to tabular or attribute data. As a result, more strategies and methods for handling and integrating both spatial and attribute data were developed, making the results more useful to planners and decision-makers. With the development of this technology, our ability to handle, manipulate, and analyze maps has improved. Therefore, using a GIS:

- Users can interrogate geographical features displayed on a computer map and retrieve associated attribute information for display or further analysis.
- Maps can be constructed by querying or analyzing attribute data.
- New sets of information can be generated by performing spatial operations (such as polygon overlay) on the integrated database.
- Different items of attribute data can be associated with one another through a shared location code.

To perform any kind of modeling or spatial analysis in a GIS, the necessary data must first be entered. Writing data to the GIS database and converting it into a computer-readable format is known as data input. There are two types of data that need to be entered into a GIS: spatial data and related non-spatial attribute data. The geographical location of features is represented by the spatial data. To depict geographic features like a street, a lake, or a forest land, one can use points, lines, and areas. These data will normally be obtained from one or more of the following sources:

- Existing maps
- Aerial photographs
- Satellite imageries
- Existing digital data
- Other GIS data bases

In turn, GIS data can be a valuable tool for image analysis, and remote sensing is a significant source of data for GIS analysis. Because of the complementary roles that GIS and remote sensing play, the need for their integration is therefore unavoidable and developing quickly.

5.6.3.1. Basic Types of Spatial Data

Four basic geometric entity types—points, lines, polygons, and continuous surfaces (area)—are intended to represent spatial data.

- **Point Data:** The most basic kind of spatial data are points. Points can be classified into two categories: those that pertain to phenomena that are discretely distributed, like water and oil wells, and those that pertain to continuous distributions, like weather stations that report soil temperature and precipitation measurements.
- **Line Data:** Line entities are segments of straight lines defined by two or more pairs of coordinates, or linear features composed of numerous points. The line entities may be dynamic (flow) or static (structural type). A set of x-y coordinate pairs that describe a continuous complex line is called an arc, chain, or string.
- **Polygon or Area Data:** The most common type of data used in GIS is polygons. These are bounded areas that are established by a multi-segment closed continuous sequence. Natural phenomena like land forms or man-made elements like forest stands or land use units can define the boundaries.
- **Continuous Surface:** Temperature, precipitation, elevation (as part of topographic data), and other variables are examples of continuous surfaces. These are essentially three-dimensional data that are handled by most GIS systems as topographic data, typically in the form of Digital Elevation Models (DEM).

Spatial data has typically been represented and stored as maps. For digitally storing geographic data, three fundamental categories of spatial data models have developed. They go by the names Vector and Raster.

5.6.3.2. Capabilities of GIS

The power of GIS lies in its ability to analyze spatial and attribute data together. The large range of analysis procedures can be divided into four categories:

i. Retrieval, reclassification and measurement,

ii. Overlay,

iii. Distance and connectivity,

iv. Neighborhood

v. Interpolation

- **Retrieval, Reclassification and Measurement Operation:** Only attribute data are modified in these functions; both spatial and attribute data are retrieved. There is no creation of new spatial elements.
- **Retrieval Operations:** These involve the selective search and manipulation and output of data. Retrieval operation includes the retrieval of data using:
  - Geometric Classifications
  - Symbolic Specifications
  - A name of code of an attribute
  - Conditional and logical statement

When working with spatial and attribute data, retrieval operations entail selective data search, manipulation, and output—all without changing the geographic location of features or generating brand-new unique entities. Retrieval operations include:

- Retrieval of data using geometric classification. Specifying the spatial domain of a point, line or area, retrieve all spatial entities and nonspatial attributes contained in the entire or in position of that spatial domain.
- Retrieval of data using symbolic specifications.
- Retrieve data using a name of code of an attribute. Retrieve using a name or code of an attribute. Example, retrieve effective depth and dominant texture of a given soil.
- Retrieval of data using conditional and logical statements. Retrieve data that satisfy alphanumeric conditions using logical expressions.
- **Reclassification Procedures:** In this process, the categories of an existing map are given new thematic values based on the initial value and the location, dimensions, and shape of the spatial configurations linked to each category. For example, a soil map may be reclassified as a permeability map. Numerical values are frequently used to denote classes in raster-based GISs. A value may be assigned to a cell to represent a class. As an illustration, a cell may be given the value 1 to represent agricultural land, 2 to represent forest land, and so forth. Both single and multiple data layers are used for classification when performing an overlay operation.

- **Measurement Functions:** A measurement function is offered by every GIS. Calculating distances, line lengths, polygon perimeters, and areas is a necessary step in measuring spatial data. Point measurements include the total number of points in a polygon, the distances between points, lines, and other points. They can also involve counting the points that fall inside a polygon.
- **Overlay Operations:** A map is produced by overlaying two or more existing maps such that the values assigned to each location are calculated as a function of the independent values associated with that location on those other maps. By joining two or more sets of distinct polygon layers together, the overlaying operation generates a new data set with newly created polygons. All GIS software packages include arithmetical and logical overlay operations.

The addition, subtraction, division, and multiplication of each value in a data layer by the value in the corresponding location in the second data layer are examples of arithmetical overlay operations. The process of selecting an area where a set of conditions is satisfied is known as logical overlay.

- **Neighborhood Operations:** Creating new data for neighborhood operations is based on taking into account a "roving window" of neighborhood points about particular target locations. They assess the features of the neighborhood around a given target location. It is essential to specify one or more target locations, the neighborhood that is being considered around each target, and the kind of function that will be performed on the neighborhood's attributes in all neighborhood operations. The search function, topographic function, and interpolation are the common neighborhood operations in most GISs.

5.6.3.3. Interpolation

- **Nearest Neighbor:** This technique gives horizontal surfaces around each point. This technique is generally used for interpolation of the rainfall data. This provides weight to be used for the rain gauge stations, for finding average rainfall over a watershed.
- **Weighted Average:** A weighted average is taken of point elevation falling within a specified radius from the interpolation point. Weight is an inverse function of a distance between the point of unknown value (interpolation point) and that of known value (data points).
- **Weighted Surface:** Weighted surface fits a surface of n-degree using the elevations at points within a specified radius from the point of interpolation.

- **Trend Surface:** Trend surface fits an overall surface with the input data. It provides the global direction in which the variable changes.
- **Kriging/Geostatistical Method:** Kriging is a mathematical technique developed by D.G. Krige for application in mining industry in South Africa. The experimental semi variogram is calculated from the data. A standard semi variogram equation is fitted to the experimental data using regression technique. This fitted curve is used in the interpolation. For groundwater data the spherical theoretical semi variogram model is best suited.

5.6.4. Applications of RS and GIS in Water Quality Studies

The process of analyzing the chemical, physical, and biological properties of water bodies and locating the source of any potential pollution or contamination that could deteriorate the quality of the water is known as water quality determination. These indicators of water quality fall into the following categories: (i) biological (algae, bacteria); (ii) physical (temperature, turbidity and clarity, color, salinity, suspended solids, dissolved solids); (iii) chemical (pH, dissolved oxygen, biological oxygen demand, nutrients (including nitrogen and phosphorus); (iv) aesthetic (odors, taints, color, and floating matter).

Surface water's backscattering properties can be considerably altered by substances used in remote sensing (Kirk, 1983). The ability to quantify these changes in the spectral signature backscattered from water and link these measured changes to a water quality parameter using empirical or analytical models is the foundation of remote sensing techniques. The substance being measured, its concentration, and the properties of the sensor all influence the best wavelength to use when measuring a water quality parameter. Suspended sediments (turbidity), algae (i.e., chlorophylls, carotenoids), chemicals (i.e., nutrients, pesticides, metals), dissolved organic matter (DOM), thermal releases, aquatic vascular plants, pathogens, and oils are major factors affecting the quality of the water in water bodies across the landscape. Remote sensing techniques can be used to measure the energy spectra of reflected solar and/or emitting thermal radiation from surface waters that are altered by suspended sediments, algae, DOM, oils, aquatic vascular plants, and thermal releases. By providing spatial and temporal views of surface water quality parameters that are not easily obtained from in situ measurements, remote sensing tools enable effective and efficient landscape monitoring, enabling the identification and quantification of water quality parameters and issues.

In freshwater surface waters, suspended sediments are the most prevalent pollutant in terms of volume and weight. The radiance emerging from surface waters in the visible and near-infrared regions of the electromagnetic spectrum is enhanced by suspended sediments. Surface water radiance is dependent on various factors such as sediment type, texture, and color, sensor view, sun angles, and water depth, as demonstrated by in situ, controlled laboratory, aircraft, and satellite measurements. The use of remote sensing for water quality monitoring began in the early 1970s. An early empirical method for estimating suspended sediments general equation was developed by Ritchie et al. in 1974. It goes like this: Y=A+BX or Y=ABX, where X is the water quality parameter of interest (i.e., suspended sediment, turbidity) and Y is the remote sensing measurement (i.e., radiance, reflectance, energy). The values of A and B, which are empirically derived factors, come from statistical relationships between the in-situ water quality parameter and the spectral reflectance value. Information regarding the band or wavelengths appropriate for that water quality parameter can be obtained from the spectral reflectance. Subsequently, other researchers used this equation to estimate the water quality parameter.

Keeping an eye on chlorophyll (algal/phytoplankton) concentrations is essential for controlling eutrophication in lakes. The spatial and temporal concentrations of chlorophyll have been measured via remote sensing. Similar to measurements of suspended sediment, the majority of remote sensing investigations into chlorophyll in water rely on empirical correlations between chlorophyll and radiance/reflectance in small bands or band ratios. As a result, field data are gathered to verify developed models or to calibrate statistical relationships.

An essential element of the best management of water resources is the routine assessment of the pattern of sediment deposition and the estimation of the reservoir's available live storage capacity. The traditional methods of quantifying sedimentation in a reservoir, such as inflow-outflow calculations and hydrographic surveys, are laborious, expensive, and time-consuming. Furthermore, a substantial amount of data is needed for the prediction of sediment deposition profiles using empirical and numerical methods, and the results are still inaccurate. Because of its spatial, spectral, and temporal characteristics, remote sensing can offer timely, repetitive, and synoptic information about the updated water spread area in a reservoir. To assess the sediment deposition pattern in a reservoir, the temporal change in water spread area is analyzed using digital analysis techniques in conjunction with a geographic information system.

Thermal pollution occurs when human activity alters the temperature of a body of water, thereby affecting biological activities. Thermal discharge from electrical power plants into streams, lakes, and coastal waters has been mapped using remotely sensed data. Remote sensing techniques can be used to accurately estimate the thermal plumes in river and coastal waters. The spatial and temporal patterns of thermal releases that are obtained from remote sensing mapping of absolute temperatures are helpful for controlling thermal releases. The outputs of mathematical models of thermal plumes can also be interpreted with the help of quantitative estimates of the temperatures of the surface waters. Because they allow for precise timing of data collection, aircraft-mounted thermal sensors are particularly helpful in research on thermal plumes. It is reasonable to anticipate seasonal variations in surface water temperatures. By routinely monitoring these seasonal variations in ocean surface temperatures with AVHRR and other satellite platforms, new insights have been gained into the function of oceans in regulating weather and climate Ritchie et al. calculated lake surface temperatures along the Mississippi River by utilizing Landsat TM thermal data. Utilizing thermal remote sensing to track temperature variations that may impact biological productivity in freshwater systems is a helpful monitoring technique. These methods enable the creation of management strategies to lessen the impact of thermal releases caused by human activity. Lake margins have been identified using satellite imagery; this method is especially helpful in remote locations or in shallow-water systems with erratic margins.

The optical characteristics of water, which result in light being scattered and absorbed rather than transmitted in straight lines, are expressed in water turbidity. For this reason, it is frequently thought of as the antithesis of clarity. Fluvial suspended sediment concentrations have frequently been determined through turbidity measurement, as suspended matter is the primary cause of turbidity in water. The spatial scale constraint may be addressed by higher-resolution satellites like IKONOS (Zhou et al., 2008).

REFERENCES

1. Bonanno, G., & Raccuia, S. A. (2018). Seagrass Halophila stipulacea: Capacity of accumulation and biomonitoring of trace elements. *Science of the Total Environment, 633,* 257–263.
2. D'costa, A. H., Shyama, S. K., MK, P. K., & Furtado, S. (2018). The Backwater Clam (*Meretrix casta*) as a bioindicator species for monitoring the pollution of an estuarine environment by genotoxic agents. *Mutation Research/Genetic Toxicology and Environmental Mutagenesis, 825,* 8–14.
3. Dixit, S. S., Smol, J. P., Kingston, J. C., & Charles, D. F. (1992). Diatoms: Powerful indicators of environmental change. *Environmental Science & Technology, 26*(1), 22–33.
4. Hamza-Chaffai, A. (2014). Usefulness of bioindicators and biomarkers in pollution biomonitoring. *International Journal of Biotechnology for Wellness Industries, 3*(1), 19–26.
5. Jain, A., Singh, B. N., Singh, S. P., Singh, H. B., & Singh, S. (2010). Exploring biodiversity as bioindicators for water pollution. In Proceedings of the National Conference on Biodiversity, Development and Poverty Alleviation (pp. 1–5). Uttar Pradesh State Biodiversity Board, Lucknow, India.
6. Kirk, J. T. O. (1983). *Light and Photosynthesis in Aquatic Ecosystems.* Cambridge University Press.
7. Omar, W. M. W. (2010). Perspectives on the use of algae as biological indicators for monitoring and protecting aquatic environments, with special reference to Malaysian freshwater ecosystems. *Tropical Life Sciences Research, 21*(2), 51.
8. Parmar, T. K., Rawtani, D., & Agrawal, Y. K. (2016). Bioindicators: The natural indicator of environmental pollution. *Frontiers in Life Science, 9*(2), 110–118.
9. Reguera, P., Couceiro, L., & Fernández, N. (2018). A review of the empirical literature on the use of limpets Patella spp. (Mollusca: Gastropoda) as bioindicators of environmental quality. *Ecotoxicology and Environmental Safety, 148,* 593–600.
10. Ritchie, J. C., & Cooper, C. M. (2001). Remote sensing techniques for determining water quality: Applications to TMDLs. In *TMDL Science Issues Conference* (pp. 367–374). Water Environment Federation.
11. Ritchie, J. C., McHenry, J. R., Schiebe, F. R., & Wilson, R. B. (1974). The relationship of reflected solar radiation and the concentration of sediment in the surface water of reservoirs. In *Remote Sensing of Earth Resources* (Vol. III, pp. 57–72). The University of Tennessee Space Institute.

12. Ritchie, J. C., Zimba, P. V., & Everitt, J. H. (2003). Remote sensing techniques to assess water quality. *Photogrammetric Engineering & Remote Sensing, 69*(6), 695–704.
13. Sivasankar, R., Ezhilarasan, P., Kumar, P. S., Naidu, S. A., Rao, G. D., Kanuri, V. V., Rao, V. R., & Ramu, K. (2018). Loricate ciliates as an indicator of eutrophication status in the estuarine and coastal waters. *Marine Pollution Bulletin, 129*(1), 207–211.
14. Sumampouw, O. J., & Risjani, Y. (2014). Bacteria as indicators of environmental pollution. *International Journal of Ecosystem, 4*(6), 251–258.
15. Wu, J. T. (1984). Phytoplankton as bioindicator for water quality in Taipei. *Botanical Bulletin of Academia Sinica, 25*(2), 205–214.
16. Zhou, Q., Zhang, J., Fu, J., Shi, J., & Jiang, G. (2008). Biomonitoring: An appealing tool for assessment of metal pollution in the aquatic ecosystem. *Analytica Chimica Acta, 606*(2), 135–150.

Chapter **6**

Restoration Techniques for Aquatic Habitats

Contents

INTRODUCTION

Aquatic habitat restoration techniques encompass a range of methods designed to restore these environments to their original state and enhance their ecological well-being. Restoring physical features like riverbanks, wetlands, and riparian zones that have been lost or degraded as a result of human activity is one of the main techniques used in habitat reconstruction. This frequently entails constructing stable, varied habitats that are able to support a variety of aquatic life by utilizing natural materials like logs and boulders. Enhancing the quality of the water is another essential component of restoring aquatic habitat. This can be accomplished by controlling industrial discharges, wastewater treatment, and agricultural runoff to lessen the amount of pollutants that enter water bodies. Putting in place vegetated buffer zones will aid in removing pollutants before they enter the water. Furthermore, reestablishing native aquatic habitats depends on managing invasive species. A decrease in biodiversity can result from native species being outcompeted by invasive plants and animals. These invasive populations are managed through mechanical removal, chemical treatments, and biological control, among other methods. Important elements of restoring aquatic habitats are also restocking and species reintroduction. To replenish depleted populations, this entails breeding native species and reintroducing them into their natural environments. To sustain these species, it is imperative to make sure that the habitat is appropriate and that there is enough food and shelter. Last but not least, the effectiveness of restoration initiatives depends on observation and flexible management. Frequent monitoring makes it possible to evaluate the ecosystem's health and the success of restoration strategies. Adaptive management is the process of continuously improving restoration outcomes by making adjustments to strategies based on new scientific knowledge and results monitoring. Aquatic habitat restoration can produce sustainable environments that meet human needs and biodiversity by combining these different approaches.

6.1. RESTORING AQUATIC ECOSYSTEMS

An "ecosystem" is a bubble of life made up of plants, animals, other organisms, the weather, and even the physical environment. An ecosystem is defined by the combined traits of all of its constituent parts. They have both biotic, or living, and abiotic, or non-living, components. Microorganisms, as well as plants and animals, are examples of living parts. The climate, geology, physical space, and other elements are examples of nonliving components. Sadly, modernization has put many ecosystems in jeopardy all over the world. Trophic cascades are being caused by species extinctions. The dynamics of the established food web are being strained by the drastic changes in the environment brought about by global warming. Considering how intricately intertwined all species are, it should come as no surprise that humans benefit from ecosystem protection. For this reason, we ought to work to preserve and rebuild ecosystems.

6.1.1. Ecosystem Restoration

Numerous ecosystems in our care are harmed. But each damaged ecosystem can be saved and returned to its original state through ecosystem restoration. Humans benefit from healthier ecosystems and increased biodiversity because these factors increase agricultural yields. Numerous techniques are available for restoring ecosystems, such as planting trees, reducing environmental stressors, and allowing nature to heal itself. While it is ideal, it is not always feasible or feasible to restore an ecosystem to its initial state. For instance, we can't just destroy things and put them back together again because that would leave some people without a place to live. Reverting farmlands to forests would not be feasible as it would result in a loss of revenue for farmers and a decrease in food production. Ecosystems should be restored as closely as possible to their natural state. But it needs to be weighed against pragmatism. Over time, restoring ecosystems has a significant positive influence on a wide range of aspects of our lives. You could consider it a long-term investment. According to research, there could be a potential income generation of up to 9 trillion US dollars in ecosystem services from the restoration of 350 million hectares of damaged aquatic and terrestrial ecosystems. Furthermore, repairing ecosystems might have a beneficial effect on our greenhouse gas issue. Restoring ecosystems has the potential to eliminate 13–26 gigatons of greenhouse gases from the atmosphere. According to estimates, the financial gains from ecosystem restoration outweigh the initial outlay by a factor of nine. Inaction can cause environmental harm, which can then result in economic harm that could be three times more expensive than ecosystem restoration. It is possible to restore any type of ecosystem, whether it be aquatic (lakes, wetlands, and oceans) or terrestrial (forests, farms, and cities). However, a restoration technique that works well in one ecosystem might not work well in another. People now have the ability to start campaigns to rebuild ecosystems. Businesses, communities, or even a single person can start their restoration initiatives in addition to governmental and developmental organizations.

6.1.1.1. Why Should We Restore Our Ecosystems?

Ecosystems require restoration for a variety of reasons. These justifications can be divided into five categories: biotic, heuristic, idealistic, pragmatic, and technocratic. The technocratic justification for conducting ecosystem restoration projects by government agencies and/or other large organizations is to fulfill specific mandates. Biologic justifications are those for ecosystem restoration that are based on protecting local biodiversity. Ecosystem restoration activities help gain a deeper comprehension of biotic expressions and ecological principles; this reasoning is categorized as heuristic. An idealistic justification for their cause can be found in the personal and cultural expression of concern or atonement for environmental degradation, re-engagement with nature, or spiritual fulfillment in relation to the restoration of our ecosystems. The practical justification for restoring ecosystems is that they offer a multitude of natural services and products. This realization and restoration of ecosystems are necessary to obtain these natural services and products. The current restoration's technocratic justification is overly limited in its application. Large organizations and/or government agency mandates ought to take the practical argument into consideration. Stated differently, they ought to see ecosystem restoration as a chance to enhance the nation's resource base. In addition, it would be extremely advantageous if individuals in positions of authority within government agencies and/or large organizations were to be instilled with the idealistic rationale. This will act as a powerful motivator for government departments and/or major corporations to carry out and oversee the success of their ecosystem restoration projects.

6.1.2. Aquatic Restoration

Many aquatic ecosystems around the world are degrading due to a variety of factors. The main causes of this ecosystem degradation are the nature of our anthropogenic activities and the increasing population. Negative shifts include homogenization of biological communities, a steep drop in the number of apex predators and ecosystem engineers, loss of biodiversity, and chemical pollution that causes an abundance of nutrients and pollutants in the aquatic environment. Recalling the practical justification for ecosystem restoration, it is imperative to stop the loss of aquatic biodiversity to protect the natural services and goods that depend on it.

The differences between freshwater and marine ecosystems are a huge factor when it comes to aquatic restoration.

- Freshwater environments tend to be smaller which exerts a greater spatial restriction on the organisms that inhabit there.
- Freshwater ecosystems include rooted vegetation and insects which exert a unique influence on the environment.
- Organisms living in marine ecosystems are better dispersed and have better connectivity with each other.

These differences should be taken into account when trying to use conservation methodologies designed for freshwater ecosystems on marine ecosystems.

6.1.3. Rivers and Lakes

Numerous people have relied on rivers, lakes, and various freshwater ecosystems as sources of food, water, and energy. They also offer defense against flooding and droughts. A multitude of flora and fauna find refuge in them. That is why we should value our freshwater ecosystems by doing the following:

6.1.3.1. Clean Up

Anything that is discarded or washed into a body of water by runoff needs to be cleaned up. Both the aesthetics and quality of the water would improve in this way. When a body of water is beautiful, people are more likely to value it and be motivated to preserve it.

6.1.3.2. Control Access

Access points that are clear and simple to use should be provided for both humans and animals, as well as machinery. The placement of these access points ought to take animal behavior into account. For instance, to prevent upsetting the fish delicate life cycle, no machinery, people, or boats should be permitted to enter an area that is known to be a fish spawning ground. This does not imply that people should be prohibited from enjoying the body of water. There would almost always be places where human activity had no effect.

6.1.3.3. Restore Vegetation

It's common knowledge that vegetation around bodies of water has advantages for those bodies of water. Thus, planting native species will help restore the thinned-out vegetation along the banks of rivers and lakes. In doing so, a buffer zone is created between the nearby source of pollution and the water through the restoration of native plant species. Eliminating invasive alien species should be part of the restoration of vegetation.

6.1.3.4. Plan with Sustainability in Mind

Fishing and harvesting should be done in such a way that doesn't deplete fish populations. This should be done on other resources as well. Check the water for sources of pollution like sewage, chemical pollutants, industrial wastes, and other kinds of effluents. Agencies that are concerned with restoring ecosystems should make agreements or pay incentives to encourage reductions in agricultural chemicals.

6.1.3.5. Current Situation

Reducing the amount of organic waste, nutrients, and pollutants that enter freshwater ecosystems has been greatly advanced in many developed nations. While more recent sewage treatment facilities have addressed this issue and discovered a means to reduce these issues, older ones introduced significant levels of organic pollution, elevated biological oxygen demand, and decreased oxygen levels in bodies of water.

Most industrialized nations have also seen a decrease in the amounts of phosphate in their water bodies. This is explained by the elimination of phosphate as a component of laundry detergents, enhanced sewage collection, and the application of tertiary treatment, which effectively eliminates phosphate. Reduced acidification of water bodies with little to no lime content is another benefit. Utilizing source-related filtering technologies and lowering fuel sulfur content helped to solve the acidification issue. In addition, serious restoration work is done to repair freshwater bodies that have been harmed. However, even in these developed nations, elevated nitrate levels remain an issue. In addition, developing nations everywhere are still finding it difficult to manage the amount of nutrients that are getting into their waterways and causing eutrophication.

6.1.3.6. Deficits in the Restoration Efforts

Despite an increase in restoration efforts, the majority of them do not adhere to a methodical and evidence-based approach. Verifiable assessments of the success of these restoration projects and whether the legally required objectives have been met are scant or nonexistent. There are restoration techniques that aim to restore the hydrogeomorphic, biogeochemical, and ecological processes that constitute a healthy river and are grounded in established theory. However, there is an issue with how these approaches are applied and evaluated; more research is required.

A case study that was done in Europe demonstrated this. This study found that adding gravel and loosening the substrate using an excavator had beneficial short-term effects that lasted no more than a year. Unfortunately, these techniques had long-term detrimental effects downstream rather than in the restoration area. Even worse, the magnitude of the adverse impacts outweighed the immediate advantages of adding gravel and loosening the substrate using an excavator. Evidence-based assessment methodologies are desperately needed for the assessment of restoration procedures. In addition, it is important to consider the taxonomic and spatial representativeness of various species groups within an aquatic ecosystem, since these groups may exhibit differing reactions to disturbances and restoration of the ecosystem.

6.1.3.7. Challenges to Restoration

Restoring ecosystems to a more favorable condition is already challenging in itself, how much more for restoring them to their original state? Here are some of the things that could make ecosystem restoration difficult:

6.1.3.7.1. Siltation and Colmation

Restoration efforts are seriously threatened by siltation and collimation of stream beds, particularly when these occur in headwater areas. Fish that spawn in gravel and freshwater mussels that are endangered may find it difficult to live and grow as a result of both siltation and colmation. Rising water temperatures, which can be caused by thermal pollution, make this issue worse. In-stream changes and land use that is dependent on the catchment are typically the causes of siltation and colmation. Having said that, when choosing the appropriate in-stream measures to employ, catchment-scale should be taken into account.

6.1.3.7.2. Naturally Fragmented Ecosystem

The majority of freshwater species that are critically endangered are found in streams. The biggest obstacles to restoration are the detrimental alterations to structural habitat quality caused by extensive alteration of stream courses and their catchments. In contrast to marine ecosystems, freshwater ecosystems are by nature more fragmented. This complicates the process of establishing connectivity in freshwater ecosystems.

However, connecting systems isn't always the best idea, particularly when they're naturally isolated from one another. It is likely that this will lead to the homogenization of biological communities. along with a diminished ability to withstand environmental stressors. This took place in Europe's extensive network of artificial canals. In naturally isolated systems, the canals created connectivity that facilitated the spread of non-native species like Ponto-Caspian gobies and zebra mussels. Native species are being threatened by the spread of alien species. Essentially, the process of restoring an ecosystem is made more difficult by the fragmented nature of freshwater ecosystems. It's unclear whether creating connectivity within a disjointed ecosystem is preferable or not.

6.1.3.7.3. The Complexity of Structural Habitat Diversity Restoration

Structural habitat diversity restoration is a key target in the restoration of streams and rivers. However, the restoration of structural habitat diversity is easier said than done. Although there are well-researched methods of achieving this kind of restoration, they aren't free from problems. These problems include inadequacies in monitoring, insufficient taxonomic representation, and underrepresentation of negative results. Such problems can lead to unsatisfactory restoration decisions and practices.

6.1.3.7.4. Incomplete Knowledge of the Autecological Requirements of Certain Species

An ecosystem's living and nonliving elements will inevitably interact with a single species. The primary focus of autecology is how these interactions would transpire. Regrettably, a great deal of information about the autecology of many different species is lacking. This ignorance stems from a lack of curiosity about these animals.

Without a thorough understanding of the interactions between all the species in an ecosystem and both living and nonliving elements, making wise restoration decisions would be a challenging task. Even closely related species can have different habitat needs despite having similar life cycles and physical characteristics, which highlights the significance of autecology. Because of their striking similarities, one species may benefit while the other suffers if one assumes otherwise. Restoration efforts that show promise in one location might not work in another because species pools can vary greatly amongst geographical regions.

6.1.3.7.5. Unknown Causes of Biodiversity Decline

Restoring ecosystems is a troubleshooting process that involves determining the underlying cause of the problem. It can be difficult to identify the issue's root cause, though. There are situations when the cause of the decline in biodiversity is not known, or there may be many environmental pressures at play, some of which may be masking the others. Furthermore, the interplay between environmental pressures exacerbates their difficulty in resolution. The primary environmental stresses that have a negative impact on freshwater biodiversity are widely known.

It is unclear how these environmental stresses interact with one another, though. It can also be challenging to determine which environmental stressors are more harmful than others and require quick attention. Not every environmental stressor has extensive documentation. Other environmental pressures are poorly understood because of their small impact. They don't really affect ecosystems, but they can still make restoration efforts more difficult. The issue of emerging contaminants also exists; as they are new, there are still many unknowns regarding them. The restoration efforts may become ineffective as a result of the uncertainties introduced by this lack of or incomplete information.

6.1.4. Oceans and Coasts

Seas and oceans cover 70% of the surface of the planet. Between 50 and 80% of the oxygen on Earth can be produced in these marine environments. They also produce fish, seaweed, and other seafood products that are used as marine food. They can also be a major source of income because they can be a tourist attraction. The application of active restoration approaches in open marine and coastal systems is limited to ecosystems that have lost their biogenic structure. The semi-terrestrial saltmarsh and mangrove vegetation, kelp beds, intertidal and subtidal seagrass beds, mollusk and polychaete reefs, and coral reefs are examples of open marine and coastal systems. Because part of their natural behavior involves altering their surroundings to create new habitats or modify the ones that already exist to suit their needs, species capable of biogenic structuring are known as "ecosystem engineers." Restoring ecosystem-engineering species improves habitat structures, which in turn affects ecosystem services and functioning.

6.1.4.1. Primary Producers

The primary producers, which include higher plants like seaweeds and seagrasses, are another crucial point to remember. They provide primary production that is immediately consumed by grazers and are typically found in salt marshes and coral reefs. Restoring terrestrial and freshwater bank habitats frequently involves planting higher plants like seagrasses, mangrove species, and saltmarsh. This method usually works well. Actually, research has indicated that growing seagrass in clusters can increase survivability. There are worries, though, that this method might result in vegetation with low genetic diversity.

6.1.4.2. Reef-Building Species

Reef-building species include corals, serpulid worms, and bivalve mollusks. The use of artificial substrata facilitates their ability to settle in a given area, particularly when they reproduce by mating with a nearby barnacle. Because of their increased capacity for settling, they can readily take the place of biogenic structures, which may result in the restoration of habitat and the restart of ecosystem processes.

6.1.4.3. Semi-enclosed Marine and Coastal Water Bodies

Restoration of semi-enclosed marine and coastal water bodies can be accomplished in many ways. One approach, and a good place to start, is to restore hydrological processes, such as the exchange between lagoons and the open sea or estuary. A step toward restoration could be as easy as altering the pace at which sediments are moved through a reservoir. Using this method, realigned coastal areas have been managed without the need for seawalls or dykes. The main disadvantage of this strategy is that it may take some time for drainage and erosional processes to start up again. Saline water would be drawn in by realigning coastal areas closer to land, destroying the terrestrial vegetation in the newly created coastal area. Mudflats can form as a result of saltmarsh plants taking over after the terrestrial vegetation dies. It would also take some time for the species composition, community structure, ecosystem function, and services of marshes made using this technique to resemble those of natural marshes. Frequently, decades may pass before ecosystem services start up again. By adding spatial heterogeneity and using some targeted restoration techniques, this process can go more quickly. Semi-enclosed marine and coastal water bodies can be successfully restored by combining two restoration techniques: managed re-alignment and altering the rate at which sediments would be transported through a reservoir. Nonetheless, it might require a while to attain complete recovery.

6.1.4.4. Restoring Ecosystems in Disused Docks

When macro-tidal ports aren't utilized for commercial shipping, they might develop an enclosed, artificially degraded ecosystem. Deeper dock basins have shown success with techniques used on lakes or reservoir management. Filter feeders, like mussels, can help improve the water quality in high-salinity docks by means of bio-filtration. A mixing airlift pump can help the mussels so that bio-filtration happens more quickly. The entire volume of the dock was able to pass through each mussel in one to two days with the help of the airlift pump. The clarity of the water significantly improved as a result of the increased bio-filtration. Human pathogens were also eliminated from the water as a result of the mussels' presence. Furthermore, there was an increase in oxygen levels. Moreover, these elevated oxygen levels were unaffected by the summer's temperature shift. A flourishing marine assemblage emerged as a result of these enhancements to the marine environment. Disused docks would give rise to unique ecosystems. These ecosystems, however, are delicate and quickly deteriorate. Within the major cities, artificial cubist marine lagoons are created as a result of the restoration of these ecosystems.

6.1.4.5. Open Systems

Owing to open systems' vast scope, eliminating acute pollution sources or ending long-term effects would hasten recovery. This is because the damaged ecosystems in the area that are starting to recover can be replenished by propagule and larval supply from unaffected areas. For example, areas that were dredged for scallops or subjected to tow fishing appear to recover quickly after the damaging activities have stopped. To allow the scallop populations to repopulate naturally, scallop dredging operations were stopped in a region off the Isle of Man. The natural recovery of the scallop population was faster than anticipated, much to everyone's surprise. Furthermore, the benthos' general nature grew increasingly diverse. Stated differently, their biodiversity increased.

6.1.4.6. Rocky Shore Systems

Rocky shorelines are resilient to oil spills, so long as harmful cleanup techniques are avoided. In one case, the cleanup efforts in Torrey Canyon, southwest Britain, ended up causing more harm than the actual oil spill. It was discovered that the first-generation dispersants used in the cleanup were more harmful than the oil itself. Dispersants work well, but they are not required. In fact, recovery occurred on shores left untreated due to seal populations within two years. Conversely, numerous species that were present in areas where dispersants were overused perished. This includes keystone species that caused biodiversity to decline. Nevertheless, limpets returned as a result of the proliferation of ephemeral and fucoid algae following the removal of the dispersants. The marine ecosystem is subject to a complex and long-lasting impact from both the oil spill and the cleanup efforts. After the oil spill cleared, seaweeds became dense, but the thriving wildlife soon ate them up. This caused multiple die-offs to occur in some areas, which in turn sparked a second algal flush. The issue was so complicated that it took the shores almost 10 years to fully recover. This indicates that the best course of action in wave-swept rocky substrates is to allow natural re-colonization processes to occur. However, by eliminating the cause of the environmental pressure, the natural recolonization process should be aided. These days, oil dispersants are less popular because of their toxicity. Today, natural recovery methods and manual cleanup are used to deal with oil spills.

6.1.4.7. Designation of Effective Marine Protected Areas

Effective marine protected area designations can eliminate human disturbance and extractive pressures from a particular habitat. This should be carried out with the proper management regimes in place to reduce human interference and promote the natural recovery of marine ecosystems. When and where marine protected areas are appropriately established, the ecosystem of the area benefits greatly. An implementation that is successful benefits invertebrate and fish populations separately. The limited coverage of marine protected areas' beneficial effects is one of their main disadvantages. Enforcement is another disadvantage. Certain marine protected areas are simply designated as such;

no stringent enforcement of protection is in place. This is not ideal because a marine protected area's efficacy greatly depends on how well it is protected.

6.1.5. Artificially Created Habitats

Certain freshwater and marine environments are man-made or artificial, and if they aren't, they have undergone significant modification. Certain rivers have undergone deepening, straightening, or both to facilitate canal construction, irrigation, and navigation. People built ponds and reservoirs to serve as sources of water for livestock and humans alike. To make navigation easier, estuaries have been straightened and dredged. To support maritime activities, other structures such as jetties, wharves, breakwaters, and enclosed dock basins were built along coastal areas. Creating hard coastal defenses to help against flooding and channeling and canalizing rivers could also result in the creation of a distinctive habitat. Numerous offshore structures, such as wind farms and oil rigs, can serve as man-made reef habitats. Many aquatic species find refuge in shipwrecks, and sinking abandoned ships on purpose is now a recognized way to create artificial reefs. Over the last half-century, diverse techniques for establishing artificial reefs have been employed to enhance marine environments, and facilitate diving tourism, recreational fishing, and commercial fishing. To increase biodiversity, artificial habitat creation is a perfectly feasible endeavor. Crevices and rock pools are examples of microhabitats that can form as a result of the structural complexity of artificial habitats. These can increase biodiversity and lessen the harsh effects of tide-out conditions. In addition, they have the potential to boost the numbers of endangered or overexploited species.

6.1.5.1. Valuing our Marine Ecosystems

Something as important as oceans and coasts should be valued by doing the following:

- **Clean Up:** Collect and remove any waste that is found on the shores and beaches. In addition, clean-up entails recycling any leftover waste or cutting back on the use of products that produce waste. The quantity of waste that ends up in our oceans is essentially decreased by the amount that is dumped in landfills. Products containing microbeads and/or microplastics are another item about which consumers should exercise caution. By using these products, microbeads and microplastics are released into the sea and oceans, endangering marine life.
- **Restore Vegetation Above and Below Marine Habitats:** Coastal ecosystems, such as sea-grass meadows, mangroves, coral reefs, and

salt marshes, are equally significant as pelagic or open ocean habitats in terms of biodiversity. They are crucial as a haven for young and tiny fish species. The fish species that depend on these shelters may see a sharp decline in population if they disappear.

6.1.5.2. Use the Ocean Wisely

Development along the coast and in the ocean, as well as fishing, can harm marine ecosystems if sustainability is not taken into consideration. Having said that, action must be taken by communities, authorities, and other stakeholders if sustainability is the desired outcome. For example, organizations can start sustainable initiatives like setting up and preserving protected areas. Communities that fish may also band together, guided by a resource person, to decide which areas should be protected and which should be fished.

6.1.6. Habitat Restoration vs. Rehabilitation

Practitioners and researchers disagree about what constitutes restoration, and there are many different definitions of what constitutes rehabilitation (Gore, 1985; Cairns, 1988; NRC, 1992; Kauffman et al., 1997). The word "restoration," which technically refers to putting an ecosystem back to its pre-disturbance state, is frequently used to describe a wide range of habitat manipulation techniques, such as mitigation, enhancement, improvement, creation of new habitat, and other scenarios (Table 6.1). In many areas where land use is predominantly agricultural, residential, urban, or industrial, true restoration is not feasible in the near future (Stanford et al., 1996). Therefore, these activities are more accurately referred to as rehabilitation. As a result, we refer to the various activities in this document as "habitat rehabilitation," using more precise language when necessary (see Table 6.1).

Table 6.1. *Commonly Used Terminology and General Definitions*

Term	Definition
Restoration	To return an aquatic system or habitat to its original, undisturbed state
Rehabilitation	To restore or improve some aspects or an ecosystem but not necessarily to fully restore all components
Habitat Enhancement or Improvement	To improve the quality of a habitat through direct manipulation (e.g., placement of instream structures, addition of nutrients)

Reclamation	To return an area to its previous habitat type but not necessarily to restore fully all functions (e.g., removal of fill to expose historic estuary, removal of a levee to allow river to periodically inundate a historic wetland)
Mitigation	Actions taken other than habitat rehabilitation to alleviate or compensate for potentially adverse effects on aquatic habitat that have been modified or lost through human activity (e.g., creation of new wetlands to replace those lost by a land development)

Source: https://www.fao.org/4/a0039e/a0039e04.htm#TopOfPage.

6.1.7. Ecological Basis for Rehabilitation

River rehabilitation projects have frequently tried to create habitats thought suitable for a single species of interest, and have done so with little understanding of the natural structure and function of rivers (Frissell and Nawa, 1992; Muhar et al., 1995; Ward et al., 2001; Roni et al., 2002, Ormerrod, 2004). Nonetheless, many scientists have noted that attempts to control specific species and habitat features rather than entire ecosystems (such as fisheries) have contributed to global decreases in fish and other aquatic species in freshwater environments (Doppelt et al., 1993; Muhar et al. Muhar, 1996; Frissell et al., 1995, 1997). Researchers and managers have also realized that when rehabilitation efforts are viewed in the context of the surrounding watershed or ecosystem, they have a higher chance of being successful in restoring one or more species and averting the extinction of others (Doppelt et al., 1993 Muhar et al., 1975 saw Reeves et al., 1995; Habersack, 2000; Beechie and Bolton, 1999).

Understanding the efficacy of different practices and the reasons behind their success or failure also depends on the watershed and ecosystem contexts (Harper et al., 1998, Jungwirth et al., 2002). The development and conservation of fisheries has also embraced an ecosystem approach by international organizations. In order to ensure the long-term survival of aquatic species, it is recognized by the Convention on Biological Diversity (CBD) (UNEP, 1998) and FAO (FAO, 1995; Sinclair and Valdimarsson, 2003) that knowledge of ecological processes, the bio-physical-chemical qualities of aquatic habitat, nutrient cycling, the significance of nontarget species, predator-prey relationships, and even the role of humans in ecosystems is essential. Therefore, some background information on watershed processes is required before discussing the efficacy of habitat rehabilitation techniques in order to help understand the variability in efficacy among techniques as well as among studies on a given rehabilitation method.

Assessment of disturbed ecosystem functions that lower river system productivity and cause declines in aquatic ecosystem integrity is necessary for the effective planning, execution, and assessment of rehabilitation initiatives (Beechie and Bolton, 1999; Buijse et al., 2002). Such assessments aim to pinpoint changes in critical processes that impact stream habitats and outline the necessary management measures to repair or revive those processes that maintain aquatic habitats and biological integrity (e.g., Muhar et al., 1995; Tockner et al., 1999; Buijse et al., 2002; Beechie et al., 2003a). This strategy places more importance on restoring the ecosystem that supports multiple species than it does on restoring individual fish populations or the populations of any other single organism. Multiple species' habitats will be restored as long as all rehabilitation efforts are in line with the main objective of restoring ecosystem processes and functions.

The scientific basis for this approach can be summarized in two important characteristics of biota and their habitats (adapted from Beechie and Bolton, 1999):

1. Biota are adapted to local environmental conditions.
2. Spatial and temporal variations in landscape processes create a dynamic mosaic of habitat conditions in a river network.

These claims imply that biota have adapted to environments that are variable in both space and time, and they also imply that the long-term survival of fish stocks or races depends on this type of environmental variability. As a result, it makes no sense to plan for the same conditions everywhere or to assume that everything will stay the same everywhere. Scientific analyses of numerous management issues over the past 10 years, including universal habitat standards, have acknowledged this (Muhar et al., 1995; Bisson et al., 1997), not managing for spatial or temporal variation in habitats (Reeves et al., 1995; Bisson et al., 1997; Habersack, 2000), and addressing symptoms of a disrupted ecosystem rather than the causes (Frissell and Nawa, 1992; Muhar et al., 1995; Spence et al., 1996; Ormerrod, 2004).

These methods typically ignore the fact that local populations are adapted to the variable natural habitat conditions that exist within their range, both in terms of time and space. In contrast, pinpointing the underlying reasons for deterioration (i.e., Rehabilitation efforts are concentrated on the processes that create and maintain habitats due to compromised ecosystem processes and functions. This helps preserve and restore the natural spatial and temporal variation of habitats to which aquatic communities are adapted, and it permits every segment of the river network to express its natural potential habitat (Muhar, 1996; Beechie and Bolton, 1999).

We emphasize that there are two key reasons why it is crucial to find the underlying causes of ecosystem degradation. First, most of the relationships between landscapes, habitats, and aquatic biota are unclear to us, and we are unable to precisely forecast how land uses will affect the conditions of aquatic habitats or how those changes will affect the biota. One could even argue that we still don't fully understand all the facets of aquatic ecosystems that have a big impact on fish populations. Previous research has shown that this ignorance has significantly deteriorated habitat. For instance, the importance of wood debris in the creation of habitats was not well recognized until the 1970s, and during the previous 200 years, river channelization and wood removal have drastically changed river habitats (Sedell and Luchessa, 1982; Muhar et al., 1995; Collins and Montgomery, 2002). Despite its significance for the composition and operation of aquatic ecosystems, scientists only suggested extensive wood removal for habitat enhancement as recently as the 1980s. Even without an understanding of the importance of wood in aquatic ecosystems, we could have prevented significant habitat loss by selecting management strategies that preserved riparian forest processes and natural wood functions in channels. Today, wood removal is far less common, though it still happens. Second, natural habitats are usually created by the movement of river channels, wood debris, and sediment; in contrast, traditional rehabilitation techniques like building boulders and log structures or placing spawning gravel attempt to create habitats that do not move in space or time. Because of this, many rehabilitation initiatives fall short of restoring habitats because they fail to acknowledge the interconnectedness of ecological and physical processes within watersheds (Frissell and Nawa, 1992; Ward et al., 2001; Ormerrod, 2004). This relative lack of knowledge leads to two main types of failure:

1. site-prescribed engineering solutions can be overwhelmed by altered watershed processes that are far removed from degraded habitats (e.g., increased sediment supply from upslope sources can bury instream structures and pools), or
2. such measures can prevent habitat formation that would otherwise naturally occur (e.g., bank protection prevents the formation of new floodplain habitats).

Avoiding these types of project failure requires that we focus on restoring ecosystem processes and functions that form and sustain aquatic habitats, rather than on the habitats themselves. Understanding these factors will also help the reader to understand project successes and failures. Beechie et al. (2003) presented a straightforward conceptual framework that outlined the connections between habitat conditions, biota, rehabilitation efforts, and ecosystem processes (Figure 6.1). We address measures that restore either ecosystem processes (e.g., restoration of floodplains, riparian planting, sediment reduction measures, or habitat connectivity (e.g., reestablishing fish passage) and compare them to rehabilitation measures that alter or directly manipulate stream environments (e.g., construction of floodplain ponds, bank hardening, and instream structures). We acknowledge that instream habitats have received the majority of attention in assessments of habitat rehabilitation, while connectivity has received less attention. The effectiveness of measures to restore the hydrologic regime or sediment supply has not received many assessments; this is partly due to the highly variable nature of these processes and the difficulty of detecting even slight changes at the watershed scale. Furthermore, it can take decades for the effects of these processes to be seen in stream channels, which makes it even harder to connect these actions to modifications in the biota and habitat features (Beechie et al., 2005).

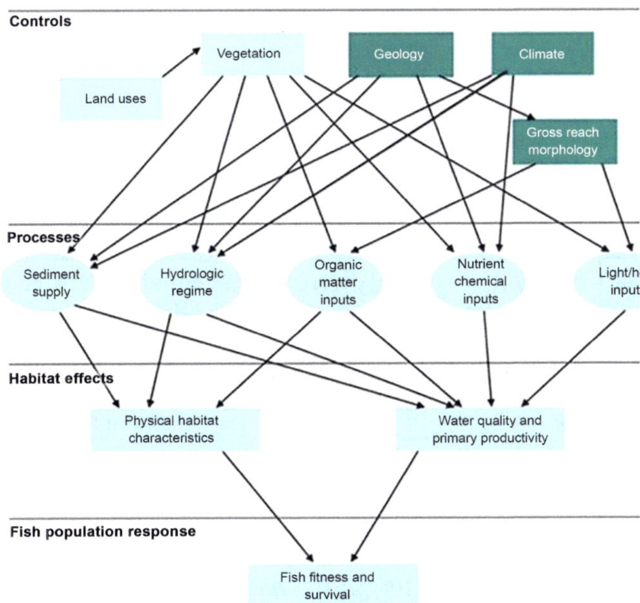

Figure 6.1. *Diagram of linkages between landscape controls, habitat-forming processes, habitat conditions and biological responses.*

Source: https://www.fao.org/4/a0039e/a0039e04.htm#TopOfPage.

River ecosystems throughout the world are driven by the same fundamental sets of processes, such as supply of water, sediment, and organic matter

(Leopold et al., 1964; Welcomme, 1985; Knighton, 1998). However, the specific mechanisms driving each of the inputs vary depending on geologic, topographic, and climatic setting (Table 6.2). Therefore, restoration or rehabilitation actions will address the same sets of processes, but techniques will vary depending on the specific mechanisms that must be addressed. For example, sediment supply to the stream network should be considered in any watershed, but certain processes of sediment supply may be emphasized depending on location.

Sediment supply is dominated by landslides in humid mountainous regions (Sidle et al., 1985), but is more commonly a function of surface erosion and gullying in gentler terrain or semiarid regions (Dunne and Leopold, 1978; Darby and Simon, 1999). Each region has different land uses that affect these erosion processes, and the methods required to address variations in the sediment supply must be tailored to the effects of particular land uses and erosion processes.

Table 6.2. *Examples of Variation in Dominant Ecosystem Processes and Human Impacts Across Ecological Settings. Assessments Should Target those Processes that are Locally Important Within Each Region or Watershed*

Watershed Process or Function	**Ecological Setting**	
	Semiarid, Gentle Terrain	**Humid Mountains**
Sediment	Gullying and surface erosion, agriculture impacts	Mass wasting and gullying, forestry impacts
Hydrology	Flashy flow regime, diversions and dams	Rainfall to snowmelt floodplain regimes, dams
Riparian functions	Grasses and shrubs retain sediment, grazing and farming impacts	Forest shades streams and supplies organic matter, logging impacts
Habitat connectivity	Culverts, dams, and dikes common	Culverts, dams, and dikes common

Source: https://www.fao.org/4/a0039e/a0039e04.htm#TopOfPage.

6.2. EFFECTIVENESS OF COMMON TECHNIQUES

Numerous strategies and tactics have been devised to revitalize freshwater environments, encompassing approaches that aim to reinstate organic processes (e.g., riparian restoration, sedimentation control) to those aiming to make quick adjustments to the physical environment (e.g., putting in place instream habitat structures) in an effort to boost target species numbers quickly. Usually, several rehabilitation projects are carried out concurrently at a single location or within a catchment.

We categorize these rehabilitation activities based on the general area where they occur or the processes they seek to restore including:

- roads and uplands,
- riparian,
- floodplain,
- dam removal,
- instream structures,
- structures in lakes and ponds,
- nutrient enrichment, and
- miscellaneous activities (e.g., habitat protection, bank protection, beaver reintroduction or removal) (Table 6.3).

We first give a brief overview of each rehabilitation method, followed by an explanation of the research on how well it works to enhance natural processes (such as the flow of wood, water, sediment, nutrients, etc.), ecological niche, and biota. When it comes to biotic responses, we mostly concentrate on fish and, to a lesser extent, macro-invertebrates. However, when available and suitable, we also report on plants and other biota. More than 700 publications on the restoration and rehabilitation of freshwater habitats were found, 334 of which presented the findings of objective assessments of the efficacy of various habitat rehabilitation strategies (Figure 6.2a).

Numerous additional books and reference materials were also looked through. There were relatively few studies from other nations, with the majority coming from the United States, Canada, and Western Europe in terms of both rehabilitation and effectiveness (Figure 6.2b). This is probably due to the fact that these developed countries have larger economies than others, as well as more funding for research initiatives and habitat restoration. While the majority of general or theoretical literature focuses on floodplain restoration or more general references on various types of restoration, effectiveness focuses on instream rehabilitation or nutrient enrichment.

Table 6.3. *Common Categories of Habitat Rehabilitation Discussed in this Section and Examples of Common Actions*

Category	Examples of Common Techniques	Typical Goals
Road rehabilitation	Removal or abandonment Resurfacing Stabilization Addition or removal of culverts	Reduce sediment supply Restore hydrology Improve water quality
Riparian rehabilitation	Fencing to exclude livestock Removal of grazing Planting of trees and vegetation Thinning or removal of understory	Restore riparian vegetation and processes Improve bank stability & instream conditions
Floodplain connectivity	Levee removal Reconnection of sloughs, lakes Excavation of new floodplain habitats Re-meandering a straightened stream	Reconnect lateral habitats Allow natural migration of channel
Dam removal and flood flows	Removal or breaching of the dam Increase instream flows Restoration of natural flood regime	Reconnect migration corridors Allow natural transport of sediment and nutrients
Instream habitat structures	Placement of log or boulder structure Engineered log jams Placement of spawning gravel Placement of brush or other cover	Improve instream habitat conditions for fish
Lakes habitat enhancement	Placement of logs and brush Artificial reefs Addition of spawning gravels	Provide cover, rearing, and spawning habitat
Nutrient enrichment	Addition of organic and inorganic nutrients	Boost productivity of system to improve biotic production Compensate for reduced nutrient levels from lack of anadromous fishes

Miscellaneous techniques	Reintroduce or remove beaver Brush removal Bank protection	Reduce or increase habitat complexity Prevent erosion or channel migration
	Habitat protection through land acquisition, conservation, easements, or legal protection (laws) Instream flows	Protect habitat from further degradation Provide adequate flows for aquatic biota and habitat

Source: https://www.fao.org/4/a0039e/a0039e04.htm#TopOfPage.

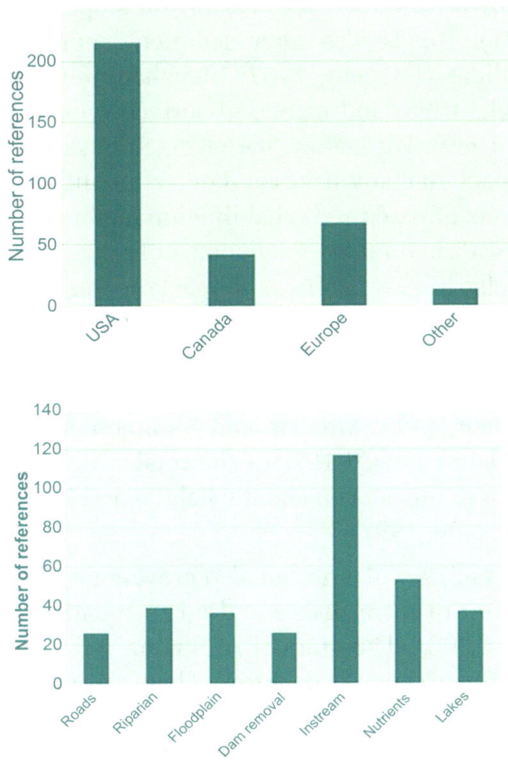

Figure 6.2. *(a) Total number of references from different regions located and used in this section; and (b) Total number of those references in each category that evaluated project effectiveness.*

Source: https://www.fao.org/4/a0039e/a0039e04.htm#TopOfPage.

6.2.1. Road Rehabilitation

Numerous detrimental effects on aquatic ecosystems can result from the construction of both paved and unpaved roads (Trombulak and Frissell, 2000; Gucinski et al., 2001). Hydrologic regimes are changed by roads (Harr et al. LaMarche and Lettenmaier, 2001; King and Tennyson, 1984) and the amount of sediment that streams receive (e.g., Sidle et al., 1985), which affect the properties of the channel and habitat (e.g., Cederholm et al., 1982; Hicks et al., 1991; Nyssen et al., and eventually have an effect on aquatic biota (Waters, 1995; Gucinski et al., 2001, 2002; Gibson et al., 2005). Drainage connections to stream channels from forest roads (e.g., ditches that drain into streams) can change how much water flows into streams, when it flows, and how much sediment washes down from road surfaces or hill slopes (Croke and Mockler, 2001; Madej, 2001). Roads also alter sediment supply through increased frequency of landslides (Dyrness, 1967; Megahan and Kidd, 1972; Sidle et al., 1985; Best et al., 1995) and increased surface erosion (Reid and Dunne, 1984; Bilby et al., 1989). Ultimately, the changes in sediment and hydrology have a negative impact on many fishes and other aquatic biota. A large body of research on the effects of roads and rehabilitation methods has concentrated on unpaved or gravel roads in rural, mountainous, or forest areas. But impermeable surfaces like roads also have a significant impact on urban watershed processes. For instance, channel incision, simplification of instream habitat, and an increase in peak flow frequency and magnitude are all caused by roads and other impermeable surface areas in urban areas (Scott et al., 1986; Booth, 1990; Wong and Chen, 1993; Moscrip and Montgomery, 1997). When roads and impervious surfaces exceed 10% of the total watershed area, changes in hydrology and negative impacts on aquatic habitat and biota will occur (Klein, 1979; Booth and Jackson, 1997).

Road crossings such as culverts can also prevent or inhibit the upstream or downstream migration of many fishes and other aquatic organisms (Larinier, 2002c; Roni et al., 2002). The amount of stream habitat made inaccessible because of human infrastructure is daunting. There are an estimated 1.4 million stream-road crossings in the United States and an estimated 2.5 million artificial barriers prevent fish passage in the United States (US Fish and Wildlife, National Fish Passage Program, Arlington, Virginia, unpublished data). While massive efforts have focused on replacing impassible culverts in streams in the United States and Canada, fish passage through culverts and other road crossings is a problem throughout North America (e.g., USGAO, 2001; Langill and Zamora, 2002; Warren and Pardew, 1998), Europe (e.g., Yanes et al., 1995; Glen, 2002; Larinier, 2002c), and elsewhere in the world.

6.2.1.1. Techniques for Reducing Road Impacts

Numerous restoration initiatives have been made in an effort to lessen the detrimental effects that roads have on aquatic environments. These fall into three categories: connectivity, hydrology restoration, and sediment reduction. Although connectivity is covered in detail in the section on floodplain rehabilitation, we will talk about it specifically in relation to small stream road crossings here. Many techniques are employed to lessen the amount of silt delivered by roads and the landslides they cause. In order to enhance the natural transport of sediment and biota, these measures include resurfacing roads, lowering traffic, adding more stream crossings, stabilizing cut and fill slopes, and replacing stream crossings. Furthermore, a popular strategy to lessen the effects of roads is the complete removal or abandonment of the road, which may involve many of the previously discussed approaches. The same strategies used to reduce sedimentation are also used to increase the number of cross drain structures, create water bars to distribute water onto the forest floor or into existing stream channels, and employ other strategies to stop roads and road ditches from acting as channel networks in order to lessen the hydrologic effects of roads in rural and forested areas. Reducing the quantity of road surface that drains into streams directly can also lessen the delivery of fine sediment. In urban areas, many novel techniques have been implemented recently to enhance storm-water infiltration and regulate increased runoff. Porous paving, removing hard surfaces and replacing them with vegetation when feasible, various seepage, retention, or detention basins or ponds, overflow wetlands, the installation of rainwater cisterns, high-flow bypass channels, alternative drainage systems, and low-impact development for new construction are a few examples of these (Sieker and Klein, 1998; Riley, 1998). Much of the work on urban systems focuses on protecting waterways in suburban or rapidly urbanizing areas by limiting impervious surface areas and protecting riparian zones and uplands. In contrast, work in existing urban centers has focused on preventing further degradation and structural improvements to channels and riparian areas when possible (Sailer, 1994; Horner and May, 1999).

At road crossings, stream reaches and channels are made more connected through the use of fish migration ladders, bridges, and gutters. The majority are made to allow adult fish to pass through roads safely, but some do not allow juvenile fish to pass through or preserve natural processes (e.g., channel morphology, including the movement of wood and sediment. While bridges can be expensive, they don't restrict the channel as much as culverts do, allowing other materials to pass through and creating a natural stream channel. Culverts with an open bottom or embedded (e.g., countersunk) pipe-arch culverts are useful for passing both young and mature salmonids because they let a natural substrate grow within the channel (Furniss et al., 1991 Clay, 1995; Larinier, 2002c). However, if the culvert size does not account for large flow events or the volume of wood and sediment transported by the stream, such culverts may restrict the stream channel (Robison et al., 1999). Additional design choices include adding baffles to the culvert to slow down flow and backwatering the culvert at the inlet or outlet. Concise summaries of culvert designs and techniques for retrofitting impassable culverts are given by Clay (1995) and Larinier (2002c). Even though the majority of culverts are made to pass adult fish, morc investigation is required to determine which kinds can pass juvenile fish at different flows.

6.2.1.2. Effectiveness of Techniques for Reducing Road Impacts

The research on the effectiveness of road rehabilitation efforts has mostly concentrated on their physical effects on the stream channel, with the exception of stream crossings like bridges and culverts, which can impede the migration of fish and other organisms. Occasionally, research has also examined processes like landslides in mountainous areas and erosion rates. Although many of the methods covered above are frequently used in mountainous regions, there aren't many published assessments of their efficacy (Switalski et al., 2004). The assessment of methods for reducing road-surface erosion has typically been restricted to contrasting the amounts of fine sediment present in road runoff across various traffic volumes and surfacing types. As an instance, Bilby et al. (1989) discovered a positive correlation between the amount of fine sediment delivered to stream channels and traffic levels. Surface erosion in the Pacific Northwest US watershed of the Clearwater River was reduced by a factor of 10 by reducing traffic (Reid and Dunne, 1984). Other studies have demonstrated that reducing traffic levels and tire pressure of trucks can reduce sediment delivery from forest roads (Foltz, 1998; Foltz and Elliot, 1998).

Reid and Dunne (1984) showed that surface erosion can be reduced by approximately 80% by using surfacing material that is 15.2 cm thick. Similarly, Kochenderfer and Helvey (1987) showed that adding 7.6 cm (3 inches) of gravel to the surface of an Appalachian Mountain forest road decreased soil loss from 47 to 6 tons per acre. Burroughs and King (1989) examined unpublished research assessing various approaches to lessen sediment erosion from roads and found that putting 7.5 to 15 cm of crushed rock on the surface significantly decreased erosion. Furthermore, paving the road surface or limiting truck traffic on unpaved roads significantly decreased erosion (Burroughs and King, 1989). Even though these studies are only available from the western United States, they indicate that these techniques, when used correctly, can lessen surface erosion. It has also been shown that removing or abandoning roads significantly lowers the amount of sediment delivered (Hickenbottom, 2000; Madej, 2001; Switalski et al., 2004). For example, Hickenbottom (2000) demonstrated in a Montana watershed that 12 months after road removal and recontouring runoff and erosion approached those of natural slopes. The type of treatment following road removal or closure (e.g., recontouring of slope, ripping of road surface, removal of stream crossing, placement of mulch, seeding, planting) can influence the sediment production and infiltration capacity of the former road bed (Cotts et al., 1991; Maynard and Hill, 1992; McNabb, 1994; Luce, 1997; Elseroad et al., 2003). In general, these studies have found that recontouring slope and site preparation (ripping and mulching of former road bed) are the most effective at inducing plant growth and reducing fine sediment production, though position of the road on the slope and time since treatment also play a role in the effectiveness of these methods. Very few evaluations of rehabilitation techniques for landslide hazard reduction have been conducted.

Anecdotal evidence suggested by Harr and Nichols (1993) suggested that removing roads in mountainous areas decreased the frequency of landslides. When Cloyd and Musser (1997) looked at a portion of more than 1,200 kilometers of stabilized or destroyed forest roads in Oregon, they discovered that untreated roads near stream crossings had higher problem ratings (erosion and mass failure severity). Studies on turbidity changes related to stream crossings and road rehabilitation have been conducted from time to time; however, the findings typically indicate transient increases in turbidity caused by construction (Brown, 2002). Lastly, it's critical to remember that the geology affects the amount of naturally occurring sediment delivered, the delivery of sediment from roads, and the overall success of efforts to reduce sediment through road resurfacing or total road removal (Bloom, 1998).

Much of the literature on urban stormwater management and hydrology has focused on modeling of potential changes (e.g., Johnson and Caldwell (1995); Sieker and Klein, 1998) and there aren't many published assessments available. Booth along with others. (2002) found that because most stormwater retention ponds are small, they have typically not been able to reduce channel erosion or restore hydrology in heavily urbanized areas. More recent techniques for lowering the effects of stormwater runoff and natural drainage systems on hydrologic and water quality (e.g., constructed wetlands, vegetated areas, and swales) have shown promise, but research and development on these concepts is still in its early phases.

Rarely are physical studies on the channel response after culvert removal carried out and published. Klein (1987) found that in a rare study on channel adjustment after culvert removal, channel adjustments were negligible in the presence of large woody debris or other channel roughness elements in a

California watershed. However, the majority of research that has been published has concentrated on biotic response or recolonization, which is typically the main goal of culvert replacement projects. Fish frequently settle in new habitats fairly quickly, and numerous studies have shown that replacing stream culverts with open channels is an efficient way to facilitate fish migration (Iversen et al., 1993; Bryant et al., 1999; Glen, 2002; see also section on floodplain rehabilitation). However, it may take several years for fish to colonize new habitats, and long-term monitoring may be required to assess success if fish populations are extremely low or if a culvert is only passable at certain water levels or seasons. Research comparing various approaches to salmon habitat rehabilitation has demonstrated that some of the biggest increases in fish production result from removing obstacles that impede fish migrations (Scully et al., 1990).

Projects to replace culverts or remove other barriers may also help invertebrates and wildlife in addition to fish. Yanes et al., Culverts both facilitated and hindered the migration of numerous mammals and reptiles, according to (1995) and Vaughan (2002), who proposed that they prevent a large number of aquatic macroinvertebrates from migrating upstream. Ultimately, the capacity of different kinds of stream crossings to permit the passage of fish and other aquatic organisms depends on a complex interplay between the physical features of the stream crossing (e.g., depth, speed, and roughness) as well as the fish's or other biota's behavior and swimming ability (Clay, 1995; Larinier, 2002b,c). It is evident from the scant published literature that culvert replacement, along with other road crossing structures that impede or prohibit fish migration, is a highly effective strategy for improving fish habitat and productivity.

6.2.2. Riparian Rehabilitation

In a watershed, riparian zones serve as the interface between terrestrial and aquatic environments. They are particularly productive zones that also serve as important habitat for both terrestrial and aquatic biota. Other important functions that riparian zones provide include processing nutrients, delivering woody debris and organic matter to a stream, providing shade, stabilizing soils, and regulating microclimate. Unfortunately, riparian areas have been and still are heavily impacted by human activities due to their proximity to water, transportation, and the productivity of the forests and soils. Common activities in riparian areas include logging, agriculture, grazing, recreation, road construction, and various residential, urban, and industrial developments. Significant efforts have been made to restore riparian areas and enhance stream habitat quality, both for ecological and cultural reasons, as well as to enhance habitat for fish and wildlife.

6.2.2.1. Techniques for Riparian Rehabilitation

Various approaches have been devised to either enhance or repair riparian state, usually with the ultimate objective of bettering aquatic environments. These can be divided into two main categories: i) fencing and grazing reduction; and ii) silviculture treatments. Since there are significant differences in the methods used to restore riparian areas, we address the two separately. Replanting trees and other vegetation with protection from additional harvesting or vegetation removal is the goal of silvicultural treatments, which aim to restore riparian areas. Efforts to reduce or eliminate grazing generally aim to relieve pressure on riparian areas by allowing vegetation to regrow naturally, a process known as passive restoration.

6.2.2.1.1. Riparian Silviculture

Riparian forests and their conditions have been enhanced or restored through the application of diverse silvicultural techniques. These include the process of sowing, planting, and taking down trees, eliminating understory vegetation that competes with the trees, eliminating disturbances that suppress the growth of nearby plants, and changing the physical environment (e.g., Riparian rehabilitation techniques can be classified according to whether they remove or kill vegetation (thinning/harvest, girdling, competitive release, pruning), or plant vegetation (seeding, planting, coppicing, staking and layering). flood regimes or sediment supply) such that desired vegetation becomes established or undesirable vegetation dies. The majority of these methods were adopted from upland forest methods, and a large portion of the literature on silviculture concentrates on methods for growing tall, straight, closely spaced, commercially significant trees, frequently as monocultures, rather than on methods for establishing varied and useful ecosystems (Smith, 1986; Burkhart et al., 1993; Curtis et al., 1998). A variety of non-silvicultural methods have also been employed to enhance riparian zones, such as managing invasive exotic plant species, stabilizing banks, and altering flow patterns. Riparian areas are comparatively vulnerable to exotic species colonization due to their high natural rates of disturbance, which result in unoccupied patches. In certain river systems, the number of exotics is quite high. As a result, the issue of managing exotic species in riparian areas is a global one (Dudgeon, 1992; Rowlinson et al., 1999; Tickner et al., 2001).

Nonetheless, the majority of these species are not regarded as noxious or invasive because they do not substantially affect ecosystem functions or outcompete native species. On the other hand, aggressive invasive species frequently create monocultures or, if allowed to spread unchecked, take over as the dominant species. When nonnative species predominate in riparian areas, restoration efforts may need removing these species. *Tamarix ramosissima*, or tamarisk, and salt cedar, or *T. chinoensis*, an invasive species common in North America and Europe, are just a few of the problematic riparian invasive species. They have light seeds that allow them to spread easily and readily, occupying recently disturbed riparian areas and degrading riparian habitat significantly throughout the American Southwest.

Numerous other invasive species exist in other parts of the world, without a doubt. Other passive techniques for riparian rehabilitation include restoring floodwaters or allowing vegetation to regrow naturally. Another way to restore riparian vegetation and river channels is to restore flooding or groundwater by removing dams or restoring normative flows.

6.2.2.2. Fencing and Grazing Reduction

Global riparian and stream habitat degradation is a result of excessive livestock grazing. Because riparian areas are closer to water and provide superior forage, livestock and native ungulates use them more frequently than upland areas, which means that grazing tends to have a greater negative impact on these areas (Armour et al., 1991).

Livestock grazing has been shown to have detrimental effects on the riparian and stream environments through altered and decreased vegetation, bank instability, increased sedimentation, channel widening, aggradation, altered hydrology, and decreased water table (Armour et al. Belsky et al., 1991; Platts, 1991, 1999). Fish habitat is often affected by changes in stream morphology, reduction in pools and depth, reduction in shade, cover, and terrestrial food supply, and increase in water temperature and quality (Armour et al. Belsky et al., 1991; Platts, 1991; 1999).

The main ways to restore grazing-affected riparian and stream channel habitats are to either completely stop grazing, use fences or other barriers to keep livestock out of certain areas of the riparian zone, or put in place a grazing reduction or management plan that allows the riparian vegetation to regrow (Elmore, 1992). Across the world, a variety of grazing management systems have been put into place; practically all of them involve controlling the length of grazing, the number of ungulates, or both. One that is especially popular is rest-rotation grazing management, which involves grazing and nongrazing intervals that are frequently rotated across several pastures. Other strategies to prevent animals from entering riparian zones include placing water sources far from the stream (i.e.supplying salt away from the stream (Platts and Nelson, 1985) or using nose pumps.

6.2.2.3. Effectiveness of Riparian Rehabilitation

6.2.2.3.1. Effectiveness of Riparian Silviculture

There aren't many long-term evaluations of riparian techniques, so their effectiveness has mostly been assessed through short-term (less than 10 years) studies of vegetation survival and growth. A variety of factors, such as herbivore grazing and understory or overstory control, affect the growth and survival of riparian plantings. Sweeney et al. (2002) investigated the effectiveness of various oak (*Querus* spp.) species over a four-year period using different riparian treatments and silvicultural methods. Survivability of oak seedlings was primarily determined by protection from herbivores. Emmingham along with others (2000) looked at more than 30 riparian projects in coastal Oregon that aimed to transform deciduous riparian areas into coniferous riparian forests. It concluded that at first, riparian silviculture treatments appear promising for establishing conifers in riparian zones that are dominated by hardwoods. On the other hand, inadequate management of the understory and overstory, defense against deer browsing (*Odocoileus* spp. the success of the project was impacted by elk (*Cervus elaphus*), beaver (*Castor canadensis*), and mountain beaver (*Aplodontia rufa*). Herbivores can thwart rehabilitation efforts, so any rehabilitation activity involving the planting and observation of vegetation must be aware of this (Emmingham et al., 2000); Merenlender and Opperman, 2000). Fencing, netting, plastic mesh cages, and chemical sprays are methods for minimizing animal damage to planted vegetation (Black, 1992).

There is a paucity of comprehensive research on fish response to riparian planting, and even fewer studies have looked at other instream biota. On the other hand, Penczak (1995) discovered that when the riparian vegetation recovered following removal, the variety of fish species in the Warta River, Poland, increased from 11 to 16. Replanted riparian buffers improved water quality and channel stability, according to a recent study on riparian fencing and replanting in New Zealand. However, responses to nutrients and fecal contaminants varied, and macroinvertebrate communities did not improve toward clean water varieties (Parkyn et al., 2003). It was suggested that in order to influence changes in water temperature and water quality, longer or larger buffers were required.

Evaluation efforts for invasive species typically focus on success of removal efforts and success of recolonization by native species (often cottonwoods and willows). Taylor and McDaniel (1998), Roelle and Gladwin (1999), Sprenger et al. (2002) and others have monitored and examined the success of various chemical, burning, mechanical, and hydrological treatments in removing tamarisk and the subsequent recolonization of native species. Application of a combination of these techniques in addition to cottonwood and willow plantings and timed irrigations have produced diverse riparian habitat (Taylor and McDaniel, 1998). Roelle and Gladwin (1999) were able to prevent the establishment of tamarisk seedlings at a rehabilitation site by regular, controlled fall flooding. For other invasive species such as Japanese knotweed (Beerling, 1991), which is known to disrupt riparian habitat throughout Europe and North America, there has been little in the way of monitoring to determine at what rate riparian habitat is being lost to the species.

According to data from the few published studies on passive riparian restoration, natural recovery rates can occasionally be high enough to eliminate the need for and possibly even be detrimental to hands-on rehabilitation. Briggs (1996) details Arivaipa Creek's comeback in Arizona, where the majority of the riparian vegetation was destroyed by a 500-year flood. To hasten the healing process, a massive recovery initiative was put into place, which involved planting thousands of cottonwood stakes. But only eight years after the flood, the riparian areas had grown back so strongly with native flora that it was impossible to find any artificial plantings. Restoring seasonally flooded forests, which are essential habitat for fish spawning and rearing, is another common practice in the natural recovery of forests. To establish fish spawning, rearing, and feeding areas in seasonally flooded forests near Tonle Sap (Great Lake) and the Mekong River, for instance, reforestation is utilized as a significant fisheries rehabilitation technique in Cambodia (Thuok, 1998). Although there isn't much long-term observation of natural recovery, it seems to be a useful strategy in the correct situations.

6.2.2.3.2. Effectiveness of Fencing and Grazing Reduction

It is challenging to draw definitive conclusions about the relative efficacy of a management strategy other than eliminating grazing because the numerous studies on the efficiency of grazing reduction techniques in restoring riparian areas have looked at a wide variety of different grazing systems. The information regarding the removal or reduction of grazing in order to enhance the physical habitat, aquatic biota, and riparian condition is compiled here. This arrangement also shows the temporal sequence in which systems recover after grazing is reduced. We attempt to give instances of the various grazing reduction techniques' efficacy.

- **Riparian Conditions:** The improvement of riparian vegetation and its functions, like shade, sediment storage, and hydrologic effects, has occurred relatively quickly (5–10 years). Numerous studies have shown that after livestock are removed from the area or that grazing intensity is drastically reduced, water storage and aquifer recharge occur. Bank stability, channel geometry, habitat complexity, and other channel characteristics also recover quickly, unless the channel is deeply incised.

Eliminating all livestock seems to be the most effective way to deal with vegetation. For instance, Robertson and Rowling (2000) discovered that understory vegetation and tree abundance were, respectively, one and three orders of magnitude higher on ungrazed sites when they compared sites that were grazed now with those where grazing had been eliminated between two and 50 years earlier. At sites where grazing was still occurring, the proportion of bare soil was lower and the amounts of fine, coarse, and fine particulate organic matter were higher. Similar to this, Platts (1991) examined grazing practices in the US, recognized 17 riparian grazing systems, and concluded that full livestock exclusion and minimal use were sufficient to protect riparian and fisheries resources. However, many variables, such as the number of grazing days, the season, the seasonal livestock dispersal behavior, and the degree of compliance, affect how well rest-rotation grazing systems allow vegetation recovery (Myers, 1989). According to a review of grazing systems conducted in Montana, USA, by Myers (1989), rest-rotation grazing systems that permitted vegetation to recover successfully typically lasted 28 days, whereas unsuccessful systems typically lasted 59 days. When Platts and Nelson (1985) studied rest-rotation grazing, they discovered that cattle grazed less in the riparian areas in the early growing season when high-quality upland vegetation was abundant and that it was difficult to achieve compliance during scheduled rest periods (Elmore, 1992). Although grazing systems that regulate the amount and timing of use of riparian and upland areas are also believed to be advantageous, historically, grazing systems have not distinguished between riparian and upland range areas (Clary and Webster, 1989). For example, spring grazing can maintain herbaceous stubble heights that sufficiently guard erodible stream banks and distribute stock equally between riparian and upland areas (Clary and Webster, 1989). When combined with rigorous monitoring, rest-rotation and other seasonal grazing techniques have demonstrated promise in preserving riparian and aquatic habitat (Myers and Swanson, 1995); however, further research is required to determine the efficacy of these techniques (Clary and Webster, 1989).

There is evidence that ungulates can contribute to the preservation of the diversity of riparian vegetation, even though native ungulate grazing may have historically damaged some riparian and stream areas (Medina et al., 2005). In a study conducted in Scotland, Humphrey and Patterson (2000) discovered that the diversity of plant communities increased when cattle grazing was reintroduced. On the other hand, riparian vegetation can also recover if native herbivores are kept from overgrazing. As an illustration, Oppermann and Merenlender (2000) discovered noticeably greater concentrations of woody plants (*Salix* spp.) along sections of a California stream where fencing had been used to keep out blacktail deer (*Odocoileus hemionus columbianus*). Additionally, the removal of livestock might result in a rise in native ungulates, which could be detrimental to attempts to restore riparian areas and the stream channel (Medina et al., 2005). These studies indicate that while grazing by wild ungulates may have some negative effects, restricted livestock grazing may occasionally have positive effects on riparian vegetation.

- **Physical Habitat and Channel Conditions:** Recovering instream habitat conditions is slower than in many riparian environments (e.g., However, many studies conducted in arid environments have noted the benefits of fencing or total grazing removal on bank stability and channel features (shade, plant growth). Overall livestock exclusion and rest-rotation grazing improved bank stability, tree cover, and pool habitat (Myers and Swanson, 1995), but additional management practices (e.g., The removal of coarse woody debris and the road crossing had an adverse effect on the instream conditions. Clary et al. (1996) in studies conducted in both Oregon and Idaho, and Clary (1999) reported improved width/depth ratios and substrate embeddedness following complete removal and reduction grazing. In the western United States, Connin (1991) examined many riparian rehabilitation projects that combined bank protection with fencing or grazing removal. Connin discovered that these initiatives generally increased bank stability, promoted the growth of riparian vegetation, and improved channel conditions. The amount of fine sediment retention is positively correlated with the width of the exclusion or buffer when livestock are kept out of only a portion of the riparian zone (Hook, 2003). These studies show that, especially in cases where past grazing was exceptionally intense, grazing systems may also contribute to the recovery of stream banks. Complete grazing removal seems to facilitate better recovery of fine sediment, bank angle, channel entrenchment, and width to depth among other in-channel factors.
- **Biotic Responses:** Relatively few published studies exist that examine the effects of grazing reduction on fishes and other aquatic biota (Rinne, 1999). Our understanding of the effects of grazing and grazing reduction on fish populations has been developed primarily through examining the effects of grazing on stream habitat characteristics (Platts, 1991; Clary, 1999). Much of the information

is based on studies comparing sites with different levels of grazing and habitat quality and fish numbers, which generally show higher stream temperatures and lower fish numbers in grazed than ungrazed sites or watersheds (e.g., Li et al., 1994; Myers and Swanson, 1991; Platts, 1991; Wu et al., 2000).

The majority of research on the effects of grazing on biota has been conducted in the western United States, where vast stretches of public land are managed for the grazing of livestock. Comprehensive reviews of grazing in the western United States were carried out by Platts (1991) and Rinne (1999), who noted that although some studies showed increases in fish populations, the studies hardly ever focused on fish or the aquatic biota. According to Rinne (1999), the majority of studies lacked the necessary length of time or experimental design to accurately identify meaningful shifts in fish populations brought about by grazing removal or reduction. Kauffman et al. (2002) conducted a study in streams in Oregon and found that after livestock removal, there were improvements in the flora, stream morphology, and densities of young-of-year rainbow trout (*Oncorhynchus mykiss*), but not in the densities of adult or juvenile trout. They proposed that the short reaches they sampled for their study were the reason why fish, both juvenile and adult, did not respond. Medina et al., (2005) examined three long-term case studies conducted in the American Southwest and discovered inconclusive findings in each of the three cases as a result of study design flaws, species interactions, upstream or watershed-scale effects, exotic species introduction, and fisheries management (stocking and regulation changes).

6.2.3. Floodplain Rehabilitation

Large-scale fishing and aquaculture operations are supported by floodplain wetlands, lakes, and other aquatic habitat in many parts of the world. Fish production has decreased in many floodplain areas as a result of rivers being channelized for industrial, urban, agricultural, and other human uses. Habitats like wetlands, sloughs, and oxbow lakes were once common in large river floodplains, but due to channelization and other human activities, they are now rare or disconnected in many river systems. In the last 10 years, large river rehabilitation—and in particular, the reconnection of isolated floodplains and their associated habitats—has become an essential part of river ecosystem rehabilitation in developed and, more recently, in developing nations (Gopal, 2004). The field of reconnecting river-floodplain systems and rehabilitating larger streams (wider than 20 meters) is still in its infancy, although research on the Danube and Rhine rivers has recently focused on these techniques (Buijse et al., 2002) and additional European rivers. Many of the river systems that are the subject of these rehabilitation initiatives are highly engineered systems that depend on human intervention to be kept in a particular condition (Bayley et al., 2000; Buijse et al., 2002). Therefore, a lot of initiatives and approaches that have come after have been and will continue to be a compromise between letting nature run its course and creating technical fixes (Cowx and Welcomme, 1998).

6.2.3.1. Overview of Floodplain Techniques

Restoring or improving lateral (connection with floodplain), longitudinal (upstream-downstream), and occasionally vertical (hyporheic) connections is the main goal of most floodplain or large river rehabilitation projects. Some of the most common rehabilitation techniques include connection of isolated floodplain habitats (e.g., Levee removal or setback, construction of a new channel, realignment or meandering of an existing channel, or creation of new floodplain ponds are examples of floodplain management practices. Other methods, which are addressed in other sections of this document, include dam removal, flood flow restoration, and riparian area rehabilitation. There are many different kinds of channels, ponds, lakes, and depressions in floodplain habitats. For fish that favor lentic habitats, these areas offer valuable overwintering habitat. Furthermore, in many instances, the rehabilitation of floodplain habitat is advised over instream habitat improvement in channels with a bankfull width of more than 12 meters and channels with a gradient of less than 2 percent (Dominquez and Cederholm, 2000; Roni et al., 2002; Pess et al., 2005). The ensuing subsections delineate diverse methodologies along with the existing knowledge regarding their efficacy. By emphasizing habitat restoration for fish, we differentiate wetland rehabilitation from floodplain rehabilitation. Wetland rehabilitation can also involve the restoration of plants, birds, mammals, and other vertebrates and invertebrates, and it frequently concentrates on habitats that are not directly connected to a stream. All of these are significant objectives of habitat rehabilitation in our opinion, but we concentrate on fish habitat in our discussion rather than just wetland science (see Kusler and Kentula, 1989, Wheeler et al., 1995, and others for more information on wetlands).

6.2.3.1.1. Connecting Isolated Habitats

The use and composition of fish species are directly impacted by the degree of connectivity between floodplain habitats (Berrebi-dit-Thomas et al., 2001; Miranda and Lucas, 2004). One of the more popular methods is to reconnect isolated habitats to the main stem by means of weir removal, levee breaching, or replacing a culvert or stream crossing to reestablish a relic channel or larger floodplain water bodies, like a pond or lake. In places where flow needs to be controlled, existing relic channels can be reconnected using notch dikes, culverts, submersible dikes, and controllable flow gates (Lister and Finnegan, 1997; Cowx and Welcomme, 1998).

Two typical issues with these kinds of projects are: i) variations in elevation between the mainstem and relic channels as a result of mainstem channel incision from the moment the floodplain channel disconnected; and ii) deterioration of the floodplain's water quality. For instance, restoration of two former meander bends that are now floodplain sloughs is one of the rehabilitation opportunities found in the Stillaguamish River in Washington State, USA (Pess et al., 1999). But between 1929 and 1991, when the bends were cut off, the river channel widened by one to two meters, making it challenging to reconnect the former meanders (Collins et al., 2003). Numerous streaming channels in Europe, North America, and other parts of the world share this issue. A method for raising the bed elevation of rivers with cut mainstem channels and reuniting them with their original channels and floodplain is the use of submersible check dams. For instance, check dams in lateral artificial channels have been used in Slovakia's Danube River to aggrade the river to the point where older channels are now reconnected, increasing the reach's water retention time (Cowx and Welcomme, 1998). Channel filling has been utilized on the Kissimmee River in Florida (USA) to reestablish floodplains and ancient meanders (Toth et al., 1993). At the places in the river where the meander crosses the main channel, levee material has been used to fill it. This approach is only effective if the floodplain or a few of its important habitats are preserved (Cowx and Welcomme, 1998).

6.2.3.1.2. Levee Breeching and Setbacks

A full or partial levee breeching or setback project increases the diversity of a floodplain's habitat by allowing a river to migrate and subsequently create and maintain different types of floodplain channels. Due to land constraints that frequently limit the relative extent of such rehabilitation actions, many of these projects combine partial and complete levee removal. One method is the "beaded approach" to floodplains, in which leveed sections of the river alternate with small sectors of the entire floodplain width (e.g., 6 to 8 km long) that are allowed to function. Wetlands and floodplain channels are two examples of the habitat features found in these sections that cover the entire width of the floodplain. The river will be able to run smoothly thanks to this family of habitats, which may also produce pulses of flow, sediment, and nutrients (Sparks, 1995). This method promotes scour, erosion, and deposition in the floodplain by allowing some areas to be submerged. This method can be applied in places like tributary junctions, where there is typically a high degree of natural channel migration.

6.2.3.1.3. Channel Reconstruction or Meandering

A channel may return to a meandering channel if it is left undisturbed for many decades, depending on the stream power, sediment load, and composition (Brookes, 1992). However, to be restored to a meandering channel connected to its floodplain, the majority of floodplain channels have a low energy and need to be directly intervened in. Making a new meandering channel is therefore one of the more popular methods for restoring rivers and reestablishing their connection to their floodplains. The majority of rivers in Denmark and northern Europe are low gradient, heavily channelized, and unlikely to revert to their original sinuous channel type absent intervention (Brookes, 1992; Hansen, 1996). For this reason, this technique is particularly popular in these regions. Restoring the sinuosity of a stream can be accomplished in one of four ways: i) by removing levees and creating a new meandering channel (instead of letting the stream naturally form its own channel as in the case of a levee set back or removal project); ii) by creating a meandering channel next to the straightened channel and then directing the river into the new channel; iii) by rerouting the straightened channel into its old channel if remnants of the original meandering channel still exist adjacent to the straightened channel; or iv) some combination of these three methods. Many smaller channels have been straightened and piped and "daylighting" or exposing these concealed watercourses and then restoring natural re-meandering pattern is an increasingly common technique in urban and agricultural landscapes (Nielsen, 1996; Riley, 1998). Sometimes in highly incised channels, it may not be feasible to aggrade the channel and reconnect it with the historical floodplain; a new floodplain is created by excavating the banks and widening the channels (Cowx and Welcomme, 1998). This is often coupled with a two-stage channel design, which incorporates a low flow channel and a new high-flow channel or new floodplain above the low flow channel but below the historic floodplain (Iversen et al., 1993). All these methods seek to reconnect the river with its floodplain, increase the diversity of habitats (mixture of slow and fast water habitats), and lengthen the stream channel (reduce the gradient).

6.2.3.1.4. Construction of New Floodplain Habitats

One method of improving habitat is to actively construct new floodplain channels, ponds, or backwaters. This process is known as the "creation of new floodplain habitats." In order to replace natural floodplain channels that have been lost as a result of flood control measures and other floodplain-isolating actions, the creation of new habitats can be a significant form of enhancement (Beechie et al., 2001). In river floodplains, two main types of habitat are usually formed: backwaters or off-channel ponds (e.g., surface and groundwater fed) and side channels (e.g., surface and groundwater fed) alcove ponds, mine dredge ponds, gravel pits, and mill ponds). The construction of side channels fed by surface or groundwater is done to give adult fish a place to spawn and to rear their juveniles (Lister and Finnegan, 1997). With an intake structure on the upstream end, they are normally connected to the main river channel at both the upstream and downstream ends. The project location can be chosen with flexibility because of the surface water source. But sedimentation issues also accompany this water source. This kind of channel might not be appropriate for systems with large suspended sediment loads due to siltation concerns. Controlling the introduction of silt and organic materials into the channel also depends on where the channel intake is located. In order to prevent bank erosion and river flooding, a surface-fed side channel usually needs a dike or control structure and bank armoring at the upstream end (Lister and Finnigan, 1997).

If the water table is low, groundwater-fed side channels can also be excavated. Groundwater-fed channels offer stable year-round water flows with little sediment, clean substrates and stable water temperatures (Sheng et al., 1990; Bonnell, 1991). Groundwater channels are excavated parallel to the main river channel, often along an existing intermittent stream or relic channel. Many of these side channels were originally constructed as channels with nearly uniform depth and substrate to serve as spawning habitats for salmonids (Sheng et al., 1990 from Giannico and Hinch; Giannico and Hinch, 2003).

A perennial or intermittent stream, pond, or wetland can be made into an off-channel pond by a variety of excavation or construction techniques, such as blasting, damming, excavation, or the installation of a flow control structure. Another common method in both North America and Europe is the creation of alcoves or backwaters that are next to and directly connected to rivers. By excavating a pond that is right next to the river channel, alcoves have been created. Additional created habitats include digging ponds and other wetlands, excavating backwaters or alcoves next to the main channel, and building new side channels that are fed by groundwater or surface water. No matter how the ponds or channels are built, it's critical to make sure that these artificial habitats are connected to the river as much as possible year-round, if not seasonally. However, in developing nations like Nigeria, floodplain ponds are occasionally intentionally dug out to separate the water and fish when the floodwaters recede, thereby offering fishing opportunities (Neiland and Ladu, 1998).

Along river channels, gravel pits, mill ponds, and mine-dredge ponds have been created as a result of floodplain mining and other floodplain excavation operations (Norman, 1998). There are more opportunities for floodplain habitats thanks to the reorganization and connection of these ponds with the river channel. For many fish species that favor slow-moving water environments, these regions can offer crucial rearing habitat; however, in certain regions, they are also believed to offer safety from predators and alien species. Many variables, such as depth, morphology, sufficient prey resources, cover, water quality, and sufficient water quality, are necessary for the effective conversion of gravel pits, mill ponds, and other artificial floodplain lakes (Norman, 1998). Large, deep ponds will offer less prey and a significantly higher risk of avulsion, whereas shallow ponds with complex shapes offer the best opportunities for juvenile fishes to successfully use them (Peterson, 1982; Norman, 1998).

6.2.3.2. Effectiveness of Floodplain Rehabilitation Techniques

6.2.3.2.1. Isolated Habitats

As with other rehabilitation methods, recovery of processes (hydrology, connectivity, sediment transport, and physical and biological response) can be used to gauge how effective floodplain rehabilitation is. While there has been much study on the restoration of connectivity and the physical and biological response, measurements of the effects on watershed processes are rare. The benefits of improving physical habitat and restoring connectivity are clear and quantifiable in terms of the amount of connected physical habitat. For example, in the Danube River the reconnection of a former side channel and canal in resulted in an additional 50 km of habitat. Reconnected floodplain ponds have proven to be effective at providing habitat for juvenile salmonids such as coho (*Oncorhynchus kisutch*) and Chinook (*O. tshawytscha*) salmon (Richards et al., 1992; Norman, 1998; Roni et al., 2002) in western North America. They are also known to provide critical rearing habitat for many other fishes including cyprinids (minnows), catastomids (suckers), and many other warm and cool water fishes. Schmutz et al. (1994) reported the colonization of a reconnected section of the Danube River and found over 40 species of fish after only one year. In the lower Rhine River, reconnection of floodplain lakes and channels lead to an increase in rheophilic cyprinids (Grift et al., 2001; Simons et al., 2001). Rheophilic cyprinid abundance and presence were positively connected with habitat connectivity and flow level to the main Rhine (Grift et al., 2001). Similar to this, the reconnection of a side channel or canal in the Danube facilitated the quick colonization of numerous species; however, because pike-perch (*Stizostedion lucioperca*), a significant sport and commercial fish, dislike bypass channels around weirs, they did not respond as favorably as other species (Schmutz et al., 1998). Given that many freshwater fish species rely entirely or partially on floodplain habitats for all or part of their life history, rapid colonization is not completely surprising (Mann, 1996; Welcomme, 1985; Cowx and Welcomme, 1998; Buijse et al., 2002).

6.2.3.2.2. Levee Breaching

The majority of research on levee breaching and setback project effectiveness has concentrated on the physical features and the lateral hydrologic connectivity of habitats. Levees and bank armoring are being removed to allow the channel to naturally migrate and regain its previous sinuosity. This is happening more and more frequently. According to preliminary findings, these initiatives successfully enhance floodplain processes like lateral erosion and nutrient transport. Projects to remove levees and restore connectivity with wetlands have been shown to alter the levels of nutrient transport (Childers et al., 1999a,b). Levee breaching on the Cosumnes River in California, as shown by Florsheim and Mount (2002), made it possible to successfully restore floodplain features like sand-splay complexes. River channels begin to migrate laterally after levee removal and fairly quickly regain some sinuosity and habitat complexity, according to data from ongoing studies in Austria (Jungwirth et al., 2002; Muhar et al., 2004). It has also been demonstrated that levee adjustments and setbacks on the Danube River benefit amphibians, dragonflies, and rheophilic fish (Chovanec et al., 2002). Additionally, Hein et al., showed that plankton production decreased as habitat connectivity decreased and that there was an increase in plankton biomass in reconnected habitats. Along with improvements in physical habitat and the restoration of natural erosional and channel migration processes, these and other Austrian studies have also shown improvements in the age structure and diversity of fish and riparian ecosystems (Jungwirth et al., 2002).

6.2.3.2.3. Re-Meandering

The amount of total stream length that increases dramatically when meanders are reinstated can be used as a proxy for the effectiveness of re-meandering. The amount of length lost as a result of straightening can be quite noticeable in large rivers. For instance, the channelization of the Mississippi River in the United States is thought to have shortened the stream by 229 kilometers (Gore and Shields, 1995). A Norwegian stream's sinuous reach was channelized, according to Hvidsten and Johnsen (1992), resulting in a reduction in stream length from 7.5 km to 2.5 km. The increase in stream length by re-meandering can be equally impressive. For example, in a review of Danish stream rehabilitation, Iversen et al. (1993), reported increases in stream length ranging from 17% to more than 60% for five river meander reinstatement projects. Clearly restoring stream and floodplain meanders dramatically increases total available habitat and habitat diversity.

One of the more comprehensive ongoing assessments of stream remeandering is the one conducted on the Brede, Cole, and Skerne rivers in Denmark and England. Evaluation of these projects showed a clear improvement in the amount of water flowing onto the floodplain (floodplain connectivity), the complexity of the habitat and the morphology of the channels, the frequency of floods, and the amount of phosphorus associated with the sediment. Research on fish fauna, aquatic vegetation, and macroinvertebrates in the River Gelså and other preserved Danish streams has yielded some encouraging findings. Streams in Austria have also shown improvements in fish species diversity and physical habitat (Jungwirth et al., 1995). It has been established that meander restoration or recreation improves stream morphology, hydrology, and sediment transport. In a similar vein, macroinvertebrates and other biota seem to recover from the effects of construction rather quickly. Long-term studies, however, have not yet been finished that definitively show a recovery or increase in fish production. As is the case with many structural modifications to physical habitat, biotic responses to floodplain habitat rehabilitation may be hampered if possible issues with water quality are not resolved.

6.2.3.2.4. Constructed Habitats

A major factor in the habitats' physical efficacy is how connected they are to the main river channel. A range of fish life stages are much more likely to use the habitats if they are kept connected by different flows. Constructed floodplain habitats have been shown to be particularly effective at both providing habitat for juvenile salmonids and increasing their survival. For instance, Solazzi et al. (2000) concluded that creation of slow-water floodplain channel habitats increased overwinter survival for coho salmon, cutthroat trout (Oncorhynchus clarki), and steelhead (*O. mykiss*). Chovanec et al. (2002) reported rapid colonization by rheophilic fish species (eg. *Chondrostoma nasus*) of constructed backwaters and side channels created on the Danube River in Austria. Similarly, Langler and Smith (2001) found age 0 coarse fish numbers and diversity higher in created backwater habitats than unmanipulated habitats in the Huntspill River, England. Habersack and Nachtnebel (1995) found that a constructed side channel of the River Drau, Austria, had a higher diversity of habitats, substrates, macroinvertebrates and higher densities of fish in the side channel than in other river reaches. One method that is particularly common for producing salmonid fish spawning habitat is the excavation of groundwater channels. As with surface-fed side channels, groundwater-fed channels serve as important rearing and spawning areas for fish, especially coho salmon. These studies show that well-designed floodplain habitats can support these fish populations.

6.2.4. Dam Removal and Flood Restoration

6.2.4.1. Techniques for Dam Removal and Flood Restoration

The primary purposes of dam removal, weir removal, fish passage or bypass channel creation, and peak flow restoration in regulated rivers are to improve fish access, maintain or restore natural riverine processes that create and maintain fish habitat, and restore longitudinal connectivity of stream and floodplain habitats (Pess et al., 2005). Many dams and diversion weirs in developed countries are nearing the end of their useful lives and are being considered for removal or restructuring due to ecological concerns, safety concerns, and the costs associated with maintaining or repairing older structures. This is in contrast to the rapidly expanding construction of dams and weirs in developing countries (Stanley and Doyle, 2003). For instance, over 500 dams have been removed in the United States in the past 20 years; the majority of these were smaller dams, standing less than 20 meters tall and having a storage capacity of under 100 acres (123 000 m^3). Similarly, one of the most popular methods for enhancing fish passage in Europe is the removal of weirs and low-head dams or the construction of bypass channels and fish passage structures.

Dams have a detrimental effect on aquatic resources because they impede fish migration, drastically alter the hydrologic, nutrient, sediment, and temperature regime downstream of the dam, and change the environment above the dam from lotic to lentic. In other regions, the majority of studies on dams have concentrated on their effects rather than their removal. Like studies on dam removal, these studies show how physical conditions or biota (e.g., Grzybkowska et al., 1990; World Commission on Dams, 2000). The consideration of these and other negative ecological consequences of dams on aquatic resources, particularly migratory fish, have made the removal of both small and large dams an increasingly common method of rehabilitation (Chisholm, 1999).

Dam removal has an impact on three areas, similar to dam construction: below a dam, in the reservoir/impoundment area, and upstream of a dam reservoir. After a dam is removed, biotic exchange between a river system below and above it increases, and aquatic migratory species (e.g., fish) that could have been obstructed by a dam (Hart and Poff, 2002). The aquatic and riparian biota, flow characteristics, sediment erosion and deposition, and the channel's instability are likely to occur in the initial years following dam removal in the former reservoir or impoundment. When a dam is removed, the instream and riparian biota usually undergo significant changes, as well as an increase in downstream sediment supplies and altered flow patterns. This implies that when floods remove or alter the composition of the vegetation and create habitat that is suitable for more flood-tolerant species, the riparian environment will undergo significant change. If there is space in the floodplain, increases in sediment are usually expected to elevate the bed and promote lateral movement. More depositional bars and other recently formed landforms as a result of these changes should favor early successional riparian species and produce a diversity of aquatic habitats.

Restoring a natural flood or flow regime is an additional strategy for restoring a hydrologic regime when dam removal is not practical. Reduction in primary productivity, colonization of riparian plants, biotic diversity, habitat connectivity, sedimentation, and nutrient cycling have all been affected by the regulation of river flows. Many riparian species, like cottonwood, become established during specific flow conditions, like floods. The natural establishment of such species becomes problematic when floods are eliminated. Furthermore, species that would not establish themselves due to floods show up in areas they do not typically visit. Flooding is also necessary for fish colonization of artificial riverine habitats. For instance, fish species diversity significantly increased after flooding in a man-made riverine wetland in the upper Mississippi River (Theiling et al., 1999). Accordingly, the practice of releasing water from dams at specific times to replicate natural flows is increasingly acknowledged as an effective rehabilitation method (Rood and Mahoney, 1993).

6.2.4.2. Effectiveness of Dam Removal and Flood Restoration

6.2.4.2.1. Effectiveness Dam Removal

Despite the fact that many of the advantages of dam removal are evident from the outset, there isn't much published research on the subject because it's a relatively new technique (e.g., fish migration and access). The studies that looked at how disconnecting dams and weirs affected connectivity, habitat, and biota below the dam, as well as processes, physical habitat, and biota directly above the dam, are summarized here. One of the effects of dam removal that can be seen the most clearly is the restoration of fish access and upstream-downstream connectivity. Hart et al., (2002) provided an overview of the outcomes of various dam removal initiatives carried out in the US and documented over 10 instances of dam removal that led to the swift influx of migratory and resident fish species into the upstream and former impoundment areas of warm- and cool-water rivers. For example, dam removal on the Clearwater River, Idaho (USA) in 1963 reconnected the main stem, increasing both habitat quality and Chinook salmon runs (Shuman, 1995). Similarly, the removal of 150-year-old Edwards

Dam on the Kennebec River in Maine (USA) resulted in large numbers of American eel (*Anguilla rostrata*), alewife (*Alosa pseudharengus*), and Atlantic and shortnose sturgeon (*Acipenser* spp.) moving upstream within the first year, as well as juvenile downstream migrants in subsequent years (Hart et al., 2002). Smith et al., (2000) reported improved fish passage following the removal of a 3 m high dam on a stream in Oregon (USA), but continued water withdrawal and other factors upstream of the dam prevented full recovery of both physical and biological conditions. Kanehl et al., (1997) also examined the effects of the removal of a low head dam in the Wisconsin (USA) and found improvement in habitat quality, biotic integrity, and smallmouth bass (*Micropterus dolomieu*) abundance and biomass five years after dam removal. While not discussed extensively in this report, the installation of bypass channels to allow fish access above diversion weirs has also proven to be highly successful rehabilitation technique for fishes in European streams (DVWK, 2002; Iversen et al., 1993). Clearly, dam removal has many benefits for migratory and lotic fishes.

A change in sediment flux causes several significant changes below a dam, the most obvious of which is a change in channel form and sediment composition. Changes in fine sediment and sediment transport have been shown in many studies; however, the kind and quantity of fine sediment that was trapped behind the former dam determines the changes in sediment. One instance is Doyle et al. (2003) investigated the erosion of fine sediment deposited in the former reservoir and the increased deposition of fine sediment downstream following the removal of low-head dams (<3 m high) in two Wisconsin rivers. Hart et al, examined 20 dam removals in the US and found that 14 of them showed increased sediment transport; however, few studies were sufficiently long to record channel changes downstream of the dam. Additional alterations include the establishment of plants, a return to a more natural temperature regime, and an increased exchange of nutrients and organic matter with the upstream regions of a watershed (Hart et al. (2002). The way reservoir-derived deposits enter and pass through downstream reaches ultimately determines the downstream effects of dam removal on ecological attributes (Stanley and Doyle, 2003). Following dam removal, changes in the macroinvertebrate community and downstream water temperature also usually follow (Hart et al., 2002). The structure of the dam and reservoir, as well as past operations, will determine whether the temperature rises or falls after the dam is removed.

Anecdotal evidence regarding the benefits of 25 dam removals in the United States was reported by Chisholm (1999). He did, however, also provide details about one dam removal that is regarded as a failure—the Fort Edward Dam in New York State, which released pollutants and heavy metals that were trapped in reservoir sediments. These pollutants and heavy metals continue to have detrimental effects on aquatic resources downstream from the former dam site. Furthermore, it is not unusual for there to be an increase in turbidity and sediment right after or soon after dam removal.

The restoration of former impoundments to river, riparian, and floodplain habitats is impacted by the removal of dams (Stanley and Doyle, 2003). This physical alteration shortens the water's residence time in a former reservoir reach, which in turn shortens the reach's capacity to hold sediment and other materials. As a result, the biota changes from being lentic to being lotic (Hart and Poff, 2002). For instance, within a year of two different dam removal projects in Wisconsin, fish and macroinvertebrates adapted to a high sediment supply reservoir environment gave way to riverine fish and macroinvertebrates. Stanley et al. conducted a related study on the Baraboo River in Wisconsin. discovered that, after a year, the macroinvertebrate assemblages in the reservoir's former impounded stream reaches resembled those in the reaches upstream and downstream. It is obvious that the riparian and aquatic vegetation in the former reservoir sites will change as a result of the dam removal. Shafroth along with others (2002) discovered that, while sites gradually became drier and were dominated by xeric species, plants that had previously colonized newly exposed soil along former reservoir bottoms were frequently composed of hydrophilic, monotypic species like stinging nettle (*Urtica dioica*) and red canary grass (*Phalaris arundinacea*). These studies show how the removal of dams causes significant changes in the physical habitat as well as the riparian and aquatic flora and fauna. Additionally, they opine that the removal of dams may have unfavorable effects, such as transient channel instability or invasive riparian species colonizing recently exposed riparian areas.

6.2.4.2.2. Effectiveness of Flood Restoration

Restoring flood flows is a relatively new technique, so data on its efficacy is a little bit scarce, although the outcomes have been excellent. The efforts to reinstate flood flows in the Grand Canyon in the United States have been the most prominent test case. In regions that were once devastated by flooding, forests and marshlands have grown along the Colorado River in the Glen Marble and Grand Canyon reaches (Stromberg, 2001). High flow test results indicate some encouraging developments in this vast river system, which had a highly dynamic flow regime before regulation. The nearshore woody vegetation that was artificially established as a result of flood control has decreased as a result of recent restoration activities that simulate floods. Researchers also observed the effects of flood simulations downstream of Glen Canyon Dam on the Colorado River and discovered that floods restored sandbar habitat and submerged patches of woody vegetation with up to one meter of sand. However, the backwater marshlands that had developed following the construction of the dam remained, and the woody vegetation was not eradicated. In a similar vein, Hill and Platts (1998) observed the impact of modified flows in the Owens River, California, through flood simulations and higher base flows. They discovered that instream habitat and fish abundance significantly improved in tandem with the changes in vegetation and that riparian vegetation quickly recovered after the altered flow regime. Speierl along with others. (2002) found that after the Main and Rodach rivers in Germany's natural flow dynamics were restored, there was an increase in species diversity and abundance. According to other research, certain riparian plant species' seedlings grow and survive better in natural flow environments than in controlled ones (Johansson and Nilsson, 2002). Likewise, riparian forests, fish populations, and bird populations can all recover as a result of restoring instream flows (Rood et al. (2003). Research on natural floods supports the use of flood restoration in controlled rivers and highlights the significance of these events for riverine ecosystems. Large-scale floods on the Missouri River in the 1980s and 1990s restored connections to a large number of wetlands and floodplain lakes that had been cut off for decades by channelization and river regulation (Galat et al., 1998). Reconnected areas had a more varied fish fauna than isolated areas, according to post-flood studies (Galat et al., 1998).

Similar to this, high water years cause the Green River in Utah, USA, to reopen its floodplain wetlands and offer rearing habitat for the endangered razorback sucker (*Xyrauchen texanus*) larvae that aren't found in the main river channel (Modde et al., 2001). Studies on natural floods and preliminary research from flood flow restoration point to advantages for riparian vegetation, aquatic biota, and sediment transport.

Adequate instream flows, also known as base flows, are necessary to preserve aquatic and riparian habitats as well as the growth of aquatic ecosystems and biota, much like increases in high flows. Rewatering stream reaches and increasing instream flows on rivers with water diversions are proven ways to lower water temperatures, enhance water quality, and generally help biota. A review of physical habitat rehabilitation techniques would not be able to adequately address the large body of literature that exists on instream flows. However, minimum instream flows must be set in addition to flood restoration in order to safeguard biota and guarantee the success of other habitat restoration initiatives.

6.2.5. Instream Habitat Structures

Physical structures have long been inserted into lotic environments to form pools, modify the morphology of channels, and offer cover and habitat to fish and other aquatic organisms (White, 1996, 2002). It is also one of the most popular and widely applied techniques in use. It was one of the first strategies to reduce habitat degradation and boost fish production in streams and rivers (Tarzwell, 1934, 1937, 1938). Over the years, numerous instream structure configurations and techniques have been employed to enhance habitat. In order to enhance fish habitat, these techniques entail inserting objects into the active stream channel, such as boulders and large woody debris (LWD), or altering the active channel itself. They are classifiable according to their goals (e.g., generate spawning gravel, dams and deflectors, cover structures, rootwads and brush bundles, gabions, logjams, and boulder or log weirs are examples of structures that can be used to produce pools and trap gravel. Because these endeavors aim to improve habitat instead of rectifying an inadequate process (e.g., Instead of being referred to as restoration, they are called instream habitat enhancement (riparian, hydrology) or returning a stream to a pre-disturbance state.

The majority of current instream enhancement methods were created in the 1930s for use in the comparatively low-gradient streams found in the Midwest of the United States (Hunt, 1993; White, 1996). Their purpose was to improve the habitat of trout that had been reduced to a simple level and deteriorated due to logging, farming, and other land uses. The main goals of early instream constructions were to decrease predation mortality, narrow channels, and offer cover for game fish such as trout. Streams across North America and Europe have subsequently used instream enhancement techniques developed in the Midwest, with differing degrees of success. Pool formation and gravel trapping have been the main objectives of the placement of instream structures in higher energy streams, such as those found in the mountainous regions of western North America. Increasing habitat heterogeneity (creating a suite of both shallow and deep water habitats) and either narrowing or widening the channel, depending on the channel and its level of degradation, are the general objectives of more recent efforts in low energy, low gradient (<1% gradient) channelized streams, such as those found in floodplains and agricultural land.

6.2.5.1. Effectiveness of Instream Structures

The majority of techniques for improving instream habitat have not been thoroughly evaluated, despite the fact that it is a well-known, widely used, and long-standing practice with over 60 published evaluations. For many years, it has been recognized that thorough monitoring and evaluation are necessary. The appropriateness of instream enhancement is hotly debated because, historically, these techniques were implemented without taking into account the factors that may have contributed to the lack of complexity in the habitat, the processes within the watershed that may have been disturbed and require correction, or the factors that may have limited the physical and biological production of a system (Frissell and Nawa, 1992; Roni et al., 2002; White, 2002).

For instance, the lack of riparian vegetation in a stream reach may be the result of land use activities like agriculture or even urbanization, which in turn may be the reason why instream wood and boulder structures fail to increase fish abundance in that reach. In these situations, instream enhancement might only offer a temporary habitat improvement if the root cause of the issue is not resolved; nevertheless, in highly altered areas, it might be the only workable solution. Thus, initiatives to improve habitat have value. In situations where certain watershed processes cannot be restored, short-term habitat enhancements may be required to safeguard rare fish species and the biota, or to offer temporary advantages. woody waste in city streams). They are especially important in conjunction with the restoration of natural processes, like sediment reduction or riparian rehabilitation (i.e, road treatments and slope stabilization), as opposed to addressing the root cause of the degradation or simplification of instream habitat, which puts these habitat enhancement strategies at risk of only treating the symptoms. It is imperative to restate that the goal of instream habitat enhancement techniques is habitat enhancement, not process restoration.

6.2.6. Nutrient Enrichment

In many lakes, streams, and other freshwater habitats, high levels of nutrients (N and P) from treated and untreated sewage, storm water inputs, and agricultural runoff (both fertilizers and animal waste) have resulted in high levels of nutrients and problems with water quality (Kiffney et al., 2005). On the other hand, the addition of organic and inorganic nutrients (N and P) has been utilized in many oligotrophic waters in temperate and arctic latitudes as a means of enhancing primary and secondary productivity and, presumably, fish growth and abundance (Benndorf, 1990; Kiffney et al., 2005). To increase fish production, nutrient additions are also frequently used in aquaculture in lakes and ponds located in floodplains.

6.2.6.1. Nutrient Enrichment Techniques

A few strategies have been developed to increase the productivity of waters and enrich nutrient levels, such as adding organic (fish carcasses) and inorganic (usually N and P) nutrients. This frequently entails redirecting nutrient-rich agricultural runoff for aquaculture purposes, adding inorganic nutrients, and adding manure or other solid organic material. A few new approaches to applying these nutrients have emerged as a result of the growing popularity of nutrient enrichment in North America. These include fish blocks, which are made from fish feed or the leftover bones and flesh from fish processing facilities, to time-release fertilizers created especially for use in streams. Although nutrient enrichment methods are sometimes referred to as restoration, they are really a type of enhancement that is typically employed in an effort to raise a water body's productivity above its historical levels. However, continuing to add nutrients is usually necessary to maintain this level of production. Western North America has seen an increase in the addition of nutrients to lakes and streams as a result of several studies showing the significance of anadromous fish to the productivity of freshwater ecosystems. According to estimates, for instance, the amount of marine-derived N and P supplied by reintroducing adult salmon to coastal streams in the Pacific Northwest has decreased to six to seven percent of historical levels (Gresh et al., 2000). Studies on fish that are not native to the area, like smelt (Osmeridae), shad (*Alosa* spp.), Atlantic salmon, and sea trout have shown comparable outcomes in North Atlantic tributaries (Durbin et al., 1979; Garman and Macko, 1998; Lyle and Elliott, 1998).

6.2.6.2. Effectiveness of Nutrient Enrichment

Lakes -Though most of the work has been done in lakes, the efficacy of these techniques has been studied in many studies conducted in lakes and streams in North America and Europe. For many years, Alaska, the midwestern and western US, British Columbia, Canada, Northwest Territories, eastern Canada, and Denmark have been manipulating N and P levels experimentally in order to boost primary and secondary production in lakes (Jeppesen et al., 1991), Japan (Ishida and Mitamura, 1986), Sweden, and Norway (Langeland, 1990). Many of these studies have shown increases in nutrients, zooplankton, benthic macroinvertebrates, phytoplankton (chlorophyll a), and primary production.

The reaction of fish to lake enrichment has proven to be more unpredictable and challenging to gauge. Sockeye salmon (*Oncorhynchus nerka*) have been the subject of the great majority of studies on fish response to lake enrichment; these studies have typically shown increased growth, juvenile survival, and occasionally adult sockeye returns. Positive findings for other species have been reported in a few studies. Johnston along with others (1999) found that rainbow trout in a lake in British Columbia responded better to inorganic fertilizers in terms of growth and reproduction, but not survival. When yellow perch (Perca flavescens) were exposed to nutrient enrichment, Essington and Houser (2003) observed that the fish grew faster and had lower mercury concentrations. Additional research has concentrated on analyzing the impact of introducing or removing fish (predators) on food webs and other elements of ecosystems. Large increases in plankton, benthos, and fish yield have also been shown in studies of nutrient additions to aquaculture ponds and reservoirs in China and Russia after the addition of manure or other organic nutrients.

Based on studies of fish as well as primary and secondary productivity, the response of a given lake to fertilization depends on a variety of factors including but not limited to: zooplankton and fish species present (Jeppesen et al., 1991; Benndorf, 1990; Carpenter et al., 1996), light levels (Bjoerk-Ramberg, 1983; Benndorf, 1990), stratification (temperature) and depth (Bjoerk-Ramberg, 1983; Jansson, 1984; Wurtsbaugh et al., 2001), whether benthos were sampled (Bjoerk-Ramberg and Anell, 1985; Vadeboncoeur et al., 2001), whether the system was P or N limited (Holmgren, 1984; Jansson et al., 2001), number of years of fertilization (Jorgenson et al., 1992), or top down versus bottom up control of biotic communities (Langeland, 1990; Stockner and Shortreed, 1994).

Streams: Although the history of adding both organic (fish tissue) and inorganic nutrients to streams and their biota is not as extensive as that of lakes, numerous studies have still looked at the effects of these two methods on stream biota. Johnston along with others. In one of the largest and longest-running studies on the addition of inorganic nutrients to a stream, (1990), Ward (1996), and McCubbing and Ward (1997, 2000) found increases in periphyton, macroinvertebrate abundance, and juvenile coho salmon and steelhead growth and condition. After analyzing case studies of inorganic nutrient additions on five distinct British Columbia watersheds, Ashley and Slaney (1997) discovered increased periphyton, invertebrate biomass, and fish growth. Wipfli et al, increased condition, growth, and production after the placement of both salmon carcasses and carcass analogs (artificial carcasses made from fish tissue) in Southeast Alaska's natural and artificial streams (2003, 2004-published studies). The growth of juvenile and adult Arctic grayling was found to increase in an Arctic tundra river after inorganic phosphorus was added, according to Deegan and Peterson (1992). All of these studies point to favorable initial reactions when inorganic nutrients are added to oligotrophic streams.

Addition of organic nutrients in the form of fish carcasses has also been examined in anadromous fish streams in Alaska, British Columbia, Washington and Minnesota. For example, Schuldt and Hershey (1995) found increased P and periphyton following addition of salmon carcasses to a Minnesota stream. In a laboratory experiment, Minakawa et al. (2002) found higher growth of caddis flies on salmon carcasses than on natural leaves. Wipfli et al. (1998; 1999) and Chaloner and Wipfli (2002) found increased chlorophyll a, macroinvertebrates, and growth rate of salmonids following additions of carcasses in artificial and natural stream channels. Studies using stable isotope analysis have demonstrated the importance of salmon carcasses in aquatic food webs for primary productivity, macroinvertebrates, fishes and even riparian vegetation and tree growth (Bilby et al., 1996; Helfield and Naiman, 2001).

While additional study is needed, these studies all suggest that the placement of salmon carcasses or an increasing in spawning fish can increase fish growth and possibly survival.

6.2.7. Lake Habitat Enhancement

Similar to the positioning of instream structures, man-made structure placement in lakes dates back to the early 1930s. Because the substrate in reservoirs lacks complexity and cover, habitat enhancement is especially common there. Similar to artificial reefs in the marine environment, artificial structures in ponds, lakes, and reservoirs serve the same functions of gathering fish, raising harvest levels, and boosting localized or total biological production in the water body. Furthermore, structures are frequently erected in natural floodplain lakes and other habitats to draw fish and enhance harvest, or in aquaculture ponds to boost fish production (Welcomme, 2002). We provide an overview of the biological efficacy of lake structures, since physical effects have not been studied. Structures that improve lake habitat can be divided into groups according to the materials they are made of and whether they are intended as spawning grounds or as cover. These include adding gravel or boulders, planting vegetation, and creating cover structures out of artificial materials or trees.

6.2.7.1. Effectiveness of Lake Habitat Enhancement

The goals and effectiveness of structures placed in lakes and ponds to improve fish habitat have been periodically reviewed and summarized. For example, in a recent survey in the United States, Tugend et al. (2002) found that 75% of the lake habitat enhancement included placement of structures, while 25% were spawning structures or substrate manipulations.

Structures designed to improve habitat can benefit wildlife by increasing spawning recruitment and survival, offering shelter or cover from predators, providing shade and locations for orientation and schooling, increasing the efficiency with which adults feed their prey, and increasing the amount of land available for harvesting and fishing. According to Prince and Maughan (1978), habitat structures in lakes and ponds enhance fish production as well as prey. The type of structure, the material used in construction, the interstitial spaces, the depth and location within the lake, the morphometric features of the lake, the substrate's complexity, the presence of aquatic macrophytes, the fish species and life stage studied, and other factors all have an impact on a structure's success. In addition, fish use of structures changes with the water temperature, season, and day (deil variation). The impact of these structures on the size of fish populations or fish production has not been investigated, in line with studies of artificial structures in the marine environment.

- **Brush Bundles and Log Cover Structures:** Made of natural materials, brush bundles, logs, and half-log structures are usually positioned on or suspended close to the bottom (Brown, 1986). Tugend et al. and Brown (1986). reviewed previous assessments of brush bundles and found that, in particular, studies on centrarchids and cyprinids had shown higher fish abundance and catches at treatment sites (brush structures) compared to control sites (no brush added), but less so for other species like yellow perch (*Perca flavescens*) (Moring et al., 1989).

The objectives of cover logs and whole logs or trees placed in lacustrine environments are usually the same as those of brush bundles. It has also been demonstrated that they react favorably to centrarchids and other warmwater fish. For instance, after half logs were placed in lakes and reservoirs, smallmouth bass (*Micropterus dolomieu*) numbers, nest density, and nest success increased (Hoff, 1991; Wills et al., 2004). Five out of 10 species found in a reservoir in Maine, US, were supported by submerged logs left over from logging operations (Moring et al., 1989). Additionally, there is some evidence that the placement of brush, vegetation, and gravel in the littoral zone can help yearling pikeperch (*Stizostedion lucioperca)* even when a predator like the Northern pike (*Esox lucius*) is present (Lapínska et al., 2001).

While many studies have examined the response in North America lakes, placement of brush and logs into lakes and floodplain habitats to enhance fish production is also utilized in the developing world (Welcomme, 2002). For example, dropping of logs to rehabilitate designated fish sanctuaries has occurred in the Tonle Sap Great Lake in Cambodia (Thuok, 1998). Similarly, Vinci et al. (2003) described three Indian variations of placing bamboo and brush into floodplain ponds in India. Brush bundles or brush parks are widely used in floodplain lakes and ponds to enhance both native and cultured fish production in developing countries (Hoggarth et al., 1999; Quasim, 2002). These are primarily installed to provide cover for fishes and many different modifications exist particularly for fish culture practices (Quasim, 2002). However, in some areas this practice is illegal or they are designed as refuges or fish sanctuaries and fishing in these areas is illegal. Studies of their effectiveness are rare, but evaluations of fish sanctuaries and brush parks in Bangladesh have shown them to be highly effective at both attracting fish and recovering endangered species (Poulsen and Hasan, 2004).

- **Cover Structures Made of Artificial Materials**: In order to add cover and complexity to the habitat in lakes, a range of artificial materials have also been employed. The materials that were most frequently mentioned in the literature were stake beds, concrete slabs or structures, plastic crates, boxes, pipes, and tires (Brown, 1986).

Hundreds of stakes hammered into the substrate or prefabricated onshore before installation are the usual components of stake beds. These artificial structures are designed to provide cover, holding, and in some cases spawning habitat.

Compared to brush bundles or logs, artificial material structures have yielded less consistent results. A few studies have documented beneficial outcomes from the erection of highly artificial structures in lakes. When Moring and Nicholson (1994) studied fish using cinder blocks, tire bundles, and brush bundles, they discovered that all species were more prevalent near artificial structures than at sites without them. Among these species were pumpkinseed sunfish (*Lepomis gibbosus*), chain pickerel (*Esox Niger*), brown bullhead (*Ameiurus nebulosus*), common shiner (*Luxilus cornutus*), and golden shiner (*Notemigonus crysoleucus*). Graham (1992) sampled Lake Anna, Virginia (USA) before and after the installation of cinderblock and log structures, and discovered more small bluegill (< 15 cm) there after treatment. In another Virginia lake, Prince and Maughan (1979a,b) discovered that tire reefs significantly concentrated bass and sunfish. In ponds treated with tires and stake beds or brush bundles compared to control ponds, Wege and Anderson (1979) reported higher growth rates of largemouth bass (*Micropterus salmoides*) and higher catch rates of both largemouth bass and bluegill.

In contrast, other studies have found no change in fish use or higher fish use of structures composed of natural materials versus artificial materials. Wills et al. (2004) examine plastic box shaped shelter devices as a habitat enhancement device and found no change in fish use before and after placement of plastic cover structures (Aqua Cribs). Pardue and Nielsen (1979) examined tilapia (*Tilapia* spp.) and bluegill (*Lepomis macrochirus*) response to pine board structures in small experimental ponds and found little evidence that structures increased fish biomass. Wilbur (1978) examined clay pipes and cement blocks with brush as fish attraction devices and found several warmwater species used the structures. However, structures that contain trees or brush were preferred by fishes over structures made of purely artificial materials (Wilbur 1978).

- **Artificial Shoals or Spawning Areas**: Typically, gravel, cobbles, boulders, or concrete rubble are placed to create suitable areas within lakes in order to create spawning areas. As was previously mentioned, a lot of cover structures, like logs and brush bundles, are made to offer cover to fishes that are spawning as well as juvenile. Much of the lake trout (*Salvelinus namaycush*) spawning areas in Lakes Ontario and Michigan in the Great Lakes of North America are artificial reefs or spawning shoals (Fitzsimons, 1996). Fitzsimons (1996) noted increased concentrations of eggs, fry, and young-of-year at artificial reefs and spawning areas in a recent review of studies on the use of artificial reefs in the Great Lakes. However, the Great Lakes' naturally occurring lake trout population has low success rates, and artificial reefs' ability to boost fish production appears to be constrained by factors other than spawning.

Many warm-water fish species, including walleye (*Stizostedion vitreum vitreum*) and sunfish (*Lepomis macrochirus*), can find spawning habitat on artificial reefs, which also increase the catch rates of other species like yellow perch. Geiling et al. (1996) examined the efficacy of building artificial reefs to create a habitat for walleye spawning in the Great Lakes. The results showed that this method was ineffective for this species and that adult walleye abundance was rarely increased. Similarly, Neuswanger and Bozek (2004) found that 85% of 20 rock placement projects intended to enhance walleye habitat in Wisconsin lakes (USA) showed no indications of increased fry production. However, the location of artificial spawning areas in lakes or rivers in relation to juvenile rearing areas determines whether or not they are successful in increasing fish populations (Jones et al. (2003). The effectiveness of enhancing spawning habitat in lakes is not well understood, but it seems to work best when the biology and ecology of the target species are taken into account during the design and placement process.

- **Shoreline Vegetation**: The installation of riparian vegetation, or lakeshore vegetation, offers lakes several advantages, such as shade, a supply of organic matter, protection from wind, which can alter the mixing, bank protection, food (terrestrial insects), and a place to find brush, logs, and other cover structures. Furthermore, vegetation is planted in lakes and reservoirs with areas that are periodically flooded, offering a suitable home for young fish like bass. After planting lakeshores, little biological assessment has been done; however, planting water-tolerant plants like Salix spp. aids in preventing erosion and stabilizing banks, which may enhance fish habitat in the water body and around structures.
- **Concentration Versus Increased Harvest:** There is concern that the placement of structures in both freshwater and marine environments is only increasing fish concentration and harvest rates, rather than increasing production (Roni and Quinn, 2001b). According to reports by Paxton and Stevenson (1979), Wege and Anderson (1979), artificial structures have higher catch rates than naturally occurring unstructured areas. These findings raise the possibility that artificial structures could result in increased fish harvesting exploitation rather than population growth. To be clear whether this is true or not, more research is required.

REFERENCES

1. Albert, S., & Trimble, T. (2000). Beavers are partners in riparian restoration on the Zuni Indian reservation. *Ecological Restoration, 18*(2), 87–92.
2. Angermeier, P. L., & Karr, J. R. (1984). Relationship between woody debris and fish habitat in a small warmwater stream. *Transactions of the American Fisheries Society, 113*, 716–726.
3. Annear, T., Chisholm, I., Beecher, H., Locke, A., Aarestad, P., Burkhart, N., Coomer, C., Estes, C., Hunt, J., Jacobson, R., Jobsis, G., Kauffman, J., Marshall, J., Mayes, K., Stalnaker, C., & Wentworth, R. (2002). *Instream Flows for Riverine Resource Stewardship*. Cheyenne, WY: Instream Flow Council.
4. Apple, L. L. (1985). Riparian habitat restoration and beavers. In R. R. Johnson, C. D. Ziebell, D. R. Paton, P. F. Ffolliott, & R. H. Hamre (Eds.), *Riparian Ecosystems and Their Management: Reconciling Conflicting Uses* (pp. 489–490). General Technical Report RM-120. Fort Collins, CO: USDA Forest Service.
5. Armantrout, N. B. (1991). Restructuring streams for anadromous salmonids. In J. Colt & R. J. White (Eds.), *Fisheries Bioengineering Symposium* (pp. 136–149). American Fisheries Society Symposium 10. Bethesda, MD: American Fisheries Society.
6. Armour, C. L., Duff, D. A., & Elmore, W. (1991). The effects of livestock grazing on riparian and stream ecosystems. *Fisheries, 16*(1), 7–11.
7. Bassett, C. E. (1994). Use and evaluation of fish habitat structures in lakes of the eastern United States by the USDA Forest Service. *Bulletin of Marine Science, 55*(2, 3), 1137–1148.
8. Bates, D. J., McBain, G. G., & Newbury, R. W. (1997). Restoration of a channelized salmonid stream, Oullette Creek, British Columbia. In J. D. Hall, P. A. Bisson, & R. E. Gresswell (Eds.), *Symposium on Sea-Run Cutthroat Trout: Biology, Management, and Future Conservation* (pp. 183–195). Corvallis, OR: American Fisheries Society, Oregon Chapter.
9. Bauer, S. B., & Ralph, S. C. (2001). Strengthening the use of aquatic habitat indicators in Clean Water Act programs. *Fisheries, 26*(6), 14–24.
10. Bayley, P. B. (1995). Understanding large river-floodplain ecosystems. *BioScience, 45*(3), 153–158.
11. Bayley, P. B., O'Hara, K., & Steel, R. (2000). Defining and achieving fish habitat rehabilitation in large, low-gradient rivers. In I. G. Cowx (Ed.), *Management and Ecology of River Fisheries* (pp. 279–287). Oxford, UK: Fishing News Books.

12. Beechie, T. J., Collins, B. D., & Pess, G. R. (2001). Holocene and recent geomorphic processes, land use, and salmonid habitat in two north Puget Sound river basins. *Water Science and Applications, 4*, 37–54.
13. Beechie, T. J., Pess, G., Beamer, E., Lucchetti, G., & Bilby, R. E. (2003a). Role of watershed assessments in recovery planning for threatened or endangered salmon. In D. Montgomery, S. Bolton, D. Booth, & L. Wall (Eds.), *Restoration of Puget Sound Rivers* (pp. 194–225). Seattle, WA: University of Washington Press.
14. Beechie, T., & Bolton, S. (1999). An approach to restoring salmonid habitat-forming processes in Pacific Northwest watersheds. *Fisheries, 24*(4), 6–15.
15. Beechie, T., Beamer, E., & Wasserman, L. (1994). Estimating coho salmon rearing habitat and smolt production losses in a large river basin, and implications for habitat restoration. *North American Journal of Fisheries Management, 14*, 797–811.
16. Beechie, T., Beamer, E., Collins, B., & Benda, L. (1996). Restoration of habitat-forming processes in Pacific Northwest watersheds: A locally adaptable approach to salmonid habitat restoration. In D. L. Peterson & C. V. Klimas (Eds.), *The Role of Restoration in Ecosystem Management* (pp. 48–67). Madison, WI: Society for Ecological Restoration.
17. Cairns, J. J. (1988). Restoration and the alternative: A research strategy. *Restoration Management Notes, 6*(2), 65–67.
18. Calow, P., & Petts, G. E. (1994). *The Rivers Handbook: Hydrological and Ecological Principles*. Oxford, UK: Blackwell Scientific.
19. Carpenter, S. R., Kitchell, J. F., Cottingham, K. L., Schindler, D. E., Christensen, D. L., Post, D. M., & Voichick, N. (1996). Chlorophyll variability, nutrient input, and grazing: Evidence from whole-lake experiments. *Ecology, 77*(3), 725–735.
20. Cederholm, C. J., Bilby, R. E., Bisson, P. A., Bumstead, T. W., Fransen, B. R., Scarlett, W. J., & Ward, J. W. (1997). Response of juvenile coho salmon and steelhead to placement of large woody debris in a coastal Washington stream. *North American Journal of Fisheries Management, 17*, 947–963.
21. Cederholm, C. J., Kunze, M. D., Murota, T., & Sibatani, A. (1999). Pacific salmon carcasses: Essential contributions of nutrients and energy for aquatic and terrestrial ecosystems. *Fisheries, 24*(10), 6–15.
22. Cederholm, C. J., Reid, L. M., Edie, B. G., & Salo, E. O. (1982). Effects of forest road erosion on salmonid spawning gravel composition and populations of the Clearwater River, Washington. In K. A. Hashagen

(Ed.), *Habitat Disturbance and Recovery: Proceedings of a Symposium* (pp. 1–17). San Francisco, CA: California Trout, Inc.

23. Cederholm, C. J., Scarlett, W. J., & Peterson, N. P. (1988). Low-cost enhancement technique for winter habitat of juvenile coho salmon. *North American Journal of Fisheries Management, 8*, 438–441.
24. Chaloner, D. T., & Wipfli, M. S. (2002). Influence of decomposing Pacific salmon carcasses on macroinvertebrate growth and standing stock in southeastern Alaska streams. *Journal of the North American Benthological Society, 21*(3), 430–442.
25. Chapman, D. W. (1996). Efficacy of structural manipulations of instream habitat in the Columbia River Basin. *Northwest Science, 5*(4), 279–293.
26. Chatwin, S. C., Howes, D. E., Schwab, J. W., & Swanston, D. N. (1994). *A Guide for Management of Landslide-Prone Terrain in the Pacific Northwest* (2nd ed.). Victoria, BC: Research Branch, BC Ministry of Forests.
27. Childers, D. L., Iwaniec, D. M., Parker, F. M., Rondeau, D., & Madden, C. (1999a). How freshwater Everglades wetlands mediate the quality of recently enhanced water inflows to the Florida Bay estuary. *Gulf Research Reports, 11*, 70–71.
28. Childers, D. L., Oehm, N. J., Parker, F., & Madden, C. (1999b). How freshwater Everglades wetlands mediate changes in water flow and nutrient loadings to the Florida Bay estuary. *Gulf Research Reports, 10.*
29. Chisholm, I. (1999). *Dam Removal Success Stories: Restoring Rivers Through Selective Removal of Dams that Don't Make Sense*. Washington, DC: American Rivers, Friends of the Earth, Trout Unlimited.
30. Chovanec, A., Schiemer, F., Waidbacher, H., & Spolwind, R. (2002). Rehabilitation of a heavily modified river section of the Danube in Vienna (Austria): Biological assessment of landscape linkages on different scales. *International Review of Hydrobiology, 87*, 2, 3.
31. Christensen, D. L., Herwig, B. R., Schindler, D. E., & Carpenter, S. R. (1996). Impacts of lakeshore residential developments on coarse woody debris in north temperate lakes. *Ecological Applications, 6*(4), 1143–1149.
32. Díaz, M., Gonzalez, E., Muñoz-Pulido, R., & Naveso, M. A. (1996). Habitat selection patterns of common cranes (*Grus grus*) wintering in holm oak (*Quercus ilex*) dehesas of central Spain: Effects of human management. *Biological Conservation, 75*(2), 119–123.
33. Dobkin, D. S., Rich, A. C., & Pyle, W. H. (1998). Habitat and avifaunal recovery from livestock grazing in a riparian meadow system of the northwestern Great Basin. *Conservation Biology, 12*(1), 209–221.

34. Domínguez, L. G., & Cederholm, C. J. (2000). Rehabilitating stream channels using large woody debris with considerations for salmonid life history and fluvial geomorphic processes. In E. E. Knudsen, C. R. Steward, D. W. MacDonald, J. E. Williams, & D. W. Reiser (Eds.), *Sustainable Fisheries Management: Pacific Salmon* (pp. 545–563). New York: Lewis Publishers.
35. Doppelt, B., Scurlock, M., Frissell, C., & Karr, J. (1993). *Entering the Watershed.* Covelo, CA: Island Press.
36. Downes, B. J., Barmuta, L. A., Fairweather, P. G., Faith, D. P., Keough, M. J., Lake, P. S., Mapstone, B. D., & Quinn, G. P. (2002). *Monitoring Ecological Impacts: Concepts and Practice in Flowing Waters*. Cambridge, UK: Cambridge University Press.
37. Doyle, E. G. (1997). *The Habitat Restoration Cost Estimation Model (HRCEM): A Cooperative Tool for Watershed Management* (M.M.A. thesis). University of Washington, Seattle, WA.
38. Doyle, M. W., Stanley, E. H., & Harbor, J. M. (2003). Channel adjustments following two dam removals in Wisconsin. *Water Resources Bulletin, 39*(1), 1011.
39. Ehlers, R. (1956). An evaluation of stream improvement project devices constructed eighteen years ago. *California Fish and Game, 42*, 203–217.
40. Ellis, L. M., Crawford, C. S., & Molles, M. C. (2001). Influence of annual flooding on terrestrial arthropod assemblages of a Rio Grande riparian forest. *Regulated Rivers: Research & Management, 17*(1), 1–20.
41. Elmore, W. (1992). Riparian responses to grazing practices. In R. J. Naiman (Ed.), *Watershed Management* (pp. 442–457). New York: Springer-Verlag.
42. Elmore, W., & Beschta, R. L. (1987). Riparian areas: Perceptions in management. *Rangelands, 9*(6), 260–265.
43. Elseroad, A. C., Fule, P. Z., & Covington, W. W. (2003). Forest road revegetation: Effects of seeding and soil amendments. *Ecological Restoration, 21*(3), 180–185.
44. Emmingham, W. H., Chan, S. S., Mikowski, D., Owston, P. W., & Bishaw, B. (2000). *Silviculture Practices for Riparian Forests in the Oregon Coast Range* (Research Contribution 24). Corvallis, OR: Forest Research Laboratory, Oregon State University.
45. Gore, J. A. (1985). *The Restoration of Rivers and Streams: Theories and Experience*. Stoneham, MA: Butterworth Publishers.
46. Kauffman, J. B., Beschta, R. L., Otting, N., & Lytjen, D. (1997). An ecological perspective of riparian and stream restoration in the western United States. *Fisheries, 22*(5), 12–24.

Chapter 7

Sustainable Use of Aquatic Resources

Contents

INTRODUCTION

All water-based ecosystems and the diverse life forms they support are considered aquatic resources. These resources are essential to economic activity, environmental health, and human survival. A crucial component of environmental management and conservation that seeks to strike a balance between human needs and the wellbeing of aquatic ecosystems is the sustainable use of aquatic resources. Fish, shellfish, seaweed, and other freshwater and marine organisms are examples of aquatic resources that offer essential services like sustenance, employment, leisure, and cultural values. Nevertheless, overfishing, pollution, habitat destruction, and climate change are putting more and more strain on these resources.

In order to manage these resources sustainably, current needs must be satisfied without endangering the capacity of future generations to satisfy their own. This necessitates a thorough comprehension of the ecological, economic, and social aspects of using aquatic resources. We can guarantee the long-term survival of aquatic ecosystems and the services they offer by implementing sustainable practices.

7.1. SUSTAINABLE AQUACULTURE AND ITS PRACTICES

Aquaculture is a sunrise industry in the current Earth food production context. Fish, mollusks, crustaceans, and aquatic plants are examples of aquatic organisms that are commercially produced through aquaculture, according to the FAO. The fishing and aquaculture industries have seen a significant shift in the last several decades due to an increase in overall production capacity, trade, and food consumption. The aquaculture industry is expanding, despite the capture fisheries experiencing a brief period of stagnation over the past few decades. New methods of aquaculture production hold great promise for improving environmental performance while reducing the world's food security issues and meeting human nutritional needs.

Finfish, crustaceans, seaweeds, mollusks, and other aquatic organisms are just a few of the many species that are cultivated in the aquaculture sector. As a result, the aquaculture industry is providing opportunities for the socioeconomic development of both rural and urban areas.

However, as the aquaculture industry grows, it also faces challenges like water scarcity, environmental degradation, a lack of available land for aquaculture, high input costs, etc. Thus, sustainability is becoming increasingly important in the aquaculture industry.

7.1.1. Sustainable Development

Despite the self-renewal of living resources, sustainability must come before resource usage. To guarantee the fulfillment and continuous satisfaction of human needs for both the present and the future generations, sustainable development places a strong emphasis on the management and conservation of natural resources as well as the direction of institutional and technological advancements. In the fields of agriculture, forestry, and fisheries, this kind of sustainable development preserves land, water, plant, and animal resources in a way that is socially acceptable, technically sound, and non-degrading of the environment.

7.1.2. Sustainable Aquaculture

The term "sustainable aquaculture" refers to aquaculture practices that prioritize environmental, economic, and social sustainability in order to enhance capacity building and efficiently utilize land for the aquaculture industry.

- **Environmental Sustainability:** Aquaculture practices should be eco-friendly. It should not impose any significant disruption to biodiversity.
- **Economic Sustainability:** The aquaculture practice should be profitable with good long-term prospects.
- **Social Sustainability:** Aquaculture practice must be socially responsible and contribute to the wellbeing of the local community.

7.1.3. Key Challenges Faced by Aquaculture Sector

With the passage of time aquaculture sector also facing some problems, which are included as follows,

- According to FAO, there are nearly 870 million people out of the 7.1 billion world population are suffering from malnutrition.
- By 2050 the food consumption will be expected to increase from 2400 kcal/day to 3000 kcal/day.
- Share of calories derived from cereals is declining in India.
- There will be around 70 million undernourished people in India by 2050.
- Over the years, there is a shrinkage of freshwater resources.
- Unavoidable intensification of aquaculture leads to environmental pollution, unprecedented disease occurrence, and heavy loss in the end.

- Input costs for aquaculture are rapidly increasing.

7.1.4. Potential of Sustainable Aquaculture

The sector's future outlook will be determined by how important it is to be able to address global issues like environmental degradation, water scarcity, population growth, food scarcity, global hunger, and other pressures placed on natural resources and aquatic ecosystems by social and economic forces.

One promising solution to address these issues both now and in the near future is sustainable aquaculture. Millions of people depend on fisheries and aquaculture for their livelihood and for the security of their food supply. It also a helping hand for millions of people as:

- A creator of employment.
- Supplier of nutritious food.
- Generator of income and economic growth.

Furthermore, scientists believe that aquaculture is a highly effective way to use resources like food, water, energy, and space while producing high-quality proteins that are vital to human health and welfare. Compared to the agriculture production sector, the modern sustainable approach of aquaculture generates less waste and carbon-nitrogen footprints. Therefore, aquaculture needs sustainable development, which could help with the problems it is currently facing.

7.1.5. Different Approaches for Making Aquaculture Sustainable

1. Focus on Environmental Impacts:
   - Development of suitable technology for environmentally feasible aquaculture practice.
   - Mitigation strategies for environmental pollution through microbial, Nanotechnology, practicing holistic methods like multi-trophic and organic aquaculture
2. Use of Renewable Energy Sources:
   - The use of renewable energy in aquaculture can be a boon as it is cost effective, environmentally friendly, and carbon neutral technology.
   - There is a wide variety of renewable energy techniques entering into aquaculture industry, such as wind-powered water pumps and solar-powered water heating systems, etc.

Investigating in these technologies reduces the long-term operating costs and reducing environmental implications.

3. Reduce Dependency on Fishmeal as Feed Ingredient:
   - As feed cost in aquaculture is skyrocketing due to the increased demand for fishmeal which also causes overexploitation of capture fisheries.
   - Shifting of dependency from animal-based feed ingredients to plant-based can be served as a way for sustainable practice.
   - Research on waste to wealth as a feed ingredient also should be promoted.
4. Promotion of Sustainable Aquaculture:
   - Rewarding sustainable farming practices in aquaculture through government policies can be done.
   - Subsidy on input costs for sustainable aquaculture farmers should be done.
   - Awareness and encouragement for sustainable aquaculture can be practiced.
5. Practice of Sustainable Organic Aquaculture

Organic aquaculture is based on the use of organic inputs in aquaculture.

- No use of inorganic fertilizer, pesticides, or antibiotics.
- It produces healthy fish which is completely devoid of any pesticide, or antibiotic residues.
- Many organic aquaculture issues still need to be resolved; steps should be taken to encourage and enhance organic aquaculture as a sustainable practice of aquaculture.

6. Investment in New Sustainable Technologies

A. Aquaponics

- Aquaponics is an integration of aquaculture and hydroponics techniques to produce both fish and plants in the same system.
- The system has high sustainable potential, such as less water requirements, minimum environmental impacts, and organic food production.
- It has the potential to produce value-added products like fish and a high-yield variety of vegetables with less labor and land, less chemical use, and not much water usage.
- It can also be adapted to diverse and changing conditions.

B. Integrated Multi-Trophic Aquaculture Systems (IMTA)

- IMTA is based on the principle of polyculture and waste

utilization from the system by culture species.

- In the system, the particulate waste materials are absorbed by filter feeder bivalves, while water-soluble waste materials absorbed by algae.
- Along with the culture of fish and crustaceans, bivalves, algae, sea weeds can be cultured, which also maximizes profit margin.

C. In-Pond Raceways (IPR)

- IPR is like cage culture techniques but can be applicable to almost any water bodies which have the advantage of controlled
- Water movement: This improves water quality and provides scope for more stocking density.
- IPR is more sustainable than cages, raceways, and intensive open pond culture.
- But still, further research is needed about the waste disposal of solid and liquid wastes.

D. Re-Circulatory Aquaculture System (RAS)

- RAS is based on the principle of water conservation and reduced waste discharge.
- It consists of a culture chamber, a settling chamber, and a biological filter. The water flows from the culture chamber to the settling chamber and then the biological chamber, which filters the water for reuse in the culture system.
- It has the potential for culturing in high stocking density, which maximizes the profit.
- RAS system conserves the water and reduces pollution.
- Fewer biological risks such as disease issues relative to farming in the natural environment.
- Lower environmental compliances.
- It has the ability to control the cultural environment, which improves FCR and reduces the need for chemicals and drug use.
- Thus RAS can be an efficient, useful tool for sustainable aquaculture.

E. Bio-Floc Technology (BFT)

- BFT technique is based on in situ remediation by which heterotrophic microbes culture is enhanced by C: N ratio and constant aeration.

- Generates a floc-like structure that is nothing but aggregates of various living and dead microbes, which provide a good nutritional value for culture organisms.
- BFT reduces feed conversion ratio (FCR) and decreases feed costs.
- This natural productivity plays an important role in recycling nutrients and maintaining water quality.
- It conserves water, utilizes waste, and reduces pollution.
- Less investment as compared to others.
- The BFT is much more sustainability potential than any other cultural system

7.2. SUSTAINABLE USE OF OCEAN RESOURCES

It is a widely accepted fact that any nation possessing marine resources ought to wish to use them responsibly. In today's world, "sustainability" can refer to nearly anything and is considered a universal "good." It therefore has no meaning. This is a significant issue since the concept of "sustainability" is crucial to the equitable and responsible use of natural resources. It is based on the idea of intergenerational equity. The majority of human activities still rely on wild species and ecosystem services in the seas, which is the final area. A fundamental tenet of ocean management, the requirement that uses of marine systems be sustainable is enshrined in both national and international agreements.

Current notions of sustainability usually include equity and economic considerations, which greatly complicates the process of establishing objective standards. There are many different measures for these socioeconomic variables, and no single set of guidelines exists for determining the ideal ratio of social to economic values in any given society. The historical development of sustainability concepts was outlined by earlier writers like Kuhlman and Farrington. It began with the simple persistence of renewable resources and progressed to the division of sustainability into social, economic, and environmental "dimensions." A significant advancement that followed was the analysis of "strong" and "weak" sustainability. While weak sustainability permits consideration of "wealth, comprising man-made assets and environmental assets," strong sustainability states that "the next generation should inherit a stock of environmental assets no less than the stock inherited by the previous generation." Weak sustainability makes it possible to trade-off material gains for declining environmental assets. When "Strong" sustainability is coupled with the requirement to make trade-offs between the various dimensions, problems can arise, particularly in situations where there are no clear guidelines for striking the right balance. To keep "sustainability from becoming an empty phrase indistinguishable from 'goodness,'" Kuhlman and Farrington explain the necessity of keeping "well-being" and "sustainability" distinct from one another. While we wholeheartedly concur with this assessment, we aim to define sustainability more precisely. Additionally, we contend that the "strong" sustainability idea of a "constancy of natural capital rule" (CNCR) is overly rigid, rejecting the idea that long-term environmental harm can be justified regardless of the positive effects on the economy and society.

The current situation has been described as the "post-Brundtland quagmire," and while we agree with their diagnosis, we would disagree with their cure, that we should "...embrace a plurality of approaches to and perspectives on sustainability, accept multiple interpretations." Ecological sustainability can be defined in a universal way, and we should at least do this, even if socio-economic factors include more substantial value judgements.

7.2.1. A Framework for Thinking About the Ecological Dimension of Sustainable Use

Our suggested method makes it possible to compare extractive and non-extractive industries fairly and measures fish stocks, the seafloor, and marine wildlife sustainably using the same units. The fundamental idea is that human activities in the marine environment shouldn't have permanent effects, meaning that when they stop, ecosystems should be able to recover to their pre-use state in a time frame that is acceptable to society.

For their uses of the marine environment to be considered sustainable, users do not need to show that the environment is currently in a natural state, or even close to one. However, users would have to prove that the system can be restored to its initial state before using it. While allowing for human-induced change, this strategy leaves open options for future generations.

Core principles of our proposed approach to the assessment of environmental sustainability.

- We do not know what future generations will need from the environment and so the onus should be on us to give them the maximum range of options. "Conservation of options" principle
- Uses with irreversible impacts remove options from future generations
- At a user level, the end of use should allow the area affected (and its ecosystem services) to return to its pre-use state.

- At an ecosystem level, the end of use should allow the ecosystem affected to return to a natural state.
- Therefore, sustainable uses are those that don't have an adverse effect on the environment that would make it impossible for the environment to revert to its original state or change into any other state that future generations might find desirable. "Conservation of equality" in terms of multiple generations.
- There is no requirement that the exploited environment be in a natural state, or even close to it, in order to demonstrate that the exploitation is sustainable.

For governments and individual users, pre-use can mean different things. In order for the ecosystem to return to its pre-use state—which might not be a natural state if other effects predate the more recent use—a user would need to persuade regulators that stopping their use would enable this to happen. Governments would have to show proof that the system could return to its natural baseline if all uses were stopped in order to claim that their management of the system as a whole represented sustainable use.

We contend that uses that result in irreversible damage are not compatible with the concept of ecological sustainability. Future generational effects could be viewed as "acceptable" losses if society finds the uses to be important enough to continue regardless. Being truthful about the distinctions between "sustainable" and "acceptable" is crucial.

The majority of human activities that involve the environment are made up of numerous smaller ones. For example, feeding, disease control, transportation, holding fish during growth, and predator protection are all necessary in finfish aquaculture. A farming operation cannot be deemed sustainable as a whole if even one of these individual activities has consequences that are unavoidable within a reasonable amount of time.

Two major categories of sustainable use apply to marine systems: maintaining commercial stocks and effectively managing the effects on non-commercial species and the larger marine environment. A use is deemed sustainable under the Marine Strategy Framework Directive if its effects reverse upon cessation, permitting a "rapid" return to a "natural" state. Therefore, just because something is being used sustainably does not mean that it is in a natural state at the moment, or even close to one. This reversibility criterion can be applied to any use, enabling a consistent comparison of various uses.

Similarly, criteria for determining whether effects on productivity, habitat provision, species richness, and ecosystem functions are "temporary" are included in the FAO guidelines on deep-sea fisheries management, the definition of Significant Adverse Impacts. Therefore, the idea of recovery as a standard for evaluating the consequences of human activity in the sea is consistent with both EU and UN policy.

The ecosystem effects of fishing could be taken into account in the same manner as stock yields if one adopted an approach based on the capacity of systems to recover. This is not usually how things have worked in the past; for example, in the past, different methods were used in the scoring systems for the effects of fishing on ecosystems and for stock sustainability in the Marine Stewardship Council (MSC) certification. The effects on biodiversity and the seabed were evaluated using qualitative metrics, whereas the sustainability of fishing on the stock has always been evaluated using quantitative metrics like MSY. We contend that this is superfluous and that, in order to identify sustainable fisheries, a system should assess the sustainability of the impacts on non-target species and the environment by applying the same standards that are used for stock species. The new MSC standards aim to restore populations to "50% of unimpacted levels within three generations or 100 years, whichever is shorter," moving towards a system where Endangered, Threatened, and Protected (ETP) species are treated more like stocks. Although we contend that this strategy should be used for ecosystem functions, biodiversity, and seafloor integrity rather than just ETP species, this is a positive step. Additionally, it is more in line with EU and UN guidelines, which contributes to the growing consensus that recoverability is a crucial metric.

Fisheries are an example where natural capital can decline over time without this being evidence of unsustainable fishing. This is contrary to the views of Ekins et al. that "...sustainability depends on the maintenance of the capital stock." When fishing begins, a stock's size normally declines; in fact, this is thought to be necessary for achieving Maximum Sustainable Yield (MSY). It is oversimplified to associate all declines with unsustainable use; rather, a decline's mechanism, rate, and duration determine whether or not it is indicative of unsustainable practices. For instance, compared to a larger, slower decline in the absence of additional damage, there may be a small but rapid decline linked to habitat damage. The idea that capital stocks should never fluctuate is the crux of the issue and the rationale behind the necessity of the concept of "weak sustainability" as a workaround. Are both unsustainable and which is worse? Under any definition of "strong sustainability," where capital stocks are fixed, fishing is not possible. The issue goes away and the need to think about the "substitution" of natural capital for financial capital vanishes if we acknowledge that under sustainable use, the natural environment can change. Daly permits varying levels of natural capital, provided that they are all at their optimal level, with a preference for Maximum Economic Yield (MEY) over MSY. Even though this is a better situation, it is not realistic to expect all competing uses of a shared environment to always have access to the resources they need at optimal efficiency. Although political decisions must be made regarding the distribution of resources and space, imbalances among users do not always indicate that a resource is being used in an environmentally unsound manner (Daly, 1990).

Hilborn with others elucidate sustainable fisheries criteria that incorporate all three of the traditional "pillars" and highlight the inconsistencies between them. Among the instances given is "sustainable overexploitation," in which harvesting can go on forever at a level that permits recovery but falls short of maximizing the fishery's economic potential. Additional conflicts arise when deciding how to allocate and concentrate quotas, as social objectives and economic efficiency are frequently at odds. There is rarely a "right" response in these situations since the optimal solution varies widely among cultures and political philosophies. Neumann and associates contend that despite their apparent sustainability, these trade-offs are often politically expedient and result in weakened ecological protections due to social and economic demands (Hilborn et al., 2015).

Determining the appropriate amount of fishing effort is one instance of this balancing act. As part of the first "National Standard" of the Magnuson-Stevens Act, the "optimum yield" of each stock has been a topic of continuous discussion in the US. It would seem that shifting from MSY to MEY (lower fishing effort, higher stock sizes, and more efficient fishing) would combine the environmental and economic aspects of sustainability. However, processors may see a decline in profits as a result of lower yields and higher unit prices, which would hurt their bottom line. Changes in management targets may have varying effects on different stakeholders, depending on the spatial and temporal scale that is being considered.

Hilborn et al. (2015) criticize the notion that evaluations of "sustainable seafood" should take into account "warm and fuzzy" feelings about fisheries. They also highlight the numerous inconsistencies that exist between the socioeconomic outcomes of ecologically sustainable fisheries and their sustainability, concluding that a return to the original concept based on stock persistence makes the most sense. We contend that this kind of strategy can be simply expanded to address environmental effects outside of the stock without undermining the idea.

7.2.2. Applying Approach to the Assessment of Sustainable Use

We look at three primary uses of marine resources. First, there are the fisheries' objectives. In this context, sustainability can be simply defined as the use of gear and levels of exploitation that do not diminish future generations of fishermen's ability to fish. Separate decisions would be made on the political and social fronts regarding which sectors of the fisheries should benefit and whether to maximize profits or total yield.

Second, some uses—like marine renewable energy—cannot deplete the resource, while others—like oil, gas, and seabed minerals—use a resource that is not replenished at appropriate intervals. Here, the only decisions are about how and when to use these resources in light of the state of the market. If environmental concerns have an impact on demand, they may also have an impact on these market conditions. These resources cannot be used at a sustainable level because they are either impractical (oil) or irrelevant (wind).

In the third and most complicated scenario, the taxa or biotopes are unintentionally the victims of an activity rather than its intended target. An example of collateral damage, also known as an economic externality, is harm done to these environmental components. Bycatch, impacts on the seafloor and benthic flora and fauna, and food web effects (such as on higher predators) would all fall under this category. However, the answer to the question of whether harm done to non-target species and the larger environment can ever be deemed sustainable is yes. Every action in the marine environment has an effect, and the fishing and energy sectors cannot exist without them.

When an activity that has an impact on the environment is done in a way that does not permanently prevent it from being used in the future, it is deemed sustainable. It does not imply that there is no current influence, nor does it imply that the system is perfect. One activity may modify a system slightly but not sustainably, whereas another may modify a system significantly but sustainably. This is due to the fact that determining whether an activity is sustainable involves more than just evaluating the here and now; it also involves making a judgment about the future. According to Hollins, a population's "instability" does not always imply a lack of resilience—in fact, it might sometimes be a sign of resilience. A pressure could have a significant impact, but its release could cause an equally quick recovery.

The idea of sustainable use is based on the idea that ecosystem use should occur within the "elastic limits" of the system. The utilization of resources in an ecosystem can result in significant changes to its natural state; however, the use of those resources can still be considered sustainable if the system can return to its pre-use state. This is undoubtedly difficult because, in many cases, we lack quantitative information regarding the pre-use condition of ecosystems where long-term activities have been occurring. However, there are methods by which individuals attempting to assert sustainability can offer proof. These consist of both historical research and current experiments.

When deciding whether a novel activity is likely to be sustainable, it would be helpful to have data from both the before and after of the proposed activity's pilot phase, which would see data collection and use occurring. When there is insufficient experience with the kind or intensity of a proposed activity, or when there is an ecological context that is less understood, such pilot phases would be required.

However, a lot of work has been done to identify such baselines and compare them to the present. At the level of the ecosystem as a whole, the situation is more challenging and will typically require assessment in the absence of complete pre-use information. There are difficulties here, of course. Similar questions arise in restoration ecology, also known as "rewilding," about which historical period serves as a suitable "target" state. Naturally, it is impossible to have complete knowledge of the past because the climatic context has changed, rendering some past states unreachable. However, there are a lot of information sources available, such as fisheries data, diaries, middens, and preserved otoliths, which show historical taxonomic abundances, sizes, and occurrences that may be used as indicators. If governments want to make statements about sustainability, we should still expect them to make their case and provide supporting documentation, despite the task's difficulty. More caution will be needed in cases where the evidence is less certain.

Fortunately, spatial management and other fisheries measures can also serve as a test of sustainable use, as they offer a means of evaluating the capacity of the seabed and natural communities to recuperate in areas where uses are eliminated or altered. Therefore, marine protected areas, or MPAs, offer more than just protection; they also present a chance to evaluate the sustainability of the sea's broader uses.

Frequently, a "counter-factual" future that might never materialize. The majority of the seabed may never revert to its pre-use condition, but if impacts were eliminated or significantly reduced, the entire region would be under sustainable use. Even if the seabed is quantifiably altered from its natural state, use does not need to cease throughout the entire region. We don't know if these activities will be discontinued by future generations to satisfy different needs.

Comparing change within areas that remain under use to change when all use is eliminated or reduced in suitably sized and replicated trial areas representative of the larger area would be the most reliable way to evaluate sustainable use. Many times, there are already spatial management policies in place that could be used as test zones for the wider sea's sustainable use. Regretfully, rather than being selected as representational, the majority of MPAs are selected because they are unique in some manner. A lack of control areas and inadequate replication are also common issues, though there are some notable exceptions, like the MPA network in California. Currently, research on MPAs offers unrealized potential to assist us in determining whether the uses we make of the oceans are sustainable. Protected areas continue to play a vital role in maintaining the larger ecosystem even though they may not have been fully utilized as tests of sustainable use. As larval and mobile adult sources, they could speed up the rate at which affected areas bounce back from stresses. It is less certain that there will be positive effects outside of MPAs, particularly outside of the immediate "halo" where spillover is typically observed, than there is evidence of change within MPAs. Even in cases when solid data are available, such as in California, where there is strong evidence of local lobster spillover, changes to larger fish populations are not supported. On the other hand, research conducted in New Zealand discovered that marine reserves significantly increased the number of snapper species. Although MPAs play a significant role, it is obvious that they cannot take the place of regulations governing the kinds and levels of human activity that are permitted in the ocean.

Our suggested framework for understanding sustainability is connected to the idea of "resilience," specifically recovery as opposed to resistance. The level of pressure was unsustainable if activities overwhelm the system's resilience to the point where it is unable to "bounce back" even when pressures are decreased. This may be the case with the Firth of Clyde cod population in Scotland, whose numbers have not rebounded despite a complete ban on directed fishing for this species. The fish population was driven to a point where it is struggling to recover due to an unsustainable level of fishing, although opinions on the role of predators in impeding recovery are divided. There is considerable disagreement over how much the effects of fishing have actually been mitigated, with some estimates of bycatch mortality being very high and spawning aggregation protection being insufficient.

The competing uses of the marine environment can be fairly compared by keeping the ability to recover as the central concept. It is important to distinguish between "acceptable" and "sustainable" uses of the marine environment in case an unsustainable activity was justified for other reasons.

7.2.3. Sustainable Use for Future Generations

Taking into account context, as well as spatial and temporal scales, is still necessary when using our concept of recoverability as the sustainability metric. If a new use of the sea is implemented briefly or over a small area of the seafloor, it may be sustainable; however, if it is implemented over an extended period of time or over a large area, it may not be.

While it is possible to come up with technical definitions of "natural" and statistical methods of detecting change towards such a state, there remains a critical societal decision. How quickly should a system be able to return to a natural state on cessation of the impacts that are thought to have altered it? Some definitions refer to a "rapid" recovery to a natural state, but give no actual timescale. Some definitions allow for different rates of recovery depending on the taxon or process in question, e.g., the FAO allows "5–20 years, taking into account the specific features of the populations and ecosystems." Damage to cold water corals or coralline algae beds in temperate waters shows little evidence of recovery after decades since the impact is thought to have occurred, but this is to be expected given the slow growth rates of the taxa involved.

It seems irrational that alternate users should have to wait longer than 10 years before the effects of prior uses have subsided, and sustainable use should permit changes in use over time. We argue that any use that degrades systems to the point where they are unable to recover in 10 years is not sustainable.

These kinds of activities are currently being explored in the deep ocean, where polymetallic nodule mining is being studied. The biology of mined nodule fields does not appear to have recovered after decades, according to recent data, which may not come as a surprise considering the biology of abyssal systems. It is not a sustainable activity because, on human timescales, the collateral damage to the local biology seems irreversible. These kinds of activities might be deemed appropriate, which brings up the question of how much irreversible harm society is willing to tolerate.

In the end, we cannot attain sustainable use unless we ensure the survival and well-being of future generations. We argue that a definition that puts an emphasis on the future is necessary because the terms "sustainability" and "sustainable" as they are currently used obscure the true problem. Children born today bear an inherited burden as a result of previous generations' unsustainable use of natural resources. Although the idea of environmental intergenerational equity is not new, the Youth Strike Movement for Climate Action shows that it is becoming more and more prevalent in public awareness. The unfairness of using Earth's resources in an unsustainable way across generations is becoming more widely recognized. Our suggested method of evaluating sustainability, which centers on resource availability in the future, implicitly recognizes intergenerational equity.

7.3. SMALL INNOVATION FOR SUSTAINABLE USE OF FISHERIES RESOURCES

Promoting sustainable use of fisheries resources can benefit greatly from small, innovative approaches.

7.3.1. Sustainable Development in Fisheries

7.3.1.1. Chronology from Enclosure of Waters to Reduction of Wastes in Environment and Resource Use

The world's fisheries saw a transition in the second half of the 20th century from coastal countries claiming expanded rights to global conservation of the aquatic environment and resources.

In 1945, the United States of America issued the "Truman Proclamation for Conservation and Protection of Fishery Resources," which established the right of nations to create conservation zones in contiguous high seas following World War II. In the 1947 Santiago Declaration, Chile, Ecuador, and Peru declared exclusive sovereignty and jurisdiction over the sea up to 200 nautical miles off their coasts. These appeared to be an attempt by nations to compete for larger sea areas.

A series of United Nations (UN) Conferences on the Law of the Sea, which held their inaugural meeting in 1958, followed this trend. Their outcome was the United Nations Convention on the Law of the Sea (UNCLOS). Adopted in 1982, UNCLOS came into force in 1994. In 1976 and 1977, the United States of America and 23 other states declared various forms of 200 nautical mile zones before UNCLOS coming into force. That marked the beginning of the 200 nautical mile rule.

In addition to the aforementioned process, there were many disagreements during the 1980s, which marked a shift in global fisheries management principles. The Bering high seas fishery and the high seas drift-net fishery were typical examples in the Pacific region. From the perspective of environmental preservation and sensible use of aquatic life resources, they were concerned. The new international pressure eventually caused the two sectors to cease. The former was criticized for overharvesting stock and negatively impacting the fishing ground's ecosystem; the latter caused injuries to unharvested fish and unintentional captures of marine mammals, marine reptiles, and seabirds. This was mirrored in the steps taken to outlaw certain high seas fisheries.

7.3.1.2. Rio Declaration

The fundamental idea of the "Rio Declaration on Environment and Development," which was adopted at the UN Conference on Environment and Development in 1992, is sustainable development. Its idea can be summed up as an attempt to achieve intergenerational equity with regard to the environment and its advantages. It aimed to break away from the traditional trend of expanding industries and production by changing both the guiding principles and modes of thought. The Rio Declaration brought up many topics that are still being discussed today, including resource exploitation, environmental preservation, ending poverty, developing nations, ecosystem health, human well-being, innovation, citizen participation in decision-making, cautionary principle, women, indigenous people, human warfare, and peace.

The majority of industrial sectors as well as scientific and technological research were impacted by the Rio Declaration. Reproductive, biological, and natural resource-based industries like fishing were among the most affected. The Sustainable Development Goals (SDGs) and Education for Sustainable Development (ESD) of today inherit the aforementioned problems.

7.3.1.3. Code of Conduct for Responsible Fisheries

The UN Food and Agriculture Organization (FAO) Department of Fisheries first proposed the idea of "responsible fishery" in 1991, and the "Code of Conduct for Responsible Fisheries (CCRF)" was released in 1995. It was a notable turning point in the global development and management of fisheries, and it has dominated global fisheries ever since.

Comparison of the texts regarding a precautionary approach, for example, clearly indicates that CCRF was prepared on the basis of the Rio Declaration; as 'In order to protect the environment, the precautionary approach shall be widely applied by States … lack of full scientific certainty shall not be used as a reason for postponing … measures to prevent environmental degradation and 'States should apply the precautionary approach widely to conservation, management and exploitation of living aquatic resources to protect them … The absence of adequate scientific information should not be used as a reason for postponing or failing to take … measures.'

The preface of the CCRF states that the organization understands that fisheries "provide a vital source of food, employment, recreation, trade, and economic wellbeing for people throughout the world, both for present and future generations" and that "conservation, management, and development of living aquatic resources, with due respect for the ecosystem and biodiversity," should therefore be conducted responsibly. The "Code sets out principles and international standards of behavior for responsible practices" is another way that the organization espouses its philosophy.

Additionally, the CCRF demonstrates its aspirational "General Principles," which cover topics like intergenerational equity in the sustainable use of resources, food security, poverty alleviation, ecological and biodiversity preservation, and responsible behavior. Furthermore, based on its central tenet of "sustainable utilization of fishery resources," CCRF concentrates on the enhanced role of fisheries as a food-supplying industry through solutions to wastes in capture fisheries, mentioning prevention of overfishing and excess fishing capacity, decisions based on the best available scientific evidence, as well as selective and environmentally safe fishing gear and practices.

Despite lacking legal force, the CCRF has stated that "conservation of the aquatic environment and sustainable utilization of living resources" should be the focus of research and development for novel technologies, policy formulation, and fisheries education. Even though the CCRF was published 25 years ago, its significance has not diminished. In light of the Sustainable Development Goals (SDGs) recent global diffusion, CCRF ought to be more significant than it was in the past.

7.3.1.4. Reduction of Wastes in Use of Fishery Resources

Following the release of the CCRF, FAO began to emphasize the need to reduce wastes in the use of fishery resources. The word "wastes" is commonly used to describe the part of fishery resources that could be used by humans but are removed from the environment due to fishing operations and only marginally or never used. Not only do wastes include common and obvious situations like the death of potential resource organisms from abandoned bycatch and ghost fishing caused by abandoned fishing gear—both of which are discussed below—but they also include a wide range of situations, such as the wastes of edible fish parts during marketing and consumption and the infrequent use of food fish for aquaculture feeding. According to the idea of "providing resources to future generations," this concept can be expanded. Today's fisheries resource management is focused on finding technological, scientific, and policy-oriented solutions to waste problems.

7.3.2. Research Toward Reduction of Wastes in Utilization of Fishery Resources

The term "innovation" describes how modern technological advancements create new business opportunities. It was defined as "creation of/transfer to new technology contributable to sustainable development" in Agenda 21, one of the Rio Declaration's annexes, but it goes beyond simple technological advancement. Instead of advocating for a game-changing invention that will alter our culture and society at large, the author has favored "small innovation," which is achievable for researchers from even the smallest institutions.

In order to minimize wastes in capture fisheries following CCRF, numerous new studies and technological advancements have been carried out since the 1980s. Ghost fishing concerns and bycatch/disposals are common topics.

7.3.2.1. Bycatch and Discards

Bycatch and discard was one of the waste problems in fisheries that was first acknowledged as resolved in the 1980s. In 1994, the FAO estimated that up to 27 million tons were discarded worldwide. Even though the overestimation of the reported value was proven to be untrue, the report served as a catalyst for the general recognition of the bycatch and discard issue, leading to the prohibition of certain fishing gear and the closure of certain fishing enterprises.

The term "bycatch" refers to catches that are not wanted in fishing operations and are either partially or completely discarded elsewhere after not being landed (utilized). Bycatch fishes are not used for many reasons, some of which vary over time following socioeconomic factors. It is not unusual for species that are discarded today to become the landed catch tomorrow as a result of shifting market and management trends. The term "bycatch" refers to the capture of non-target species on occasion; however, this definition is not always accurate and has led to a lot of misconceptions. Another way that discards are referred to is as "discards at sea," but this is also inaccurate. A place for discards is not a contentious place. The main issue is the bycatch and disposal of species and/or sizes of organisms that should be protected due to the goal of resource and environmental conservation. In this situation, landing page promotion is not the answer. Avoiding being captured by those organisms is the only solution.

7.3.2.1.1. Bycatch and Discard in Prawn Trawl Fishery

One of the fishing industries where bycatch and discards first raised concerns was the prawn trawl fishery. In the 1980s, we began field research that included a survey of the bycatch and discard situation as well as the development of a bycatch exclusion device in the prawn trawl fishery in the Gulf of Papua, Papua New Guinea (PNG). According to observations made while on board, the sector's discard ratio for finfish to prawn retention was roughly 9~14. Even though only a tiny percentage of bycatch finfish were landed in the sector, the ratio was roughly equal to 10~15, which was a commonly accepted approximation for tropical prawn trawling. The ratio was multiplied by the amount of prawn production reported in the PNG government statistics for the sector and the annual discard of 11~20,000 tons. This quantity was roughly equal to the estimated total production of Papua New Guinea's coastal fisheries. Furthermore, the majority of the discarded species were grunts and ponyfish, which are important sources of food for coastal villagers. For these reasons, the author concluded that the studied sector had caused a significant waste in the use of coastal fishery resources. In the South Pacific, this was the first study on bycatch and discard.

7.3.2.1.2. Development of Bycatch-Reductive Trammel-Nets

Researchers and technicians worldwide have developed a wide range of gear designs and devices for the purpose of reducing bycatch. This section contains a brief case study that the author conducted for the creation of fishing gear that is focused on excluding bycatch.

In the 1980s, the discarding of small flounder bycatch in bottom trammel-nets intended for cuttlefish raised concerns in Kagoshima prefecture, Japan. The prefectural regulations imposed a dilemma whereby flounder with a total length of less than 25 cm were prohibited from landing, forcing fishermen to discard them.

The release of artificially hatched juvenile flounder in mariculture resulted in twice as much waste being discarded as small individuals. Matsuoka along with others observed the vertical positions of entanglement for both species in five outer meshes. It was discovered that cuttlefish are caught in the lowest mesh convergently, but they flounder relatively evenly (Matsuoka et al., 1995).

The observation that the trial fishing did not result in a decrease in cuttlefish catches led to the hypothesis that the thick brown twine used to suspend the net webbing was the attraction for cuttlefish. As a result, a new design was created that added more visible twines to the gear to improve catch efficiency. The next hypothesis examined whether the distance a swimming layer of flounder has from the seabed is proportional to fish size. This was prompted by another experimental result showing that large flounder individuals were caught in the semi-demersal net, while small ones were excluded. It prompted new research that combined hydrodynamics and fish behavior science.

7.3.2.2. Ghost Fishing

Ghost fishing is the practice of leaving abandoned or lost fishing gear in the water where it continues to kill aquatic life without human intervention or harvesting. In the science of capture fisheries, we only consider mortality resulting from abandoned fishing gear; we do not include mortality resulting from other man-made materials, like plastic debris, in our definition.

In the middle of the 1970s, ghost fishing was said to be one of the most significant problems facing capture fisheries. However, some people questioned it because the phenomenon's invisible process was apparent to regular fishing operators. Scientists and managers involved in fisheries started to worry about it in the late 1980s. Ghost fishing using cage traps was demonstrated in Japan in the early 1990s.

7.4. STRATEGIES FOR SUSTAINABLE USE OF LIVING AQUATIC RESOURCES

These strategies seek to protect fisheries stocks, preserve biodiversity, involve fishing communities in management, and increase fishery productivity through the use of environmentally friendly technologies.

7.4.1. Freshwater Ecosystems

In order for freshwater habitats to support human life, threats to their water quality must be reduced, if not eliminated. In the uplands, forest denudation is typically the cause of flooding and sedimentation (a type of pollution); however, excessive sedimentation can be an effect of mining. Waste effluents from industries, aquaculture, and agriculture are examples of additional pollutants. Reforestation of deforested areas and initiatives for pollution and flood control are some of the solutions to these issues.

Large-scale reforestation brings back the regular flow of water in rivers that once supported a variety of prawn species, food fish (such as gobies and mullets), and other creatures that produced goods. River systems that have dried up may be restored to yield again.

Overuse poses a further risk to freshwater resources. Chemical poisoning and dynamiting have both been used on rivers. Fish cage overcrowding is a common sight in lakes that have been used for fish farming. Luzon's Laguna de Bay is a prime example of how overfishing and declining water quality have led to a decline in species richness and fish yields. Another example is Lake Lanao, where the majority of the 12 species of endemic cyprinid fishes in eight genera disappeared before researchers could investigate them. Another example is Lake Buhi in the Bicol area, which is losing its sinarapan, or gobies.

In addition to protection, research is crucial. The potential of endemic fish species for aquarium use, aquaculture, and ranching is one study area. An additional factor is the genetic enhancement of introduced species that are currently employed in aquaculture, such as tilapias and African and Thai catfish.

7.4.2. Shallow-water Marine Ecosystems

Estuaries, mangroves, seagrass beds, coral reefs, soft-bottom communities, and open waters are some of these productive ecosystems. Shallow waters are home to the majority of marine production that is accessible to the majority of the human population. For instance, coral reefs produce more than 30 tons of fish annually per square kilometer. Mangroves play a significant role in fisheries by providing 13–47% of the carbon needed by demersal food chains through their leaf litter.

7.4.3. Protective Management

To address environmental issues, many management techniques have been put into practice recently. Protective management of shallow-water ecosystems is one tactic. Throughout Southeast Asia, reserves, sanctuaries, and marine parks have been established. Therefore, fishing has been prohibited temporarily or permanently in certain areas of coral reefs, or even in entire reef areas. Additionally, coral reefs have been put to a variety of uses.

When it comes to coral reefs, the idea of marine fishery reserves (MFRs) is beginning to be recognized as a practical solution for coral reef fisheries that are sustainable. MFRs are reef areas that are off-limits to human exploitation forever. They safeguard reef fish populations that act as spawning grounds and guarantee the recruitment of juvenile fish to both other reefs and the entire reef system. Experiments conducted in central Visayas, Philippines, have shown that an MFR is effective in maintaining species richness, abundance, and community structure within the reserve as well as high fish yields outside the reserve. Fish yields in Sumilon Island, Cebu, were higher during the protected period than during the unprotected period. It was evident that fish were being exported from the MFR to the fishing area.

MFRs may provide fish larvae to coral reefs that are located downstream of the main currents. This may be the case for the fisheries on Palawan Island, where fish larvae from the Tubbataha National Marine Park may replenish the stocks by migrating westward along the Sulu Sea's dominant currents.

Mangrove ecosystem protection seems to be working in certain Southeast Asian nations, like Brunei and Malaysia. However, mangroves in the Philippines are protected on paper but not in practice. In the entire nation, there are most likely very few tiny mangrove patches that have received some level of protection.

Despite the valiant efforts of scientists like Miguel Fortes, the seagrass beds are still ignored. Seagrass beds that are protected do not exist in the nation. Likewise, no regions with a soft bottom are strictly protected. The soft-bottom benthos are, nevertheless, automatically protected when specific bays are periodically closed to trawlers.

Estuaries function as fish nurseries and are typically productive places. However, estuaries are primarily affected by pollution, particularly from household waste, chemicals from industry, and sediment from upstream areas. Manila Bay's estuarine areas are contaminated. Other examples are the river mouths where mine tailings end up. This nation does not have any protected estuaries. Small island ecosystems that are nearly pristine should also be protected. Small, heavily forested islands in Malaysia, like Pulau Tioman and Pulau Redang, have been developed as resorts or leisure areas. To ensure the survival of both the marine and terrestrial components of these island ecosystems, it is crucial to maintain the proper balance between development and conservation.

7.4.4. Rehabilitation Schemes

Restoration of the original associated biota and rehabilitation of degraded environments are the other strategies to make our coastal ecosystems work for people. These programs include the establishment of artificial reefs, coral transplantation, and mangrove reforestation. A nationwide program for mangrove reforestation has been started by the Department of Environment and Natural Resources (DENR). One effect of this activity is an increase in organic matter in coastal areas, which provides food for various organisms. The dependent human population has access to protein food because of the increased species diversity in coastal ecosystems. Certain island communities in the Visayas had begun planting mangrove trees long before the DENR program was implemented, both for construction purposes and as nurseries to draw in fishery species. Since they still contain sizable tracts of primary mangrove swamps, Southeast Asian nations like Malaysia, Indonesia, and Brunei most likely do not require the replanting of mangroves.

It has been demonstrated that transplanting corals is possible, and certain economically significant corals—like blue coral, for example—have been seen to grow well. Fish and other organisms linked to reefs are drawn to areas of depleted bottom that are restored through coral transplantation. Coral transplantation is required due to the extensive loss of coral reefs in Southeast Asia, as well as their significance as recreational sites and a source of valuable products. It has been suggested that artificial reefs be established, albeit cautiously, to improve habitat and serve as a management tool for artisanal fisheries. Experience in the Philippines has demonstrated that, with comparatively little work, artificial reefs can produce a respectable biomass of fish. A transplantation process for seagrass is being used in the experiment stage to speed up the consolidation of loose bottom. A few seagrass species have been transplanted by the UP-MSI near Marinduque Island.

7.5. PROMOTING SUSTAINABLE AQUACULTURE THROUGH ECONOMIC AND OTHER INCENTIVES

Many agencies (local, national, regional or international) have provided incentives for aquaculture development. These include subsidies for production inputs or outputs, e.g.,

- subsidies to fixed capital or to operating capital by direct grants, soft loans or debt-equity swaps;
- income and profit tax rebates (tax holidays);
- public investment in collective infrastructure (e.g., water channels and bunds, electricity supply);
- market promotion of aquaculture products;
- investment in human capital (skill development through training and extension) and public investment into innovation (research and development); and
- technology transfer.

These organizations' primary focus has always been to hasten the expansion of the aquaculture industry in order to achieve economic and social goals, including job creation and income generation, foreign exchange earnings, and rural development to reduce poverty and supply food. These policy goals and instruments are not unique to aquaculture, but in the past, developing aquaculture sectors in both developed and developing nations have profited from this kind of public support. The main arguments for this support have been:

- the risky character of aquaculture production for which knowledge is still rather limited;
- the relatively high-tech content of some aquaculture practices; and
- the significant potential of aquaculture to provide socio-economic development opportunities, especially in deprived regions and poor countries.

Another argument is that because scale economies do not yet apply and learning-by-doing benefits are not yet realized, "infant" industries suffer from higher costs. As a result, without government support, a late starter may never become globally competitive, or may do so only after a very long time.

The kind of public policy and support required to guarantee that aquaculture development proceeds in a sustainable manner and is not adversely affected by other economic activities has received more attention in recent years.

The destruction or removal of ecologically significant habitats, waste production levels that surpass the assimilation capacity of nearshore waters and freshwater aquifers, the capturing of wild post larvae for stocking and high by-catch mortalities, and the transfer of pathogens and genes between cultured and wild stocks are among the issues regarding the environmental sustainability of some aquaculture operations. Concerns about economic, social, and cultural sustainability include the sharp declines in output brought on by disease outbreaks; the hindrance of local communities' customary access to shared resources; and the socioeconomic, cultural, and nutritional effects of switching from multicrop agriculture to pond culture (Ridler, 1998).

The frequently unsatisfactory results of command and control measures—the establishment of rules and regulations that either permit or prohibit specific behaviors—have led to a growing interest in using economic incentives to achieve sustainability goals, particularly when growth incentives are also present. The general goal of command and control measures is to prevent private operators from being incentivized to overuse or contaminate natural resources by different forms of market failure. On the other hand, economic incentives make an effort to match sustainability goals with the incentive structure.

The supply of sustainability incentives is justified economically by many different kinds of market failures. Important inputs for aquaculture production, like land, water, and certain valuable environmental resources and functions, usually have no market or a poorly defined one. Consequently, these inputs are either not priced at all or underpriced in relation to their social value. This consequently has the effect of adding a "subsidy" component to the price of the cultivated products, the ultimate beneficiaries of which are the consumers.

Incentives should ideally establish "win-win" scenarios that accomplish goals for environmental preservation, social and economic development, and both. However, it is a known fact that growth-oriented incentives either created or aided in the development of unsustainable production systems. In a similar vein, sustainability incentives may, at least initially, make it more difficult to meet growth goals. Furthermore, domestic producers may face a competitive disadvantage when compared to foreign producers who are exempt from internalizing environmental costs due to proper social costing of production inputs.

7.5.1. Command and Control Versus Economic Incentive Schemes

Standards for command and control (C&C) are typically designed to govern the execution of a particular task or group of tasks. Effective C&C typically requires both the eventual sanctioning of trespasses and the monitoring of compliance. The main drawbacks of the C&C approach are its excessive restriction, lack of flexibility, inability to adjust on a case-by-case basis, and tendency to impede technological advancement. Furthermore, producers are not incentivized by the regulations supporting the C&C approach to meet standards that are higher than those mandated by law. Although C&C is frequently criticized for these reasons, government organizations frequently use it, and occasionally the industry even requests it. Regulations are created in accordance with public administration logic, frequently without regard to their enforceability. Politically speaking, "something has been done," and since everyone is held to the same standard or norm, it creates a feeling of justice. However, implementation is frequently hidden by the political arena's persistent weakness in results monitoring and accountability.

An alternative method is to use incentives. The purpose of incentives is to signal public objectives while allowing for some latitude in individual and group decision-making to address them, rather than to outright prohibit or permit. Incentives operate indirectly through factors that influence individual and group decision-making, like the pursuit of profit or adherence to norms. Market or social forces can be highly effective means of influencing the overall result of individual actions in the direction of goals that have been jointly established.

Different kinds of incentives can be developed in isolation or in combination:

- improving the institutional framework (definition of rights and participatory processes);
- developing collective values (education, information, training);
- creating nonmarket economic incentives (taxes and subsidies); and
- establishing market incentives (tradable property/access rights; eco-labelling).

Each of these instruments depends on command and control in one way or another. Properly defined and practically enforced property rights are necessary to create the conditions for an efficient market over them. Likewise, certification standards must be set and followed in order to create a market-based incentive for eco-friendly production practices through product eco-labelling. Therefore, incentives plus command and control should be viewed as a continuum of policy means, with relative advantages or disadvantages depending on what they are intended to achieve, rather than as two distinct approaches. The variety of incentive tools that are currently available is most likely underutilized, with a persistent bias toward command and control.

7.5.2. Crafting Institutions to Improve the Structure of Rights and Stakeholders Participation

The first arena to provide incentives is to define and enforce rights. The enforcement of human rights is as important for social sustainability as the definition and enforcement of rights of nature for environmental sustainability. Natural resource abundance has given way to scarcity over the past few decades as a result of numerous resources becoming irreversibly lost, depleted, or deteriorated. An important contributing factor to the unsustainable use patterns of many natural resources has been free and unrestricted access, or simply the absence of enforcement of access rights. The main points of contention brought up by opponents of specific aquaculture practices are the preservation of valuable ecosystems and landscapes, the preservation of water quantity and quality, the accessibility of the shore, and the prevention of land erosion. Numerous conflicts between potentially conflicting interests might be avoided if fundamental guidelines for impartially resolving disputes between them were defined beforehand.

Private and public promoters of aquaculture development would benefit from legislation on integrated coastal area management that would define access rights and limitations to different types of activities, as well as recognize fundamental individual rights (like the right to access the shore or water with certain properties). This would enable them to plan their activities in a more secure and informed manner. Well-defined individual or collective rights serve as a motivator because their holders, whether they be on the side of the aquaculture promoter or another interested party, can assert their claims in court or utilize them as a tool for persuasion. How to balance the recognition of the historical rights of the first entrant or user, the rights of the best economic offer, and the rights of future generations is a crucial and frequently contentious issue when defining and allocating use rights to natural resources.

The establishment of private property regimes has limitations for many types of environmental resources. These restrictions are brought about by the interdependence of several natural resources' functions as well as the high expense of enforcement. Thus, under a regime of private property, nature is frequently difficult to appropriate. For the purpose of protecting land and landscape, managing fresh, brackish, and marine waters, and preserving biodiversity and important ecosystems, new approaches to defining common property rights are required. In addition to the legal definition of rights, other factors that must be taken into account include the design of institutions and decision-making procedures, the organization of interest representation, and the creation of enforcement and monitoring programs. The ability to enforce rights regimes is the primary challenge in their design and implementation. However, this challenge shouldn't stop deliberate attempts to create, openly debate, and enact these rights and obligations. Although it sounds good in theory, private property rights depend on a few key elements to be met in order for the market to distribute them.

7.5.3. Fees, Charges, Taxes and Subsidies

The goal of economic tools such as taxes, fees, charges, and subsidies is to "get prices right" for the inputs and/or outputs used in production. Real social and environmental costs (or benefits) of goods and services may be distorted by government price controls and price support programs, as well as by insufficient or nonexistent markets, or by market failures. Economic tools directly impact the costs and benefits of private enterprise. They aim to persuade people and businesses to adopt more environmentally and socially responsible behaviors.

Applying economic instruments can be challenging because it requires careful design of the tax or levy to communicate to producers and consumers the goal of the measure. The fundamental idea behind a pollution tax or charge is to make up for the harm caused to the victim. The 1992 Rio Declaration on Environment and Development and the "polluters pay principle," which was embraced by Organization for Economic Cooperation and Development (OECD) nations in 1972, are both based on this logic. Its roots can be found in the arguments made by Pigou, who suggested enacting a tax to carry it out. The "Pigouvian tax," as it is aptly named, ought to be established at the pecuniary value of the marginal harm that pollution causes at the "optimal" pollution point.

Optimal pollution is typically defined as a level at which the costs of further pollution reduction exceed the benefits to the environment, society, and economy. It is not always zero pollution. Implementing such a tax has its difficulties:

- it is often next to impossible to determine the optimal pollution level because of measurement and valuation problems of the damage caused and cost of mitigation and clean-up; and
- it is also difficult to calculate and apply the tax in proportion to the value of the damage caused.

In actuality, environmental levies and taxes may be imposed on closely related inputs or outputs rather than the polluting material released or natural resources utilized.

Like all instruments, economic instruments require some capacity for oversight and enforcement. In order to function effectively, the authority must be able to keep an eye on the phenomena that serve as the basis for tax computation in addition to collecting the tax or fee. Since individual farms are highly decentralized, it is difficult to implement such control. As a result, taxes can be more easily collected at the level of production outputs, where they can be imposed directly at the source of pollution (pesticides, antibiotics, and feed). But as a result, regardless of the actual harm done, all types of economic activity utilizing these inputs will be subject to a uniform tax. This makes the tax a far less effective tool for enacting policy. In the worst case scenario, a tax might have unforeseen consequences that make environmental damage worse rather than better. To encourage farmers to implement more effective feeding practices that lead to low feed conversion ratios and low organic effluent loading per unit of production, think about taxing pellet feed. While some farmers might be motivated to cut costs on pellet feed, others might decide to switch to low-quality, non-taxable alternatives with high feed conversion ratios.

Apart from the previously mentioned implementation issues, environmental taxes are criticized in aquaculture, as they are in any other industry, due to their impact on competitiveness and their tendency to discourage innovation by increasing production costs. The following reasons make these arguments potentially untenable. A more stable and higher production level for all farms can be achieved by a well-thought-out and executed environmental tax that, for example, lowers water pollution.

Additionally, farmers will look to technological advancements to minimize their usage of the taxed production input. There isn't much evidence to refute the use of economic instruments theoretically. Creating affordable plans that are both practically implementable and accomplish the desired goals is where the challenges lie.

Grants and subsidies have been extensively utilized to offset the high degree of risk associated with the establishment of aquaculture farms. They have been reasonably successful in many countries when assessed based on whether or not they have accelerated production growth. When assessing their effectiveness in promoting sustainable development, however, their record is less impressive for some varieties of aquaculture. If subsidy programs are to be kept in place at all, they seem to need to be reoriented toward achieving a more rational and efficient use of the environment and other resources. Targeted research to create more ecologically friendly production methods is one obvious area. Clearly addressing poverty alleviation is important, not only for humanitarian and social reasons but also because poverty contributes significantly to environmental degradation in developing nations.

7.5.4. Information Dissemination, Development of Normative Values and Other Tools of Persuasion

Shared cultural norms and values that govern behavior are a useful source of incentives that direct both individual and group decision-making. The issue is that when social groups experience rapid change and are exposed to powerful external influences, traditional value systems—which are grounded in shared knowledge and sustained by moral obligations and social sanctions—may disintegrate. These are frequent cases where traditional norms and values that are perceived as impediments to modernization are undermined by short-term economic and political interests. Maintaining or even reviving traditional sources of behavioral norms that can help to foster sustainability seems to require careful balance, as does not limiting the dynamic changes required for social and economic advancement. Establishing and promoting shared values is undoubtedly necessary to cultivate human resource development and build communal concerns about sustainability.

Education and, more broadly speaking, persuasion, are important factors. It is important to raise public awareness of potential hazards, train aquaculturists on the environmental and social aspects of their industry, and acquaint legislators and other decision-makers with these issues.

Producer and industry associations have taken on a more significant role in the development of production and product standards, codes of conduct, codes of best practices, and awareness-raising, education, and training in recent years. These initiatives, while occasionally extending the reach of established institutions and groups, are primarily a contemporary reaction to the well-known environmental, socioeconomic, and commercial challenges that certain kinds of aquaculture systems face. Industry codes that are voluntary can cut down on the amount of money that the government has to spend on regulatory actions. Furthermore, it is well known that assuring customers and middlemen—who include wholesalers and retailers—that a product is produced without harming the environment and in accordance with labor laws and human rights could boost demand for it in both domestic and foreign markets.

Coordination of the actions of decision-making entities, like individual farmers or cooperatives, is a key role of a code. Common issues can be handled by agreeing to act in a certain way. For example, keeping the water quality in shared water bodies that are used by multiple farmers intact. Compared to command and control, codes have the advantage of being more adaptable and easily "amendable." On the other hand, regulations that begin as suggestions or voluntary guidelines may have an extraordinary "legal career" in which they quickly transform into legally binding laws. This could occur because the vast majority of people who are impacted by the recommendations or guidelines agree with them.

7.5.5. Eco-Labeling

Governments, businesses, and nongovernmental organizations (NGOs) have all implemented eco-labeling programs for a variety of purposes and sectors. One thing they have in common is that consumers are being directed to consider product attributes other than price and required health and quality standards.

These characteristics can be linked to environmental and ecological goals as well as economic and social ones, such as fair trade, helping small farmers, and discouraging child labor, as well as goals relating to the quality of health products, like organic production.

At the 1992 United Nations Conference on Environment and Development (UNCED) in Rio de Janeiro, Brazil, governments agreed to "encourage expansion of environmental labeling and other environmentally related product information programs designed to assist consumers to make informed choices," recognizing the potential value of eco-labeling schemes in creating market-based incentives for environmentally friendly products and production processes. Customers are afforded the chance to articulate their ecological and environmental concerns via the products they select.

Price and/or market share differences between products with an eco-label and those whose producers choose not to pursue such labeling, or between products that meet the requirements to be eco-labeled, are predicted to arise from consumer preferences. A certification process based on a set of requirements is used to obtain the label (i,e., the desired standard). Potential differences in price and/or market share give businesses an economic incentive to certify their product(s).

According to their design and administration, eco-labeling schemes are often classified into:

- First party labeling schemes, emanating from individual companies based on their own product standards;
- Second party labeling schemes, established by industry associations for their members' products. Certification criteria are elaborated by the members, sometimes by drawing upon external expertise from academia and environmental organizations, and verification of compliance is achieved through internal certification procedures within the industry or employment of external certifying companies; and
- Third party labeling schemes, usually independent from the producers, distributors and sellers of the labelled products; while the criteria may be established through a negotiation process among the various interested parties, they are often motivated by the environmental objectives of the private initiators of such schemes.

Because of the increased confidence that private commercial interests won't compromise the criteria applied to the schemes and that strict compliance with them will be based on verifiable and impartial certification procedures, environmental organizations typically support eco-labeling schemes that are based on third party certification. Right now, the most alluring opportunity is provided by the quickly growing markets for organic food products, which have price premiums ranging from 30 to over 100%. These comparatively high premiums might not be solely—or even primarily—related to consumers' desire to purchase goods made using environmentally friendly technologies; rather, they might also be related to their desire to consume wholesome, uncontaminated food. There are currently no international guidelines that are specifically designed for the production of organic aquaculture; however, many of the requirements that are established for agricultural production can also be easily applied to aquaculture. It is anticipated that considerable modifications to production procedures and methods will be needed in order to meet the current international standards for organic food when it comes to fish and shellfish raised in aquaculture, particularly in systems that are intensive and semi-intensive.

There are worries that eco-labelling is a product-differentiation and marketing tactic that will eventually help large corporations with production, distribution, and marketing. Small-scale producers would not be able to take advantage of economies of scale in production as well as in product certification, labeling, and verification because they would have less access to expensive "best available technologies," particularly those in developing nations. Small farmers can compete in both domestic and foreign markets, according to experiences with organic farming. However, to take advantage of scale economies, they might need the right institutional and organizational arrangements (such as group certification, cooperative input purchases, and product marketing, among others).

7.6. DEVELOPING MARINE ECOTOURISM FOR A SUSTAINABLE BLUE ECONOMY

In addition to more general tourism activities, many of which are coastal, marine ecotourism is linked to and dependent upon other facets of the blue economy.

7.6.1. Defining Marine Ecotourism

Marine ecotourism and marine tourism share a common pool of natural capital, making them closely related. In 2014, it was estimated that the global marine tourism industry would be valued at USD 390 billion. Prior to COVID-19, it is estimated that 80% of all tourism took place in coastal regions. Coastal and marine tourism was expected to account for 26% of the global blue economy by 2030, having grown at a rate of 3 percent annually prior to 2020.

Determining the exact nature of marine tourism is difficult due to its complexity, which has evolved in tandem with technological advancements (such as helmet diving tours and submersible rides) and public interest. Products catering to specific markets are constantly being created; these range from art trails and underwater sculptures to shark diving and seafood festivals. Many people consider ecotourism to be a subset of nature-based tourism, which uses natural resources but typically gives conservation and education less of an emphasis.

Ecotourism aims to achieve economic, ecological, and social objectives. Learning, interpretation, low-impact activities that reduce adverse effects on the environment and society, community involvement and control, support for conservation, and adherence to indigenous people's rights and beliefs are typical features of ecotourism.

120 million tourists were served by marine ecotourism in 2016, with a global value of USD 50 billion. Sometimes the financial advantages of marine ecotourism are measured; see, for instance, Spalding et al. An estimated $19 billion is made annually worldwide by coral reef ecosystems, which draw highly paid ecotourists for their diving and snorkeling experiences. An additional $16 billion is made by "reef-adjacent" tourism, which includes beach and seafood-related activities. The expansion of ecotourism models is expected to be a future trend in marine tourism, according to experts in the blue economy (Spalding et al., 2017) (Table 7.1).

Table 7.1. *Marine Ecotourism Publications*

WOS					
Author	**Center/Institution**	**Number of Articles**	**Citations**	**Average**	**h-Index**
Kovačić, M.	Rijeka University, Rijeka, Croatia	13	67	5.15	5
Luković, T.	Dubrovnik University, Croatia	9	27	3	3
Saayman, M.	North-West University, Potchefstroom, South Africa	6	77	12.83	4
Favro, S.	Sveučilište u Splitu, Split, Croatia	5	22	4.4	4
Kizielewicz, J.	Gdynia Maritime University, Poland	4	13	3.25	3
Gračan, D.	Rijeka University, Rijeka, Croatia	4	2	0.5	1
SCOPUS					

Author	Center/Institution	No. of Articles	Citations	Average	h-index
Kovačić, M.	Rijeka University, Rijeka, Croatia	15	50	4.16	4
Luković, T.	Dubrovnik University, Croatia	11	22	2	3
Favro, S.	Sveučilište u Splitu, Split, Croatia	9	20	2.22	3
Saayman, M.	North-West University, Potchefstroom, South Africa	7	120	17.14	5
Gračan, D.	Rijeka University, Rijeka, Croatia	6	8	1.33	2
Gržetić, Z.	Hydrographic Institute of the Republic of Croatia, Split, Croatia	5	20	4	2

Note: Authors on WoS and Scopus by number of articles published.

Source: Martínez Vázquez RM, Milán García J, De Pablo Valenciano J. (2021). Analysis and trends of global research on nautical, maritime and marine tourism. Journal of Marine Science and Engineering, 9(1):93. https://doi.org/10.3390/jmse9010093.

7.6.1.1. Marine Ecotourism Impacts

In comparison to its terrestrial counterpart, the marine ecotourism industry is still poorly researched, and the supply side of the industry is noticeably understudied. However, a variety of impact-related studies have been carried out on the industry. It's common knowledge that marine ecotourism is a high-value, low-impact tourism industry that can boost financial incentives for marine conservation. It has been discovered that tourists who behave sustainably and show concern for the local community spend more than typical tourists. Shark divers, or marine ecotourists, spent 21% more in the area than the typical visitor, according to research by Topelko and Dearden. Studies reveal an increasing desire for environmentally friendly travel options and visitor experiences. Among 2,000 Western consumers surveyed, 80% expressed interest in eco-friendly or sustainable vacations.

Visitors pay to interact with the fauna in marine and aquatic ecosystems through the use of marine ecotourism. Many times, the money raised from this source goes directly toward protecting marine areas, advancing conservation science, and promoting research. Visitors' appreciation of conservation is further increased by up-close experiences with the marine environment made possible by ecotourism. This is becoming more and more crucial since humans need a stronger connection to the natural world in order to think about how their actions affect the environment.

7.6.2. How Does Marine Ecotourism Support a Blue Economy?

Many people believe that the blue economy concept embodies the sustainable use of ocean resources for ecological and social well-being. Government priorities for the tourism industry to restore and strengthen natural capital and support "thriving and sustainable regions" are directly supported by the development of marine ecotourism in Aotearoa New Zealand which is linked to the blue economy. In addition to supporting a blue economy that benefits the local community, marine ecotourism offers financial incentives for bettering ocean management. Despite this, policymakers in New Zealand undervalue and ignore marine ecotourism, which has not reached its full potential.

Ecosystem services are all of the socioeconomic advantages that come from the natural world. Intending to inform stakeholders about the benefits that nature offers, the Ecosystem Services Approach (ESA) considers a comprehensive range of possible advantages that humans may derive from a particular ecosystem. A better grasp of the benefits and trade-offs involved, as well as the investments required, can be obtained by using an ESA approach to model scenarios and cumulative effects. In actuality, however, decision-making related to tourism is dispersed among several departments and does not take the ecosystem as a whole into account when making decisions.

Due to the many feedback loops and connections that bind ecosystems together, decisions are frequently made that could result in the loss of a single habitat and have an impact on many stakeholder groups. The Environmental Science Association (ESA) argues that all decisions should take into account the benefits that ecosystems offer, including to human welfare.

Marine ecosystem services include those provided by mangroves, coral reefs, and sea grass. The socioeconomic value of larger coastal and marine areas is increased by conserving the abundant marine ecosystem services that are often concentrated in small geographic areas.

Activities that take place on the ocean and along the coast are inextricably linked to other sectors of the blue economy, including marine ecotourism. Therefore, it is crucial to adopt a comprehensive disciplinary perspective in order to comprehend the dynamics and difficulties the industry faces. For instance, detrimental practices like Navy sonar, runoff, and algal blooms brought on by extractive industries all have an adverse effect on the marine ecosystem and deteriorate the "product" for marine ecotourism. Similarly, because fisheries development has cumulative and circular effects on the ocean, it affects marine ecotourism. For instance, a decrease in bird populations and fewer whale sightings were observed in the Scottish Isles as a result of rising sea temperatures brought on by climate change and industrial sand eel fishing, which is a major source of food for pelagic birds and mink whales. These factors also had an impact on the region's coastal and marine ecotourism industry. On the other hand, tourism can supplement declining fishing revenue and serve as a means of preserving fishing customs and cultural heritage. Perhaps the most well-known and lucrative marine ecotourism destination in the world, Australia's Great Barrier Reef, is in danger due to the connections between climate change and coral reef bleaching.

7.6.3. Marine Ecotourism and Ecosystem Management

Access to marine resources is necessary for marine ecotourism, which is becoming more and more acknowledged as a key component of managing marine ecosystems. Since it offers a comprehensive perspective of all effects, including those caused by humans on the marine environment, ecosystem-based management (EBM), which has been pushed as a strategy for both healthy seas and sustainable marine tourism, is a widely used tool for ocean management.

A recent assessment of the state of ocean governance, while excluding tourism specifically, identifies the main obstacles to ocean governance as the variety of human interests in the ocean and the uncertainties in scientific understanding. They contend further that in addition to established institutions, rules, and knowledge sources, a more sustainable form of ocean governance in the future will require the inclusion of community interests and stakeholder engagement. Among these varied interests, marine tourism is acknowledged as playing a significant role. According to Cánovas-Molina and García-Frapolli, tourism plays a significant role in the conflicts surrounding MPAs. It has the ability to both increase ecological conservation awareness and endanger it by exposing MPAs to more human interaction.

Although summaries of the potential for enhanced marine management stress the importance of interacting with a variety of stakeholders, Lachapelle et al., alert natural resource planners to the fact that changes in the use of human resources will increase the pressure on them to accomplish adaptive management in the face of goal-related conflict. "Sectoral interplay," or the difficulties in balancing the goals and directives of regulatory bodies with the more limited interests and focuses of other stakeholders in EBM, environmental justice, the disparities in the costs and effects of nature-based tourism on communities and other stakeholders, and the necessity of empowering community voices, are some of the specific obstacles to successful ecosystem management that are mentioned (Phelan et al., 2020).

The problem of interactions between the various interest groups involved is the most commonly addressed challenge to ecosystem management involving tourism; this issue is also brought up in the development of marine ecotourism policies. It is reported that social exploitation, disempowerment, and environmental degradation result from low or poorly managed stakeholder engagement in ecosystem management. In these situations, the rights, well-being, or other components of the ecosystem are impacted by the interests of more powerful actors. Aiming to lessen this imbalance, proposed frameworks also acknowledge this problem (Dimmock, et al., 2014).

The specific conflict between economic and other wellbeing objectives has been found to have significant implications for ecosystem management that involves tourism, which is related to the contested interests of participants in ecosystem management. Bogadottir records the detrimental effects on local tourism and communities of government initiatives to support Blue Growth in the Faroe Islands. Bogadottir's argument is predicated on the effects of an expanding social metabolism, which is fueled by an emphasis on economic expansion and leads to increased consumption and disruption of the marine ecosystem. Furthermore, Llausas et al., examine why policies based on the idea of carrying capacity have failed to stop tourists from overusing an MPA because of the financial interests of stronger players and the way that current tourism practices are ingrained. However, tourism is also thought to offer a way to put people's well-being first while giving people and communities a financial boost. Ateljevic offers the possibility of combining regenerative principles with tourism to create a more sustainable transition, whereby tourism guarantees a market for products originating from regenerative agriculture, even though he does not provide a marine example. According to Knowles, adventure tourism has a similar chance to attract well-off travelers who care about the environment and are willing to spend money on eco-friendly activities that boost the local economy (Bogadottir, 2020).

REFERENCES

1. Bogadottir, R. (2020). Blue growth and its discontents in the Faroe Islands: An island perspective on blue (de)growth, sustainability, and environmental justice. *Sustainability Science, 15*(1), 103–115.
2. Daly, H. E. (1990). Toward some operational principles of sustainable development. *Ecological Economics, 2*(1), 1–6.
3. Dimmock, K., Hawkins, E. R., & Tiyce, M. (2014). Stakeholders, industry knowledge and adaptive management in the Australian whale-watching industry. *Journal of Sustainable Tourism, 22*(7), 1108–1121.
4. Ekins, P., Simon, S., Deutsch, L., Folke, C., & De Groot, R. (2003). A framework for the practical application of the concepts of critical natural capital and strong sustainability. *Ecological Economics, 44*(2, 3), 165–185.
5. Hilborn, R., Fulton, E. A., Green, B. S., Hartmann, K., Tracey, S. R., & Watson, R. A. (2015). When is a fishery sustainable? *Canadian Journal of Fisheries and Aquatic Sciences, 72*(9), 1433–1441.
6. Matsuoka, T., Kasu, J., & Nagaleta, H. (1995). Catch and fish-loss model for vertical longline fishing with particular emphasis on injury of unharvested fish. *Fisheries Science, 61*, 760–765.
7. Matsuoka, T., Ono, T., Kawamura, G., & Ishida, H. (2004). Development of species-selective semidemersal trammel-net for cuttlefish towards reduction of incidental catch of sub-legal flounder. *Memorandum of the Faculty of Fisheries, Kagoshima University, 53*, 19–25.
8. Nakashima, T., & Matsuoka, T. (2004). Ghost-fishing ability decreasing over time for lost bottom-gillnet and estimation of total number of mortality. *Nippon Suisan Gakkaishi, 70*, 728–737.
9. Nakashima, T., & Matsuoka, T. (2005). Ghost-fishing mortality and fish aggregation by lost bottom-gillnet tangled around fish aggregation device. *Nippon Suisan Gakkaishi, 71*, 178–187.
10. Neumann, B., Ott, K., & Kenchington, R. (2017). Strong sustainability in coastal areas: A conceptual interpretation of SDG 14. *Sustainability Science, 12*(6), 1019–1035.
11. Phelan, A., Ruhanen, L., & Mair, J. (2020). Ecosystem services approach for community-based ecotourism: Towards an equitable and sustainable blue economy. *Journal of Sustainable Tourism, 28*(10), 1665–1685.
12. Piedad-Pascual, F., Cruz, E., & Sumalangcay, A. Jr. (1990). Supplemental feeding on *Penaeus monodon* juveniles with diets containing various levels of defatted soybean meal. *Aquaculture, 89*, 183–191.

13. Pimentel, D., & Pimentel, M. (2003). Sustainability of meat-based and plant-based diets and the environment. *American Journal of Clinical Nutrition, 78*(3), 660–663.
14. Rice, J., Arvanitidis, C., Borja, A., Frid, C., Hiddink, J. G., Krause, J., Lorance, P., Ragnarsson, S. Á., Sköld, M., Trabucco, B., Enserink, L., & Norkko, A. (2012). Indicators for sea-floor integrity under the European Marine Strategy Framework Directive. *Ecological Indicators, 12*(1), 174–184.
15. Ridler, N. (1998). Aquaculture and the role of government. *Aquaculture Europe, 22*(4), 6–12.
16. Roxas, F. M. Y., Rivera, J. P. R., & Gutierrez, E. L. M. (2020). Framework for creating sustainable tourism using systems thinking. *Current Issues in Tourism, 23*(3), 280–296.
17. Shreck, A., Getz, C., & Feenstra, G. (2006). Social sustainability, farm labor, and organic agriculture: Findings from an exploratory analysis. *Agricultural and Human Values, 24*(4), 439–449.
18. Smythe, T., Thompson, R., & Garcia-Quijano, C. (2019). The importance of brokers in collaborative marine ecosystem-based management of social-ecological systems. *Coastal Management, 47*(5), 473–491.
19. Sneddon, C., Howarth, R. B., & Norgaard, R. B. (2006). Sustainable development in a post-Brundtland world. *Ecological Economics, 57*(2), 253–268.
20. Spalding, M., Burke, L., Wood, S. A., Ashpole, J., Hutchison, J., & Ermgassen, P. (2017). Mapping the global value and distribution of coral reef tourism. *Marine Policy, 82*, 104–113.
21. Tacon, A., & Metian, M. (2008). Global overview on the use of fish meal and fish oil in industrially compounded aquafeeds: Trends and future prospects. *Aquaculture, 285*, 146–158.
22. Verdegem, M., & Bosma, R. (2009). Water withdrawal for brackish and inland aquaculture, and options to produce more fish in ponds with present water use. *Water Policy, 11*, 52–68.
23. Verdegem, M., Bosma, R., & Verreth, A. (2006). Reducing water use for animal production through aquaculture. *Water Resources Development, 22*(1), 101–113.
24. Watanabe, H., & Ono, S. (2000). Analytical review of the elaboration of FAO Code of Conduct for Responsible Fisheries. *Report of Tokyo University of Fisheries, 35*, 153–176.

Chapter 8

Conservation Strategies and Management Approaches

Contents

INTRODUCTION

Aquatic biodiversity is threatened by many factors. Among the main offenders are invasive species, pollution, overfishing, habitat destruction, and climate change. Aquatic ecosystems are disrupted by human activities like overfishing, coastal development, and pollution release, which results in a decline in biodiversity and the degradation of important habitats. It is the responsibility of all people, groups, and governments around the world to work together to preserve and promote biodiversity in our aquatic environments. We can become true stewards of the blue planet by appreciating the importance of aquatic biodiversity, addressing the threats it faces, and taking proactive measures to preserve and restore our water ecosystems.

The practice of safeguarding and maintaining ocean ecosystems and biodiversity is known as marine conservation, one of the many marine sciences. It encompasses not just the ocean but also mangroves, estuaries, and other threatened coastal habitats that are vital to marine life. Conservation strategies are essential for ensuring the health and vitality of oceans, which are vital to the survival of many species and the overall well-being of the planet. These tactics center on lessening the effects of human activity on marine ecosystems, encouraging the sustainable use of marine resources, and defending important species and habitats. Strategies for effective marine conservation can combine a variety of techniques and are customized to the unique requirements and environmental circumstances of the affected area. For marine conservation efforts to be successful, cooperation between the government, non-governmental organizations, business community, and local communities is frequently essential. Based on the best available science, conservation efforts ought to be incorporated with those of other industries, like tourism and energy.

8.1. IN-SITU AND EX-SITU CONSERVATION METHODS

Preservation of biodiversity is essential to preserving the ecological equilibrium and guaranteeing the long-term viability of life on Earth. Two main strategies have been identified for this endeavor: ex-situ and in-situ conservation. Although they both seek to protect species and ecosystems, they do so through various approaches and plans of action.

8.1.1. In-Situ Conservation

The preservation of species and ecosystems in their native environments is referred to as "in-situ conservation." Its main objective is to preserve and manage entire ecosystems, along with the variety of plants and animals that they support.

This method highlights the significance of preserving the complex interactions that exist between organisms and their surroundings while also acknowledging the intrinsic value of biodiversity. The creation and administration of protected areas, such as national parks, wildlife sanctuaries, and marine reserves, are among the tactics used in in-situ conservation. Numerous species find refuge in these places, which offer them sufficient habitat and defense against human activities like pollution, poaching, and habitat destruction. Furthermore, community-based conservation initiatives—in which local communities actively participate in the management and protection of natural resources, promoting sustainable practices that benefit both people and the environment—are frequently included in in-situ conservation efforts.

The capacity of in-situ conservation to maintain the organic evolutionary processes that gradually sculpt ecosystems is one of its main benefits. In-situ conservation contributes to the preservation of genetic diversity and permits species to evolve in response to shifting environmental circumstances by allowing them to interact and adapt within their natural environments.

8.1.2. Ex-Situ Conservation

Contrarily, ex-situ conservation refers to the preservation of species away from their native environments. This strategy is usually used when there are immediate threats to the species, such as habitat loss, overexploitation, or disease outbreaks, which cannot be sufficiently addressed by in-situ conservation alone. Programs for captive breeding, botanical gardens, seed banks, and gene banks are examples of ex-situ conservation techniques. These establishments function as safeguards against extinction by acting as archives for genetic material or endangered species. For example, captive breeding programs raise threatened species in controlled settings in preparation for their eventual release back into the wild when threats are reduced or their natural habitats are restored.

Ex-situ conservation is essential in keeping species from going extinct, but it is not without its difficulties. In captive breeding programs, preserving genetic diversity and avoiding inbreeding depression are major concerns. Furthermore, it can be difficult to release captive-bred animals back into the wild because they might not have the survival instincts needed to thrive there. Integration of In-Situ and Ex-Situ Conservation: The complementary nature of in-situ and ex-situ conservation approaches has gained more attention in recent years. It has been shown that integrated conservation strategies—which incorporate aspects of both approaches—are more successful in preserving biodiversity.

By providing individuals for reintroduction programs, genetic material for breeding programs, and scientific research to guide management decisions, for instance, ex-situ conservation efforts can support in-situ conservation efforts.

On the other hand, in-situ conservation guarantees the long-term survival of species within their native ecosystems, serving as the ultimate objective for ex-situ efforts.

8.2. MARINE PROTECTED AREAS (MPAS) AND FRESHWATER RESERVES

An area of the ocean where human activity is restricted by the government is known as a marine protected area (MPA). Numerous MPAs permit uses of the land that do not harm the environment. Some have fishing bans. Some don't even permit people to enter the area. Because of the numerous risks that the ocean and its inhabitants face, MPAs have been established. Overfishing, littering, water pollution, and climate change are some of the threats facing the ocean. Numerous fish, marine mammals, and other sea creatures have seen population declines as a result of these threats. Marine parks, marine conservation zones, marine reserves, marine sanctuaries, and no-take zones are just a few of the many names that can be applied to marine protected areas. Globally, more than 5,000 MPAs have been created. As of 2023, they collectively occupy slightly over 8% of the ocean.

Numerous aquatic environments can have marine protected areas established. A few MPAs are located in open waters. Numerous MPAs guard the coast. Others discuss estuaries, which are rivers' entry points into the ocean. Freshwater and saltwater coexist in estuaries. MPAs also include some freshwater habitats, such as Great Lakes protected areas.

8.2.1. Goals of MPAs

The objectives of different MPAs vary. Many MPAs are primarily concerned with safeguarding marine environments and the diversity of life they support. For instance, many small islands and the surrounding waters are protected by the Galápagos Marine Reserve, which is located about 1,000 kilometers (600 miles) off the west coast of South America. This reserve has an amazing diversity of habitats, including mangrove swamps, where trees grow directly in salty seawater, coral reefs, and chilly ocean currents. 3,000 distinct plant and animal species can be found in the waters surrounding the Galápagos Islands, including uncommon species like the marine iguana (*Amblyrhynchus cristatus*), which is the only seagoing lizard in the world.

Shipwrecks and other historic sites are the focus of some MPAs. During the American Civil War in the 1860s, a warship named USS Monitor sank in a storm off the coast of North Carolina. To preserve the ship's remnants, the

USS Monitor National Marine Sanctuary was founded in 1975. It was the first national marine sanctuary in the country.

In order to guarantee that resources are sustainable and won't run out, other MPAs have been established. These MPAs guarantee that fish can reproduce and sustain healthy populations by imposing restrictions that prohibit overfishing. People are able to continue their way of life by fishing year after year thanks to this. Georges Bank, off the coast of New England (a region in the northeast of the U.S.) and Nova Scotia, Canada, was once one of the world's greatest fisheries. However, because of centuries of intensive fishing, the numbers of flounder, haddock (*Melanogrammus aeglefinus*), cod, and other species fell. Fish populations started to rise and fishing got better after the US and Canada created many MPAs.

8.2.2. Levels of Protection

The degree of protection offered by various MPAs varies. The most stringent kind of MPA forbids any human entry. This stops people from disturbing sensitive habitats in addition to stopping them from fishing. No-entry MPAs are frequently used for research and are typically small. The enormous Seaflower Reserve off the Caribbean coast of Colombia prohibits human access in certain areas. There are less stringent MPAs. While fishing and collecting are prohibited in no-take MPAs, visitors are still permitted to travel through the area and enjoy recreational activities like swimming and snorkeling. A small island located 18 kilometers (11 miles) off the coast of Belize, Central America, is home to Laughing Bird Caye National Park, which is entirely a no-take MPA.

While the area is protected in multiple-use MPAs, some fishing is permitted. Many national parks, such as Acadia National Park in the U.S. state of Maine, are multiple-use MPAs. Numerous MPAs are separated into various zones. Fishing is permitted in some areas, but entry may be completely prohibited in others. The Great Barrier Reef in Australia is home to one of the biggest MPAs in the world. There are zones within the Great Barrier Reef Marine Park. Both commercial and recreational fishing is permitted in some of these zones. Roughly one-third of the park is off-limits to fishing. Fish and coral populations have grown since these zones were established.

8.2.3. Establishing an MPA

Many MPAs are established by national governments. MPAs are also created by tribal, state, and local governments. For example, the U.S. state of California has established the Point Lobos State Marine Reserve to protect underwater canyons and kelp forests. The Quileute Tribe of the U.S. state of Washington works with the federal government to keep the Olympic Coast National Marine Sanctuary a sustainable fishery.

Sometimes, national governments collaborate to create a transboundary MPA. The Mediterranean Marine Mammal Sanctuary of the Pelagos was founded by France, Monaco, and Italy collectively. It includes both international waters and portions of the sea that are within each country's borders. In certain MPAs, the degree of protection is constant throughout the year. Others only forbid people from entering a region during specific times of the year, usually when important species are reproducing. In the Irish Sea, for instance, fishing is prohibited during the cod spawning season when the fish lay and fertilize their eggs. This aids in the population conservation of cod.

8.2.4. Benefits of Marine Protected Areas (MPAs)

The ecosystems within and surrounding Marine Protected Areas (MPAs) can greatly benefit ecologically from MPAs and MPA networks. According to studies, MPAs can increase the size of individual marine organisms as well as the variety and abundance of marine life, which in turn increases the resilience and productivity of the ecosystem as a whole.

8.2.4.1. Safeguard Biodiversity and Protect At-Risk Species and Habitats

The ecosystems they safeguard as well as other ecosystems can greatly benefit ecologically from the establishment of Marine Protected Areas (MPAs) and MPA networks. According to studies, MPAs can increase the size of individual marine organisms as well as the variety and abundance of marine life, which in turn increases the resilience and productivity of the ecosystem as a whole. Well-designed MPAs protect habitats essential to life cycle stages, including spawning, juvenile rearing, and feeding, and they also provide a haven for harvested species. These actions help to preserve the structure, function, integrity, richness, and resilience of ecosystems. Furthermore, species that depend on freshwater and marine environments frequently supplement nearby terrestrial protected areas (e.g., salmon.) and aid in the recovery and restoration of vulnerable species, environments, and ecosystems, such as glass sponge reefs. Additionally, mobile or migratory species may find a "stress free" haven and safe haven in MPAs. Marine life in MPAs has, on average, 4:5 times more biomass, 28% larger body size, and 21% higher species diversity than in other non-protected areas, according to a global synthesis of 124 fully-protected MPAs. Ecosystems that are more diverse have been demonstrated to be better able to endure stressful events and bounce back from them rather quickly. Biodiversity can act as a significant barrier against the effects of climate change.

8.2.4.2. MPAs Can Help Mitigate Against Climate Change

The current climate and biodiversity crises can be mitigated and better adapted to with the support of Marine Protected Areas (MPAs). By preserving biodiversity and safeguarding areas like salt marshes, seagrass meadows, and kelp forests that absorb carbon dioxide, neutralize storm surges, and protect coastlines from rising sea levels, MPAs serve as an ecological "insurance policy" against ecosystem collapse and contribute to climate change mitigation. Seagrasses, such as eelgrass, cover only 0.2 percent of the world's sea floor, but studies show that they contribute roughly 10% of the ocean's capacity to store carbon. In addition to protecting coastlines from erosion and contributing to the reduction of sediment in the water, seagrass meadows serve as vital habitat space for numerous species. It is astounding that seagrasses can absorb carbon from the atmosphere up to 35 times faster than terrestrial forests. When implemented, the Northern Shelf Bioregion MPA Network will protect large swaths of eelgrass meadows and kelp forests across the coast of B.C., helping to preserve the invaluable ecosystem services they provide such as mitigating climate change.

8.2.4.3. MPAs Help to Support Healthy Coastal Communities

Marine protected areas (MPAs) are an extremely effective tool for the conservation of marine biodiversity and ecosystems as well as for the sustainable management of marine resources, according to decades' worth of scientific research. The vital ecosystem services that human communities depend on and receive from the marine environment are preserved and improved by MPAs. Marine ecosystem services include: provisioning services (e.g., harvestable goods—fish, shellfish and seaweed for food, minerals, health products), cultural (e.g., non-material benefits—recreation, education, research, spiritual, cultural heritage and identity), regulating (e.g., climate regulation, waste treatment, erosion protection, oxygen production), and supporting services (e.g., biodiversity and gene pool protection, ecosystem process and lifecycle maintenance). The well-being of the individual and the community is substantially enhanced by each of these services.

The coastal communities that MPAs are established around are unavoidably impacted by their implementation. As a result, a significant amount of research has been done to determine the value and economic effects of MPAs on coastal communities. Although MPAs can have both advantages and disadvantages, the research is unequivocal: overall, well-designed and well-managed MPAs yield greater benefits and provide significant returns on investment.

From an economic perspective, MPAs contribute to B.C.'s "blue economy" that encompasses industries such as fisheries, tourism, shipping, and more.

According to research, MPAs can yield benefits of up to $20 for every dollar "invested" in them. Moreover, these benefits could materialize within five years. Tourism and recreation are B.C.'s second largest ($5.8 billion/year) and second fastest growing sector in our blue economy, after transportation. A wealth of wildlife and well-preserved ecosystems are extremely beneficial to this industry. Ecotourism and recreation-based businesses, when run sustainably, can support conservation efforts by creating revenue (e.g., for study purposes) and educating the general public. Recent studies on the non-financial advantages of MPAs indicate that they offer people a healing and restorative environment that can enhance wellbeing and have positive effects on mental health.

8.2.4.4. Increase The Commercial and Recreational Fishing Catches in Surrounding Areas

According to estimates from the Food and Agriculture Organization of the United Nations, 85% of marine fish stocks are either overfished or completely exploited. The main causes of these decreases are overfishing and poor management of fish stocks, and the effects of climate change are now being felt at a never-before-seen pace. This poses a serious threat to food security and social justice in addition to causing ecological disasters. For 3 billion people worldwide, the ocean serves as their main source of protein. In addition, access to marine resources and fisheries support the cultural fabric of many coastal communities. As reliable fish stocks become ever more inaccessible in B.C., the "devastating effects for…cultures and food security" pose a threat for many coastal communities.

Case studies from the literature on global conservation show how marine protected areas (MPAs) can benefit commercial and recreational fisheries. They achieve this by greatly boosting the quality and quantity of specific fish and shellfish species. Highly protected MPAs offer a safe haven for harvested species to grow larger and live longer by lowering or eliminating fishing pressure within their borders. Over their lives, larger, older fish can give birth to hundreds or even thousands more offspring, some of which will eventually find their way into unprotected waters. The "spillover effect" is a well-researched phenomenon that may aid in stock recovery.

According to a case study from the Pacific Northwest, lingcod laid 20 times more eggs and copper rockfish laid 100 times more eggs in an MPA boundary than they did in an unprotected "no-take" area. MPAs are an investment in sustainable fisheries and contribute to the preservation of food and cultural security for coastal communities and future generations by serving as a provisioning service for neighboring fisheries. MPAs can be an effective tool for conservation management by supporting the restoration and sustainability of fisheries, particularly when paired with improved fisheries management techniques.

The advantages of well-managed MPAs frequently outweigh the preliminary or estimated costs to coastal communities, fisheries, and other stakeholder groups. Depending on the lifespan and distribution rates of target species, the economic benefits of MPAs to fisheries—such as higher yields, higher revenues, and lower effort-related costs—may materialize in less than 10 years and be felt up to 100 kilometers from the MPA's borders. The World Economic Forum declared the following in 2020.

"As evidence of the benefits of MPAs continues to emerge, the people and organizations governing them at local, national and international scales need to learn and evolve."

8.2.4.5. Provide Opportunity for Co-Governance and Protect Culturally Important Sites

Over the course of more than 10 years, 17 First Nations, the B. C. and the Canadian government have been collaborating to create the Northern Shelf Bioregion Marine Protected Area Network, Canada's first marine protected area (MPA) network. When put into practice, this network of MPAs will uphold Indigenous legal traditions within the territories that they have been governing and sustainably managing for thousands of years, offering a chance for co-governance and reconciliation. Equal power sharing between the Canadian and Indigenous governments is ideal in a co-governed MPA. It may be possible to move to more exclusively Indigenous-governed protected areas in the future if management is based on both Canadian and Indigenous law. Indigenous nations have the inherent right to govern over both marine and terrestrial areas, and they are able to designate MPAs and other protected areas under their own laws.

MPAs can be established to support traditional harvesting practices and conserve and protect areas that hold spiritual or cultural significance for coastal First Nations, in addition to safeguarding species and ecosystems. MPAs can be a useful tool to guarantee long-term food security for coastal communities by safeguarding culturally significant features and First Nations' harvesting areas.

8.2.5. Economic and Social Implications

MPAs are viewed by society as a public investment in marine resources. A benefit-cost framework, which evaluates whether the potential benefits of protection—adjusted to account for risks—outweigh the potential costs, can serve as a guide when deciding whether to commit resources. The possible advantages and disadvantages of MPAs are realized over the long and short terms, just like other public investments. For instance, blocking off a region that has typically produced a sizable catch would likely result in a decrease in the overall catch, at least initially. Total catch levels may rise as fish populations in the MPA start to rebound and spillover to the remaining fishable waters grows. How long might it take for this response to manifest? This will depend, partly, on the fish's biological makeup (e.g., quickly expanding or slowly expanding stocks) and the degree to which the fishermen adapt to the biological and economic circumstances.

The potential advantages and expenses of MPAs are split into the following extractive (e.g., fishing (both commercial and recreational) and non-extractive users (e.g., eco-tourists) and administration. Table 8.1 provides examples of the cost and benefit categories for this group of stakeholders.

Table 8.1. *Comparison of law-making details of the CAMLR Convention, the OSPAR Convention, and the BBNJ Agreement*

Instruments	Purposes	Entered Into Force	Initial Parties	MPAs	Management Body
The CAMLR Convention	Conservation of Antarctic marine living resources	1982	Argentina, Australia, Belgium, Chile, the French Republic, the German Democratic Republic, the Federal Republic of Germany, Japan, New Zealand, Norway, Poland, the Republic of South Africa, the Soviet Union, UK, and USA	The South Orkney Islands South Shelf MPA; the Ross Sea region MPA	CCAMLR

The OSPAR Convention	Prevention and elimination of pollution of the marine environment, or with respect to the protection of the marine environment against the adverse effects of human activities	1998	Belgium, Denmark, Finland, France, Germany, Iceland, Ireland, Luxembourg, Netherlands, Norway, Portugal, Spain, Sweden, Switzerland, UK, and EU	The OSPAR network of MPAs	OSPAR Commission
The BBNJ Agreement	Conservation and sustainable use of marine biological diversity of areas beyond national jurisdiction	No in force	This agreement is now not open to signature, but many states and organizations participated in Conferences [32]	no	COP

Source: Jiang R, Guo P. (2023). Sustainable Management of Marine Protected Areas in the High Seas: From Regional Treaties to a Global New Agreement on Biodiversity in Areas beyond National Jurisdiction. Sustainability, 15(15), 11575. https://doi.org/10.3390/su151511575.

8.2.5.1. Benefits to Extractive Users

If there is a substantial spillover effect, MPAs that stabilize or grow fish populations inside their borders may also serve a similar purpose outside the protected area. Consequently, there may be less fluctuation in the overall catch levels or a rise in the total catch over the long term. Ultimately, the net effect on catch is determined by the fish stock's condition and the area set aside's relative importance in the overall catch prior to the closure. For instance, there is less chance that the overall catch will rise if the area contributed a sizable portion of it. This is because there is a higher threshold level that the biological spillover effects must cross. Conversely, there is a greater chance of increasing the total catch if the catch level represents a negligible portion of the total catch, perhaps as a result of overexploitation prior to closure. It has been demonstrated that, in an overfished system where fish stocks are neither too sessile nor too mobile, long-run total catch levels should rise after a region is set aside from harvesting. By altering the make-up of the catch, MPAs can also raise a fishery's market value. Traditional area and time closures have been used, among other things, to protect stocks and increase market value by changing selectivity. For example, if a closure is imposed in a region where a significant fraction of the stock is young, selectivity may be affected. This will cause the fishing effort to be diverted to other locations or to a different time of day, which should result in a larger catch and possibly higher market prices. If the shift in catch composition from smaller to larger fish is accompanied by a move to a more valuable product form (such as frozen to fresh product), there is still another potential source of increased revenue.

8.2.5.2. Costs to Extractive Users

The biological advantages of MPAs have been extensively studied in the literature, but the potential costs have not gotten as much attention. Fishing is a complicated activity that depends on a variety of variables, such as the kind of boat and equipment used, the species being targeted, the density of the stock, the time of day, and the areas fished. The inputs used in this process have an effect on fishing expenses, which in turn has an effect on the amount of activity. For example, one could imagine that the cost per unit of catch could be reduced if there was less congestion at a particular fishing ground. Geographical features that forbid the use of specific types of gear or stronger currents that increase fuel consumption could also cause costs to vary amongst fishing grounds. How the introduction of MPAs might impact these kinds of costs is a crucial question in the ongoing discussion.

Lowering the area available for fishing suggests that there may be more vessel congestion on the remaining grounds, at least initially. The effects of congestion may lead to higher fuel consumption and capital costs (e.g., fish-finding apparatus). Significantly lowering the number of fishable waters may also result in a rise in disputes over resource allocation and entanglements in gear amongst resource users. For instance, a trawler that has been forced to relocate by an MPA may encounter trouble if it enters a region that is usually reserved for fixed-gear fishermen. In this case, the fixed-gear fishermen—who might not have been directly impacted by the closures—as well as the displaced trawler incur higher harvesting costs. The implementation of an MPA may cause fishing pressure to shift from one species to another, increasing competition for the catch of the second species, so that congestion effects are not limited to the fishery for which the closure was imposed.

When choosing a fishing spot, a fisherman considers a variety of factors, such as the season, species to target, anticipated time at sea, anticipated catch rates, costs associated with transportation, search and rescue, location-specific expenses, dockside prices, and weather-related events. We talk about transportation costs, including time spent traveling and fishing, to get an idea of how these factors might alter following a closure. Fuel consumption will increase and vessels' time spent in the water with fishing gear will decrease after an MPA is sat. This is especially true if vessels spend more time steaming to and from the fishing grounds. Skippers may be encouraged to invest in more capacity, such as bigger holds, better fishing equipment, and more horsepower, by the increases in steam-time. Thus, an MPA may unintentionally increase capital expenditures in the fishery at a time when managers are looking for ways to reduce capacity—for example, by implementing a vessel buy-back program.

The time and fuel used to search for fish account for a sizable portion of the total variable harvesting costs in many fisheries. A fundamental tenet of harvesting cost modeling is that, all other things being equal, larger stock levels imply lower search costs. The term "stock effect" is frequently used to describe this phenomenon. MPAs may be able to lower this portion of harvesting costs in fisheries where there is a substantial stock effect. For example, there is a chance that stock levels will rise across the fishery if the biological spillover effects are significant and widespread. One would anticipate a drop in searching expenses in this situation. If fishermen only focus in the areas surrounding the MPAs, searching costs may also go down if the MPAs have a significant spillover. However, after MPAs are established, searching costs may rise if the spillover effects are minimal. For instance, fishermen who have been displaced and are familiar with fish concentrations in their area may need to invest more time and energy in learning about stock concentrations and oceanographic conditions in the remaining unprotected areas.

Given that fishing is one of the riskiest professions in the world, fishery managers should be aware that some regulations may make the industry even riskier. If a sheltered lagoon or other near-shore habitat is closed, fishermen may be forced to fish in farther-off waters. Fishermen will be more vulnerable to storms if they operate farther offshore because it will take longer to return to port. If inshore fishermen, who are affected by the MPA, are unable to obtain the funding required to upgrade their boats and equipment before going offshore, these risks may worsen. The potential for increased occupational risks and higher costs from an increase in search and rescue missions stems from the combination of inadequate vessels and inexperience of the displaced fishermen who are "forced" to operate in new, riskier environments.

8.2.5.3. Benefits to Non-Extractive Users

Studies reveal that marine protected areas (MPAs) can boost biodiversity and enable a marine ecosystem to revert to its "natural" state (e.g., Parrish and Boersma, 1999). These modifications to the marine ecosystem might be appreciated by non-extractive users. Its utility may be increased, for instance, by designating a specific marine area, such as a park or sanctuary, to preserve biodiversity. This area could be utilized for diving and photography. Even though they might not use the area, people who value the ocean's existence might find it appealing to improve its health. If an MPA does draw more tourists, the local economy may benefit from more jobs, income, and tax income. It is even possible that potential losses from fewer catches for commercial or recreational purposes as a result of the closure could be offset by potential increases in tourism-related revenue. It goes without saying that an MPA's location and environment would have a significant impact on the size of these costs and benefits. For instance, a protected area offshore that is primarily home to species that live on the ocean floor is unlikely to have much of a tourism potential, but a closure of a coral reef might. Generally speaking, an MPA can offer protection as well as the potential for financial gains to industries not directly related to commercial fishing. Actually, research conducted in Caribbean marine parks revealed that both could be achieved with the right management.

8.2.5.4. Costs to Non-Extractive Users

Critical questions about sustainable use are raised when the long-term carrying capacity of a designated area, like a marine park, is unknown. The very biological benefits—such as abundant biodiversity and more fish—and cultural benefits that closing the area to extractive users was meant to protect could be diminished if tourism is allowed to continue unchecked. The popularity of the site may also have a negative effect on the utility that visitors derive from it, since increased traffic leads to higher levels of congestion on the site. If there is a direct relationship between the demand to visit a site and site quality (e.g., the probability of seeing large fauna and flora, or a special coral reef ecosystem) then congestion and the detrimental impacts associated with it could lead to fewer visitors and lower economic benefits. In the end, unrestrained tourism may diminish all values obtained from the MPA, suggesting that siting choices should be made in conjunction with management strategies intended to minimize the influence of non-extractive users.

The possible changes brought about by cutting off a sizable portion of economically viable fishable waters may also have a negative impact on people who cherish traditional coastal fishing communities. A tiny, remote fishing village that concentrates on a fishery that is found on a near-shore coral reef could serve as an example. A sizable area may be protected if the near-shore reef's in situ biodiversity is to be preserved by the MPA. Local fishermen who might not have the financial means to operate in waters further offshore could be forced out of business as a result of the closure. Working in the tourism industry may be the best option for certain communities, but as the economy transitions from fishing to tourism, traditional ways of life may be lost. Additionally, if increased tourism and negative effects like traffic are linked to the MPA, community members who do not actively engage in fishing may feel as though they have also lost their traditional way of life.

8.2.5.5. Management Benefits

MPAs might offer "undisturbed" regions that open up new avenues for scientific study and act as a safeguard against mistakes made by managers. Researchers contend that these regions can serve as controls for tracking and analyzing fish population recovery, which will enhance estimations of population parameters (e.g., more accurate growth and natural mortality rate estimates) and stock evaluations, which currently depend too much on data from commercial catches. Fisheries management in the long run can only be improved by lowering the current degree of uncertainty and advancing the science behind stock abundance forecasts. If the main causes of overfishing are inadequate stock assessments or a lack of political will to set total allowable catches at sustainable levels, then MPAs may be a good safeguard against such errors.

Fisheries responses to MPAs and the existence of external threats, such as nutrient pollution and weather disturbances, will determine how well the MPAs accomplish their management goals. According to behavioral models of the fishing industry used in two studies examining the implications of MPAs, there will be an increase in harvesting pressure outside of MPA boundaries, especially along those lines. Anecdotal evidence suggests that this has happened outside of a marine protected area (MPA) off the coast of Kenya, where the density of traps increases significantly in the immediate vicinity of the MPA. The benefits of spillover have dissipated quickly due to the increased effort along the MPA's edge. Furthermore, the costs of greater competition may push fishermen to use techniques that, given the new MPA constraint, maximize private return while also increasing the amount of habitat destruction in the water that is still fishable.

Because a marine ecosystem is open to wide oceanographic shocks and is vulnerable to external threats, it may provide less protection against such threats and have less value as a hedge for MPAs. MPAs are not shielded from outside threats such as nutrient pollution, coastal erosion, freshwater diversion, and large-scale environmental shocks like El Nino. The "dead zone" in the

Gulf of Mexico and other environmental harm are examples of how upstream nutrient runoff poses a threat to a protected area's health. However, the presence of a unique location, such as the Flower Gardens National Marine Sanctuary off the coast of Texas and Louisiana, may aid in building support for efforts to lessen environmental harm and upstream nutrient pollution.

According to certain research, MPAs could be a practical substitute for traditional fishery management techniques when attempting to manage intricate, multispecies fisheries. Because market demands or regulations may cause effort to be diverted from one species to another or increase discards, multispecies fisheries are frequently challenging to manage. MPAs may provide a second-best option to stop overfishing in multispecies fisheries, like the reef fisheries in the Gulf of Mexico and the Southeastern United States, if they are appropriately conceived and positioned.

8.2.5.6. Management Costs

Due to the increased need for enforcement and monitoring, MPAs are likely to raise management costs in the absence of any changes to the fishery's management. The anticipated biological benefits might not materialize if the protected area is not adequately enforced or monitored. Because fishing grounds are remote, there is a strong incentive to poach in protected areas, and enforcement resources are limited, some seasoned fishermen feel that MPA boundaries are simply unenforceable. On the other hand, the cost of monitoring and enforcement is probably going to be influenced by things like MPA size, location, and use restrictions, fishery management laws, regional fishing customs and practices, and the technology that is available. If fishery managers use cutting-edge technology, like a global positioning satellite network that can track fishing vessels automatically and notify authorities when one enters a closed area, the cost increase might not be as great. The Great Barrier Reef in Australia and Georges Bank off the coast of Maine are two locations where this technology is being successfully applied, and it has decreased the need for fishery patrol boats.

8.2.6. Political Economy of Implementation Decisions

A public investment occurs when marine resources, which are federally and state-owned and owned by all Americans, are set aside for a specific purpose, such as an MPA. Weighing the above-described possible costs and benefits against equally significant moral, ethical, and cultural values will determinc whether or not such an investment is desirable.

8.2.6.1. Benefit-Cost Analysis

MPAs have advantages and disadvantages that can be methodically determined and explained. But it's frequently not possible to precisely calculate the expected net benefits by putting all costs and benefits into monetary terms. Similar to other public investments, closing off areas implies an intertemporal tradeoff, possibly even spanning generations, as many of the costs are incurred immediately while the potential benefits are often realized later. Another source of difficulty is the degree of imprecision and complexity involved in estimating how a new management tool will affect biological and economic systems.

The task of estimating and measuring the non-extractive use values connected to the closure presents another challenge for managers. It is widely acknowledged that these values play a significant role in determining how resources are allocated. Actually, a lot of the recent arguments in favor of MPAs have focused on the potential for significant economic gains from ecotourism as well as the conservation benefits of preserving biodiversity. However, it will be challenging in the near future to fully incorporate them in decision making processes given the current dearth of research on the magnitude of these benefits for marine resources.

8.2.6.2. Equity and Fairness Concerns

MPAs are not an exception to the rule that all public investments have equity or fairness implications. Due to the likelihood that MPAs will disproportionately impact user groups, equity concerns among the process's stakeholders may quickly surface. For instance, if an MPA is located in a near-shore area, the offshore fleet may benefit most while the inshore fleet may bear the greatest costs (such as the direct loss of fishable waters). It should come as no surprise that distributional effects play a significant role in political decision-making.

Both commercial and recreational fishermen frequently worry that while closures will save money, they won't reap the rewards. Any shift in the way benefits and costs are distributed, whether actual or perceived, is bound to cause a great deal of disagreement. Gaining a deeper comprehension of the advantages and disadvantages could point to strategies for making up for the losers and advancing the MPA proposal. If not, it would be reasonable to anticipate political interference meant to limit management alternatives. For instance, legislation requiring scientific proof of damages prior to MPA adoption may be called for if recreational fishermen fear their interests will be harmed by arbitrary closures.

In certain situations, the equity issue may be resolved by altering an MPA's design. This is a choice that is worthy of thought. But adjusting conservation instruments might not be the best way to deal with inequality. Although the public nature of marine resources suggests that the establishment of an MPA does not amount to a "taking" (though fishermen may not see it that way) necessitating payment, paying those who actually incur the costs directly when MPAs are put into place is another option.

Fairness concerns highlight the significance of a public outreach plan to determine which stakeholders must agree on management goals, MPA location and design, and MPA use. Resource managers face the danger of prioritizing where to locate MPAs and how much of the resource to protect when choosing MPAs, and they also run the risk of undervaluing socioeconomic factors. As a result, those who will be most impacted by the choice are not offered the chance to weigh in until much later, when it might seem as though crucial choices have already been made. This strategy frequently stifles attempts to create MPAs because the people with the greatest stakes will work to thwart the initiative. Resource managers should identify all stakeholders, including commercial and recreational fishers, divers, environmentalists, and other concerned citizens, at the outset of the process and involve them at every stage of decision-making, rather than taking this risk. In this manner, the ecological and socioeconomic aspects of creating MPAs can be taken into account simultaneously, as these factors frequently determine whether MPAs succeed or fail.

The Tortugas Ecological Reserve, located in the Florida Keys National Marine Sanctuary about 70 miles west of Key West, was successfully established as a result of this strategy. After the proposal was made, a working group comprised of stakeholders within the Sanctuary Advisory Council was established to develop the reserve's boundary alternatives and draft an additional environmental impact statement. Among the technical advisory teams was a physical scientist (e.g., biologists and oceanographers) and a socioeconomic group (e.g., sociologists, and economists) whose job it was to counsel the working group as it developed potential reserve boundary options. The socioeconomics team produced maps of the commercial and recreational fisheries, created user group profiles, identified all current users of the study area, and quantified the spatial distribution of economic use and activity. The stakeholders settled on a preferred option for the location of the ecological reserve after two years and numerous discussions.

Fishery managers are looking to MPAs as crucial instruments in the recovery of numerous fisheries that are in danger of collapsing. Strong scientific evidence actually supports the effectiveness of marine protected areas (MPAs) in maintaining special marine habitats, reviving fish populations that live there, and safeguarding priceless artifacts. However, depending on the size, scope, and location, they may play a variety of roles in supplying biodiversity, acting as a buffer against careless management and natural disasters, presenting research opportunities, and serving as emigration sources for the surrounding areas. For instance, the enhancement of marine biodiversity may require the preservation of a sizable number of representative habitats. If the intention is to improve fisheries, MPAs might be positioned to maximize spillover. Regardless of the objective, the planning, placement, and administration of MPAs must be guided by scientific research. It will also require careful planning and financial resources to enforce boundaries, make necessary design changes, and keep an eye on MPAs to determine whether they are successful or not.

A potential strategy to raise the likelihood that MPAs will be successful in improving fisheries is to combine them with other complementary management tools, like individual transferable quotas, that deal with the root causes of overfishing. Otherwise, as fishing activities increase in response to recovering fish stocks, the economic and possibly biological benefits could vanish quickly. In order to better integrate the preservation of fish habitat with the economic well-being of fishermen, it would be beneficial to combine MPAs with other, more comprehensive strategies that can address the root causes of overfishing and overcapacity. Anyone interested in the sustainable use of our ocean resources should find it appealing to improve the fishery's biological and economic health, as this will help win fishermen's support for marine protected areas (MPAs). However, it's crucial to remember that protected areas don't deal with the overfishing issues that plague many fisheries. In general, the specifics of the MPA under consideration, the fishermen's behavioral response, the pre-existing biological and economic conditions, and the institutional environment will determine whether the use of MPAs has any appreciable effect on excess effort or on catch levels.

Of course, MPAs frequently offer significant ancillary benefits in addition to improving fisheries. Protecting coral reefs, for instance, will have benefits unrelated to the improvement of fisheries because they store carbon and offer special recreational opportunities, like scuba diving or underwater photography. Furthermore, people who may never visit the location but gain from knowing that rare habitats and endangered species are safeguarded may find value in MPAs. These non-extractive values may be more valuable than the value of lost fishing opportunities, despite their difficulty in measurement. However, comparing these values to the more easily quantifiable effects of fishing remains challenging until further data are gathered and research on the magnitude of such values is conducted.

Naturally, one would want to know under what circumstances one might anticipate an MPA's benefits to exceed its expenses. Regretfully, the answer is contingent upon the objectives, as well as the extent and placement of the area selected for closure. For instance, the likelihood that a person will see aspects of the "pristine" habitat that are significant to them increases with the benefits to non-extractive users and the site's access costs decrease. It makes sense that the near-shore coral reefs, rich in plants and animals, would provide the greatest non-extractive benefits. The possibility of beneficial short-term effects may be increased in areas set aside to protect fish stocks where older and larger fish either fetch a higher price in commercial markets or are more alluring to recreational users. Lastly, areas that contribute relatively less to the livelihood of fishermen may be set aside if lowering potential costs to them is a priority. One reason why a particular potential MPA site might not contribute as much as another could be that the species is widely distributed throughout the fishing grounds, or that fishing in the set-aside area is among the least lucrative. Another explanation might be that the MPA's current size prevents it from having any appreciable effect on fishing revenues.

The benefit-cost framework enables decision-makers to assess the net economic benefits to society from funding MPAs, regardless of the specifics of the current design. Fairness concerns among those impacted by MPAs may be resolved if the analysis accounts for the likely intertemporal nature of the returns and provides information on the timing and distribution of benefits and costs. Nonetheless, political barriers to MPA adoption are probably not going to go away. When a protected area is set aside, fishermen initially bear the majority of the costs, and improvements in catch size and composition might not materialize for some time. Fishermen, like many others, may be inclined to take political action to block the siting of an MPA due to the uncertainty surrounding the costs and benefits of MPAs and how they will be divided. This is especially true when there is a belief that the current management regime is sustainable.

Stronger stakeholder decision-making is probably going to result in MPA decisions. Fishery managers should include fishermen, environmentalists, and other interested parties in the process of establishing MPA goals and accepting accountability for their attainment. For instance, fishermen could assist biologists and fishery managers in choosing locations for reserves that will maximize fishery benefits while posing the fewest financial obstacles to the fishermen. However, decentralization of the decision-making process and local or regional information sharing are necessary for fishermen and other stakeholders to engage effectively. Thus, an endeavor that is intrinsically scientific can be shaped by the values and beliefs of local resource users. Additionally, it might make implementation more accountable and raise the possibility that fishermen will gain from it.

8.3. SUSTAINABLE FISHERIES MANAGEMENT

A sustainable fishing method is one that minimizes its negative effects on habitats and ecosystems while maintaining a sufficient stock of fish in the oceans. Fisheries need to be managed well in order for this to occur. Fishing is an essential source of food and nutrition for communities all over the world, and it also supports their way of life. A third of the world's population gets their protein from seafood, and 38 million people work in wild capture fisheries. By practicing sustainable fishing, we can help end hunger and poverty while securing food for the future.

But excessive bycatch, unrestricted fishing, and overfishing are examples of unsustainable fishing methods that endanger our oceans. The demand for seafood is rising worldwide, and over a third of the world's fisheries have been overfished. Sustainable fishing has the potential to stop this decrease and guarantee that there are enough fish in the ocean for fishing to continue for a very long time. For the purpose of providing a growing population with a nutrient-rich food source, the long-term health of fish stocks is also essential. If fisheries are managed sustainably, 16 million more tonnes of seafood could be produced annually, enough protein to meet the needs of 72 million people worldwide. As a source of protein, certified sustainable wild-capture fishing can also lessen the strain on land-based agriculture. In comparison to animal proteins found on land, seafood generally has a lower carbon footprint.

Researchers determine the maximum number of fish that can be harvested without endangering the stock's long-term health. This entails gathering information on the size of the stock, the location and timing of the species' spawning, and the proportion of juveniles who are expected to mature into adults. They also evaluate environmental elements like predation by other species that could have an impact on the stock. A variety of management techniques, such as banning fishing during spawning seasons and imposing size restrictions to safeguard juvenile fish, can be used to safeguard stocks against overfishing. Adopting preventive measures known as harvest control rules, which mandate reducing catches in the event of a stock population decline, is another essential component of sustainable fishing. This is especially crucial when multiple nations share stocks and coordinated action is required to stop overfishing.

Sustainable fishing minimizes its negative effects on threatened, endangered, and protected species while promoting the health and diversity of ocean ecosystems. Every species plays a distinct role in the ecosystems of the oceans and is a part of a healthy predator-prey food web. The entire food web may be affected if one species goes extinct as a result of overfishing or excessive bycatch. A healthy ecosystem benefits more than just fish stocks; it can also aid in the ocean's ability to control the temperature. Seagrass, mollusc shells, and plant-like plankton are among the components of the ocean ecosystem that retain carbon dioxide from the atmosphere after it dissolves in seawater. Sustainable fishing is more crucial than ever because climate change is already having a big impact on our oceans and fish stocks.

8.3.1. Sustainable Fishing Minimize Impacts on the Environment

Not just the species that are being targeted by fishing can be affected. The marine environment may be endangered by problems like habitat degradation, misplaced fishing gear, and undesired catch if practices are not sustainable. Bycatch, which includes non-target fish and shellfish as well as marine mammals, reptiles, amphibians, and seabirds, should be reduced in fisheries by putting measures in place. This could entail making adjustments to equipment, staying away from fishing in regions where undesirable or endangered species are known to inhabit or migrate through, and adjusting methods to reduce the possibility of encounters. Seabird entanglements, for instance, can be minimized by fishing at night rather than during the day and lowering lines and nets farther into the ocean. Additionally, gear can be changed to let non-target species get away or discourage them from approaching nets too closely. This entails including exclusionary devices, such as bird-scaring lines or acoustic "pingers" that repel marine mammals, to allow larger marine animals to escape from nets.

Seafloor-interacting fishing techniques, like bottom trawling and dredging, are controversial because they run the risk of destroying seafloor habitats. However, with careful management, these techniques can be maintained. Studying seafloor habitats can help identify areas that are sensitive and should be avoided, such as those that are home to slow-growing, delicate species like corals. In certain situations, fishermen can set nets at a depth that completely avoids contact with the seabed, and they can also use lighter gear to lessen the impact of contact. Additionally, actions can be taken to lessen the impact of any lost or abandoned fishing gear as well as to prevent fishing gear from ending up at sea. This can involve keeping an eye out for misplaced equipment, recovering it, and utilizing biodegradable locks or panels on equipment like lobster pots to let species get away.

8.3.2. Sustainable Fishing Practices

It is possible to fish sustainably and still enjoy seafood while preserving population levels for future generations. For thousands of years, people have fished sustainably in many indigenous cultures. Some of the lessons from these cultures are reflected in the sustainable fishing methods used today.

Fishing methods that simultaneously maintain and increase fish populations have long been used by the Tagbanua people of the Philippines. Even now, they still adhere to these procedures. In order to allow fish stocks to replenish themselves, Tagbanua's only fish for particular species during specific times of the year that are determined by the moon and tides. They designate specific regions as protected areas where fishing is forbidden, such as coral reefs. These traditional fishermen mainly use hook-and-line techniques when fishing, only taking in enough fish to feed their families and themselves. The local Irrawaddy dolphins, who get tangled in more contemporary fishing gear like nets and traps, can suffer harm or even die. A 2007 study praised traditional Tagbanua practices as a means of preventing this from happening.

The South Pacific's traditional Polynesian cultures have also always been dependent on the resources of the ocean. Cast nets, spearfishing, and hook-and-line fishing were their most popular fishing techniques in the past. Hooks made of stone, shell, or bone were intended to capture particular animals. Additionally, fishermen would make 6-foot, 2-meter spears. Once more focusing on particular animals, they would spear fish from above or dive beneath. Fishermen working alone or in groups utilized cast nets. Fish clusters could be captured by casting nets from the coast or from canoes. These strategies were all focused on the fish that local communities and fishermen's families needed.

Even now, some of these environmentally friendly fishing techniques are in use. Spearfishing and cast-net fishing are popular among native Hawaiians. Spearfishing is a modern sport that is popular throughout the world, especially in Asia, Australia, Africa, and South America. Nowadays, spears are frequently propelled underwater by spear guns. In some parts of the United States, like Florida and Hawaii, spearfishing is a well-liked recreational activity. Because it only catches a small amount of bycatch and only targets one fish at a time, this fishing technique is regarded as sustainable.

It's likely that if you've ever gone fishing, you utilized a rod and reel. The classic hook-and-line method of fishing has been updated with rod-and-reel fishing. Different sizes and shapes of rods and reels enable both commercial and recreational fishermen to target a broad range of freshwater and saltwater fish species. Fishermen can catch pelagic species like sailfish, bottom-dwelling species like flounder, and freshwater species like catfish and trout thanks to the various types of rods and reels, as well as the various locations and baits. Rod-and-reel fishing allows non-targeted species to be released right away, reducing bycatch. Additionally, overfishing is avoided because only one fish is taken at a time. Rod-and-reel fishing is a more environmentally friendly option for commercial fishermen than longlining. Refusing to eat fish and other seafood is another way to stop overfishing and bycatch. Prominent marine scientist and National Geographic Explorer-in-Residence Dr. Sylvia Earle advises that until we have more knowledge about how to preserve healthy fish and wildlife populations, people should refrain from consuming seafood.

"I personally have stopped eating seafood," she explained to *National Geographic*. "I know too much. I know that every fish counts at this point. Some more than others, but I can no longer bear the thought of eating tuna knowing in what dire straits they currently are. If we value the ocean and the ocean's health at all, we have to understand that fish are critical to maintaining the integrity of ocean systems, which in turn make the planet work."

8.3.3. Fisheries Management

Fish and other aquatic life continue to be important sources of food and raw materials for a large number of people, communities, and nations. Through fisheries management, we must lessen overfishing and bycatch in order to preserve fish stocks. Fish population management is a difficult endeavor. Collaboration is necessary at all governmental levels, from local communities to international governments. Governments are in charge of controlling fishing in their coastal waters. The management of fisheries in US waters that are five to 321 kilometers (three to 200 miles) from land is under the purview of NOAA Fisheries. Nearer to the coast, local governments are in charge of the ocean.

Naturally, the opinions of various stakeholders regarding fishing regulations vary. The fishermen themselves are concerned about preserving fish populations for future generations as well as their own means of subsistence. In their efforts to preserve freshwater and marine habitats, conservationists frequently try to stop fishing and other activities that force animals from their natural homes. Ordinary people want to be able to buy the seafood they adore. Scientists work to maintain the wellbeing of ecosystems in fresh and saltwater.

Large portions of the vast ocean are not included in a country's territorial waters. The "high seas," or international regions that are not a part of any one country, make up the majority of Earth's waters. It is difficult to regulate fishing in international waters because different countries have conflicting interests and management strategies that need to be agreed upon.

On the other hand, numerous international agreements are in force. Nations with common economic interests in a given region comprise the 17 Regional Fisheries Management Organizations (RFMOs). Member states are obligated to abide by RFMO regulations, which may include catch limits and requirements for the kinds of gear that can be used. Although there is evidence that these regulations have reduced bycatch (like dolphins in tuna nets), it has proven difficult to keep fish stocks healthy. Although it is very difficult to enforce fishing regulations on the high seas, member nations have made efforts to address the issue of illegal fishing and stop the importation of seafood that has been caught illegally. Preserving salmon stocks is the main purpose of the North Pacific Anadromous Fish Commission (NPAFC), one agency that has shown success in enforcement. The United States, South Korea, Japan, Canada, and Russia are the members. The commission forbids fishing for salmon on the open ocean, which is mostly done with drift nets. Usually close to the surface of the ocean, drift nets float freely in the currents. They are employed in the capture of sardines and salmon that are in schools. Sadly, a lot of bycatch is produced by these nets, entwining marine mammals, seabirds, and other non-targeted species. Regulating fisheries is about creating laws based on data from science. These rules might be predicated on information about the life histories, migration patterns, or other aspects of certain species.

In the US, for example, bluefin tuna fishing is heavily regulated. This species can only be caught by hand-thrown harpoons or rod and reels. They are only allowed to catch one fish at a time thanks to this regulation. A fish must measure at least 185 centimeters (73 inches) in order to be taken from the ocean. This regulation aims to protect fish by allowing them to spawn before being harvested. Furthermore, the amount of fish that can be caught annually is limited. When that quota is reached, the season's fishing is over. To preserve fish populations for future generations, regulations such as these take into account the biology and natural history of the species (Alnafissa et al., 2021).

8.3.4. Consumers

As buyers, we have the option to select seafood from sustainably managed fisheries. To achieve this, we ought to inform ourselves about the origins and methods of fishing for fish. The ongoing challenge is getting policymakers to look ahead and take into account consumer needs, fishermen's livelihoods, and scientific data.

8.4. RESTORATION ECOLOGY AND HABITAT REHABILITATION

Aquatic ecosystems comprise a wide range of habitats, from highly oligotrophic mountain lakes and streams to chemo-autotrophic black smokers at volcanic deep-sea ridges. These habitats are all threatened in different ways, creating a range of conservation opportunities as well as challenges. Fishes are a taxonomic group that makes up between 25,000 and 40,000 species, which is estimated to account for more than 50% of all vertebrate diversity. This highlights the significance of aquatic biodiversity for the planet's overall biodiversity, which is defined by water. The greatest phyletic diversity is found in the oceans and seas, where life first originated. Only a small number of these phyla (e.g., echinoderms, chaetognaths, and priapulids) are specific to the ocean and do not exist in freshwater environments. When it comes to conservation, freshwater species are sometimes highly restricted in space, whereas marine species are generally more widely distributed. The benefits that aquatic ecosystems and biodiversity—both in freshwater and oceanic environments—offer to humankind are becoming more widely acknowledged. Human pressures on the oceans and fresh waters range from global change (i.e., human-induced climate change, ocean acidification, global homogenization of biodiversity owing to invasive non-native species), through regional (e.g., eutrophication, overfishing) and local (e.g., point source pollution, habitat degradation and loss) scale impacts. They all work together, usually in a synergistic way, to exacerbate the situation. Local effects can spread across long stretches of coastline, entire freshwater catchments, and even regional seas. Modifications in freshwater catchments have the potential to transfer issues to coastal and estuarine systems, especially in semi-enclosed bays, seas, and estuaries (e.g., The North, Irish, Baltic, and Adriatic seas). Consequently, issues at sea, like overfishing, may have an impact on migratory fish populations that are significant for freshwater ecosystems.

Current projections for aquatic ecosystems, habitats, communities, and species are dominated by the global decline in conservation status and trends. Terrestrial ecosystems in Europe have a higher overall percentage of favorable conditions than freshwater and marine ecosystems, for both species and habitats. For example, the effects of ocean acidification and climate change add to the cumulative regional and local stressors, such that only 7% of marine species assessments indicate ′favorable conservation status′. Policymakers and the general public have high expectations that deteriorating systems must be repaired or at the very least stopped from declining further. Europe serves as an example of how difficult this is, even in developed nations where meeting environmental goals is a top political priority. The majority of freshwater bodies in Europe cannot be expected to reach the desired ′good ecological status or potential′ in the near future, despite declining water pollution and the formulation of explicit restoration targets in the European Water Framework Directive and the Habitats Directive (Hering et al., 2010; Geist, 2014). The Marine Strategy Framework Directive, which aims to create productive and healthy seas with a "good environmental status," was inspired by similar concerns for marine systems.

The aforementioned instances highlight the necessity of habitat restoration and recovery. This is a complex and difficult challenge that includes physical (e.g., natural hydrodynamics, sediment dynamics, de-linearizing coasts and inland waterways), chemical (contaminants, nutrients) as well as broader connectivity issues (need for habitat patch complexity connected across landscapes, catchments, and seascapes) to ensure ecosystem functioning and the maintenance of resulting ecosystem services. Targeted actions to stop the loss of biodiversity are framed by growing policy concerns and public opinion. Open aquatic ecosystems with high connectivity have a lot of potential for natural recovery and recolonization, but oftentimes some natural intervention is still required nature through restoration-based intervention. However, some ecological and biological restoration goals may clash with other goals related to the use, like flood control (which affects both freshwater and marine habitats), hydropower production (which occurs mostly in stream and river ecosystems), or resource extraction that poses a risk of contamination (e.g., fish harvested from marine environments near oil rigs) (Hawkins et al., 1999a; 2002a).

8.4.1. Restoration Targets and Options

Any degraded ecosystem's recovery process begins with removing or minimizing current impacts. Diagnosis and system knowledge are necessary in order to focus on the actions that will have the greatest impact. If possible, reference sites that are unaltered and undisturbed should be used as models for action and as benchmarks for success. Additionally, recovery must be addressed at the proper scale, which is typically the coastal cell and, on occasion, the regional scale in fully marine systems, as well as the catchment in freshwater and estuarine systems, or even smaller local scales. At the local site-specific scale, smaller enclosed systems like lakes, ponds, and brackish and fully saline lagoons can be addressed. There are many different types of restoration projects in both freshwater and marine environments; Figure 8.1 is an example.

Figure 8.1. *Example of a habitat-reconstruction project to address the perceived absence of large woody material and pools as key limiting factors in an urban stream (Longfellow Creek, Seattle, WA, USA). Other factors, however (largely water chemistry and flow), have precluded any meaningful biological success here Source: Booth DB, Scholz JG, Beechie TJ, Ralph SC. (2016). Integrating limiting-factors analysis with process-based restoration to improve recovery of endangered salmonids in the Pacific Northwest, USA. Water, 8(5):174. https://doi.org/10.3390/w8050174.*

8.4.1.1. Fresh Waters

Most industrialized nations have made significant strides in lowering the amounts of pollutants, nutrients, and organic waste that enter streams, rivers, and lakes, and subsequently enter estuaries and coastal waters, in freshwater ecosystems. Sewage treatment plants from earlier generations addressed low oxygen, biological oxygen demand, and organic loading. The loads of phosphate in most surface water bodies have significantly decreased in less than 30 years due to its replacement in laundry detergents, better sewage collection, and tertiary treatment, while elevated levels of nitrate remain a concern even in industrialized countries. Similarly, the use of source-related filtering technologies, fuel sulfur reduction, and active restoration of affected lakes and rivers by liming greatly reduced the problem of airborne acidification of water bodies in poorly buffered, non-calcareous areas (Hultberg and Andersson, 1982). While many developing nations continue to confront significant obstacles in managing the nutrient and eutrophication levels in their waterways (Gleick et al., 2011), it is important to recognize the significant advancements made in this area in the majority of developed nations.

In contrast, deficiencies in structural habitat quality caused by the severe alteration of stream courses and their catchments remain the main obstacle to future restoration, especially in stream ecosystems that support the most critically endangered species of all freshwater habitats. Freshwater ecosystems, as opposed to ocean ecosystems, are typically far more fragmented both naturally (lakes and streams separated by land) and artificially (streams with dams and weirs), with detrimental effects on biodiversity and ecosystem functioning (Mueller et al., 2011). As a consequence, establishing connectivity, particularly in stream and river ecosystems, has become a major priority, to date mostly targeting charismatic fish species such as salmonids (e.g., Atlantic salmon, *Salmo salar*) or European eel (*Anguilla anguilla*), while migration of smaller fishes and of other species, in general, is much less valued.

According to recent research, if fish bypass channels are built with a hydraulic gradient that is absent from highly flow-regulated and dammed systems, they can serve as significant substitute habitats in degraded streams while also restoring fish migration (Pander et al., 2013). Therefore, choosing a technical fish pass over a nature-like bypass channel can have very different restoration effects that go beyond simply restoring fish passage. Fish bypass channels that mimic nature can also be used to restore environmental flow dynamics, at least locally. On the other hand, since this can lead to the homogenization of biological communities and decreased resilience to stressors, connectivity should not be established between naturally isolated systems. A prime example is the intricate network of artificial canals that connect all of Europe for navigation purposes. These canals allow non-native species, like the Ponto-Caspian gobies and the zebra mussel *Dreissena polymorpha*, to spread throughout the continent, endangering the native biota.

The siltation and colmation of stream beds is another major deficiency in habitat quality, particularly in headwater areas. Particularly in light of rising water temperatures, excessive siltation and clogging of the stream bed frequently serves as the primary barrier to the survival and growth of endangered species of freshwater mussels and gravel-spawning fish. It is unlikely that this issue can be resolved with cosmetic in-stream measures without taking the catchment scale into proper consideration, as this problem stems from both catchment-dependent land use and in-stream modifications of geomorphic structure and flow patterns. The preservation of functional stream beds and the repair of structural deficiencies have emerged as top concerns in the framework of numerous European policies, including the Water Framework Directive and the Habitats Directive. It would be ideal if stream bed restoration followed a methodical, evidence-based approach, but this is not always the case. Although the field of river restoration has experienced exponential growth in the last few decades, there hasn't been much empirical research done to determine whether individual or combined restoration projects (Figure 8.1) meet the legally required objectives of enhancing the composition and functionality of streams and rivers.

The goal of functional perspectives on river restoration is to restore all of the hydrogeomorphic, biogeochemical, and ecological processes that constitute a healthy river. While this goal is supported by widely accepted theory, research on how to carry out and evaluate functional restoration projects is still in its early stages. The majority of commonly used in-stream restorations, like adding gravel and using excavators to loosen the substrate, have only been shown to have short-term positive effects in basins with intense agricultural land use, often lasting less than a year, according to case studies of several European streams (Sternecker et al., 2013a; Mueller et al., 2014a; Pander et al., 2015). Furthermore, these studies demonstrate that certain methodological strategies, like substrate loosening, can even outweigh the short-term benefits of these measures in the restored sites in terms of their detrimental effects on downstream habitats and biota. As such, there is a pressing need to demand evidence-based methods for stream bed restoration and to evaluate the application of some of these measures critically. Given that various organismal groups (e.g., Fish, macrophytes, periphyton, and macroinvertebrates) can all react differently to disturbance and restoration, so it's important to take both taxonomic and spatial representativeness into account when evaluating the effects. As demonstrated in fishes, changing habitat preferences along different life stages can strongly determine the success of habitat restoration. Streams' unidirectional environmental flows, in addition to these geomorphological features, are a crucial component that has controlled biotic adaptation. It is well known that the habitat patchiness and diversity in streams are determined by structural features of the stream, such as stream sinuosity and in-stream structural elements, such as dead wood or macrophytes. These are the first features to go in order to prevent flooding. In stream ecology, restoration of environmental flows—which is linked to the natural flow regime as a crucial driver not only to initiate fish spawning and connections with floodplains—has gained widespread acceptance. Numerous recent examples specifically address this aspect of restoration.

One of the main concerns in freshwater habitat restoration is whether achieving a "good ecological state or potential" is actually the best restoration goal for all surface water bodies, or if it would be more effective to take a more differentiated approach, restoring some sites to excellent condition while leaving some of the more degraded systems in place. The main finding was that while some structural measures, like adding dead wood, were helpful for common species, none of the applied measures were able to restore functional populations of specialized endangered fishes, as demonstrated by a case study that attempted to restore fish populations in a highly modified river system in Germany. Therefore, it is debatable whether such investments can withstand a close examination of the effectiveness of the restoration. Benefits of restored reaches for streams have been suggested to spill over into nearby, unrestored areas. Even though it is possible that species, particularly those that are mobile, will have effects that are observed outside of the restoration areas that are spatially limited, many of the strategies that have been suggested and those that have been put into practice to restore streams are still in need of thorough scientific validation.

As some of the theories and methods for rehabilitating freshwater environments (e.g., re-oligotrophication and acidification mitigation) are well-established and supported by numerous examples, the restoration of structural habitat diversity—a crucial goal in stream and river restoration—is much more difficult to achieve and has been the subject of fewer well-designed studies that apply rigorous before-after control-impact (BACI) monitoring with adequate taxonomic representation and, when feasible, spatial replication. Additionally, there is a glaring underrepresentation of unfavorable outcomes, which are equally important for future research to analyze as favorable ones. These issues can, among other things, result in ambivalent restoration decisions and practices, as well as uncertainty about whether funding for such restoration will improve the targeted biological status as expected, e.g., in the European

Water Framework Directive. Making decisions about restoration is further complicated by our incomplete understanding of the autecological requirements of the great majority of species that have not been used or studied by humans. It is not unusual for closely related species to have quite different habitat needs, despite sharing similar life cycles and physical characteristics. For example, in freshwater amphipods and other macroinvertebrate taxa, native and non-native species exhibited varying degrees of success in colonizing newly created and restored habitats. This variation was largely due to the physical properties of the habitat, such as flow current, rather than just connectivity with source habitats. Findings from specific restoration projects are frequently hard to generalize because species pools can vary greatly depending on the geographic location and possible local species inventory.

When trying to address a combination of pressures with little information on their relative importance or locally unknown causes of biodiversity decline, restoration becomes even more difficult. Unknown interactions between pressures are another frequent problem that can impede restoration from proceeding as intended. Even though the primary global pressures and the documented decline in freshwater biodiversity are well known, it can often be challenging to determine the relative importance of each factor to the scale of operation of a restoration action due to the interaction of multiple factors and the occasionally lengthy lag times between impact and observed effect. This is made even more difficult by the incomplete understanding of the effects of other variables that may also be significant and need to be taken into consideration in the restoration of aquatic ecosystems (such as recently identified toxic compounds, or "emerging contaminants"), for which there is currently insufficient data due to a lack of routine monitoring to provide a complete picture of their significance and interactions. A certain amount of uncertainty must also be taken into account in restoration efforts due to incomplete knowledge about the precise factors that lead to species declines.

8.4.1.2. Marine Ecosystems

The extent of ecosystem enclosure in marine environments influences the potential for active restoration. As a result, systems with restricted exchange, like brackish or salinity lagoons—which are comparable to freshwater lakes and ponds—show the most potential. There are fewer options for restoration as systems get more open and experience large-scale material (mainly detritus) and propagule flow (which affects population dynamics and genetic exchange) exchange. On the other hand, the possibility of a natural recovery via recolonization from distant sources rises (Figure 8.2).

Figure 8.2. *View of a typical LWM-depauperate river in the Pacific Northwest (Nisqually River, western Washington State).*

Source: Booth DB, Scholz JG, Beechie TJ, & Ralph SC. (2016). Integrating limiting-factors analysis with process-based restoration to improve recovery of endangered salmonids in the Pacific Northwest, USA. Water, 8(5):174. https://doi.org/10.3390/w8050174.

Where the biogenic structure has been lost in open marine and coastal systems, active restoration techniques can be used. These include coral reefs, mollusk and polychaete reefs, kelp beds, semi-terrestrial saltmarsh and mangrove vegetation, and intertidal and subtidal seagrass beds. Biogenic structuring species act as ′ecosystem engineers′ (Jones et al., 1994). The restoration of such species affects ecosystem functioning and, eventually, services, in addition to giving habitat more structure. Other significant primary producers include higher plants like seagrasses, those found in saltmarshes, seaweeds, and the symbiotic algae that live with corals. In the case of macrophytic plants and algae, these sources contribute significantly to the primary production that is either directly or indirectly consumed by grazers in detrital food webs.

Higher plants, like seagrasses, mangrove species, and saltmarsh, can be planted using methods commonly employed in freshwater and terrestrial habitat restoration, or what is known as ecologically conscious gardening. According to recent research, seagrass clump planting increases survival through facilitation (Silliman et al., 2015). These ideas probably hold true for pioneer plants that are found low on saltmarshes, where it has been demonstrated that facilitation both intraspecifically and interspecifically increases survival. However, maintaining genetic diversity has drawn criticism, particularly when it comes to vegetatively reproducing clonal plants like seagrasses.

Artificial substrata have been shown to improve settlement in reef-building species, such as bivalve mollusks, serpulid worms, and corals, especially in gregarious settlements. When biogenic structure is replaced, ecosystem processes can eventually resume in restored habitats, resulting in functional equivalency.

There are numerous approaches to restoring semi-enclosed marine and coastal water bodies. Restoring hydrological processes, such as the exchange between lagoons and the open sea or estuary, is a straightforward place to start. Changing the sluicing regime could be an easy solution for this. This method has also been used in managed coastal re-alignment projects, in which arable land or reclaimed pasture is submerged, turning it back into saltmarsh, and dykes and seawalls have been removed or neglected. However, it may take some time for drainage and erosional processes to restore the physical pattern of branching creeks and gullies. Pioneer saltmarsh plants can quickly take over the mud flats that result from the terrestrial vegetation dying off first. It takes many decades for the features of these artificial marshes to fully recover from those of natural marshes, including species composition, community structure, ecosystem functioning, and services. It has been demonstrated that hundreds of years after early colonists in New England once again drained and reclaimed marshes for agricultural use, the effects of their actions are still visible today. Things can move more quickly if spatial heterogeneity and some focused restoration techniques are added. The marsh restores some ecosystem services quickly through controlled realignment (e.g., flood defense); however, it will take a long time for the full range of benefits and anything resembling the biodiversity of old marshes to materialize.

Work on closed, locked dock basins typical of macro-tidal ports—which are no longer in use for commercial shipping—is an extreme example of restoring degraded artificial ecosystems (Figure 8.1(D)–(F)). When the size of shipping increased and containerization became the norm, these docks were no longer necessary. In deeper dock basins, techniques from reservoir or lake management have proven effective. Helixor air lift pumps prevented the accumulation of an anoxic water layer below the thermocline and destratified the dock. Dense phytoplankton blooms in the nutrient-rich surface layer that subsequently used oxygen in the bottom layer to decompose were the cause of this anoxic zone.

One further benefit of docks with a moderate salinity level is the possibility of a dense colonization of filter feeders, like mussels. This was improved in

Sandon Dock on the Mersey docks by the rope culture of mussels, which were later sold for human consumption after being purified with ultraviolet light. A dense natural settlement of mussels developed in the Albert Docks. The airlift pump's mixing made it easier for mussels to biofilter. Every one to two days, the entire volume of the dock was thought to pass through a mussel. This resulted in incredibly clear water throughout the summer with constant oxygen levels. In the end, there was a healthy marine community in the docks. Human pathogens were also eliminated through bio-filtration, bringing the water up to EU bathing standards. In the end, the wider cleanup of the Mersey estuary and catchment resulted in the 1990s sea trout (Salmo trutta) migration being caught in the docks. Similar de-stratification methods were applied to the inland freshwater dock systems at Manchester (Salford Quays, the Manchester Ship Canal's terminus), in a cross-over back to freshwater.

Perhaps a better term to describe the creation of these fascinating "lagoonoids" in the center of cities would be ecosystem rehabilitation or remediation. Restoring ecosystems in abandoned docks produced healthy but entirely artificial cubist marine lagoons. Ignoring semantics, Liverpool's clean docks water plays a significant role in the revitalization of the inner city. The excellent water is used for leisure and creates a premium environment for residential buildings, hotels, museums (such as the Beatles Museum, National Maritime Museum, and Tate Northern), eateries, bars, and retail establishments.

Because propagule and larval supply come from unaffected source populations, recovery from acute pollution incidents or the cessation of chronic impacts, like habitat destruction by bottom fishing, can occur very quickly in more open systems. In areas off-limits to towed fishing gear, particularly scallop dredging, which removes a lot of bycatch and degrades the seabed by destroying habitat, the recovery of seabed communities has happened fairly quickly. As part of a fisheries management plan to test scallop reseeding, a section off the Isle of Man was closed to scallop dredging in the 1990s. Despite the sporadic encroachment of illegal fishing, the natural recovery was faster than that of the nearby fished areas. In contrast to the area that was continuously dredged, the benthos' general nature became much more heterogeneous and scallop abundance increased.

In addition, rocky shore systems can recover from oil spills very quickly if the wrong cleanup techniques are avoided. More toxic first-generation dispersants were used to treat many shorelines after the Torrey Canyon oil spill in southwest Britain than the oil itself. Recovery occurred in about two years on untreated shores like Godrevy, where worries about seal populations prevented the use of dispersants. In sharp contrast, patellan limpet grazers died in areas where dispersants were applied excessively and liberally. On the rocky shores of Europe, these species are keystones. When they were gone, ephemeral and fucoid algae proliferated in vast numbers until they were eventually eaten by recolonizing limpets. Under the thick seaweeds, these flourished until the food was consumed. In some areas, there was a subsequent die-off, which resulted in another algae flush. The shores that were treated took about 10 years to recover. It may have taken 15 years for the most severely affected sites to return to normal, patchy conditions with sporadic algal escapes. This story serves as a warning about the wisdom of allowing natural recolonization processes to take place in some habitats, especially wave-swept rocky substrates. However, it takes determination to resist the need to appear as though one is "doing something" in the event of an oil spill. Thankfully, reactions to subsequent oil spills have been far more measured, relying primarily on manual cleanup and natural recovery processes and applying far less toxic dispersants onshore.

By designating effective marine protected areas (MPAs) and implementing suitable management regimes, such as highly protected no-take zones that minimize human disturbance, the global removal of extractive pressures and human disturbance should maximize the capacity of marine ecosystems to recover. When and where this has occurred, the outcomes have been excellent. When put into practice, they can significantly improve the populations of fish and invertebrates; even a small amount of protection can be beneficial. Sadly, these regions are only slightly covered and are expanding slowly. Another significant problem is enforcement, as many MPAs exist only in name. However, the effectiveness of these fishery closure measures varies depending on the situation. There were no advantages to lessening dredge-disturbance in regions with high levels of natural disturbance, such as shallow coarse sediments in Cardigan Bay.

Throughout the world, artificially produced or heavily altered freshwater and marine environments are a common sight. Since the dawn of civilization, rivers have been deepened and straightened to facilitate navigation, and canals and irrigation channels have been dug. Ponds and reservoirs were built to supply water for aquaculture, livestock, and human consumption. Similarly, many estuaries are dredged and straightened to aid in navigation; as a result, jetties, wharves, breakwaters, and enclosed dock basins have long been features of coastal regions. Fears of flooding have resulted in the canalization and channelization of rivers as well as the growth of strong coastal defenses. In developed nations, an artificial weir is frequently used to transition between freshwater and marine environments. Nowadays there are many offshore structures, from wind farms to oil rigs. Next up will be tidal and wave energy. The seabed is also covered in a large number of shipwrecks, some of which were intentionally sunk to form reefs. Over the past half-century, numerous initiatives have been made to improve habitats through the construction of artificial reefs, mainly to facilitate diving tourism, recreational fishing, and commercial fishing. Thus, a large amount of artificial habitat—mostly made of hard materials—has been produced, capable of supporting high densities of sessile invertebrates and fouling algae. Artificial habitat designed for a single use can be carefully altered to improve biodiversity and serve a secondary purpose. Recent construction projects on seawalls in Australia, many of which were constructed in reaction to stormier and rising seas, provide a great example. Incorporating structural complexity during the design phase, retrofitting it, and installing it during routine maintenance can result in a variety of microhabitats, such as rock pools and crevices, that can improve harsh tide-out conditions and boost biodiversity. These methods can also increase the number of endangered or overfished species. While it is generally preferable to avoid engineering interventions and instead rely on natural defenses or managed coastal re-alignment, in situations where property and infrastructure must be defended, environmentally sensitive design can help to lessen the negative effects on the environment. Even though artificial structures contribute to ocean sprawl and this is not recovery or restoration, the creation of such habitats can at least partially restore ecosystem services and biodiversity, particularly in urban areas, and counteract the loss of biodiversity in nearby sedimentary habitats.

8.4.2. Commonalities and Differences Between Marine and Freshwater Systems

There are similarities and differences between the restoration of freshwater and marine systems with regard to the primary pressures that need to be addressed, the success stories that can direct future restoration work, the challenges that still need to be addressed, and suggestions for prioritizing work (Figure 8.3).

Figure 8.3. Example of bank armoring that locally precludes maintenance of active channel-migration processes along the Twisp to Carlton reach of the Methow River. Flow is from the upper left to the bottom right of the images; base imagery from Google Earth. Left panel: red lines indicate field-mapped extent of bank armoring; numbers along the channel are river miles. Right panel: air photo-mapped channel

centerlines through this reach: blue = 1893, yellow = 1940s, orange = 1960s–1970s, red = 2004, air photo base = 2013. The extent of channel migration has been significantly reduced post-1940s, in large measure from bank armoring.

Source: Booth DB, Scholz JG, Beechie TJ, Ralph SC. (2016). Integrating limiting-factors analysis with process-based restoration to improve recovery of endangered salmonids in the Pacific Northwest, USA. Water, 8(5):174. https://doi.org/10.3390/w8050174.

When considering aquatic ecosystems on a global scale, it becomes clear that they are all subject to common pressures such as pollution, overexploitation, nonnative species, climate change, and habitat loss or degradation, even though descriptions of these pressures occasionally only cover freshwater or marine habitats. Disparities emerge when particular elements within these groups are taken into account: freshwater ecosystems are primarily affected by changes in flow regimes, an increase in the demand for hydropower and water, as well as the building of substantial infrastructure, intensification of catchment land use, and the issue of increased fine sedimentation and colmation. On the other hand, stresses associated with habitat loss from fishing, diving tourism, desalination plant effluents, microplastics, mishaps involving oil exploration and transportation, and disposal at sea are unique to marine ecosystems.

Successful restoration initiatives include the structural habitat improvements that come from the construction of some artificial reefs in the seas, but not all of them, and the structural improvements along previously channelized rivers that are realized in urban settings. Examples such as the re-oligotrophication of many European lakes, as well as the reversal of the effects of acidification in previously acidified lakes (e.g., in southern Sweden), and in central-European lignite mining sites can provide guidance to similar restoration challenges elsewhere.

However, this should not be taken as a diversion from the fact that common issues pertaining to global warming and change are probably only going to get worse. It is difficult but necessary to balance ecological and societal needs, which typically prevents quick decisions on restoration. In freshwater habitats, the ongoing loss of biodiversity may also be caused by factors and interactions of factors that are currently underestimated. The many uses and services of freshwater ecosystems include drinking water supply, irrigation, flood protection, hydropower, and recreational uses of water bodies. While the issues facing coastal systems are similar, they are not as significant in the open ocean. Along with massive overfishing, especially of higher trophic levels, the threat of ocean acidification—a rise in atmospheric CO_2 concentrations and subsequent effects on primary production and carbon sequestration—in the open ocean is a serious concern. In contrast to marine systems outside of the 200 mile zone, which are regarded as international, the restoration effects in freshwater systems are typically more spatially limited and have clearly defined national responsibilities. This can be a barrier to effective restoration engagement.

The majority of oceanic systems are far more open and connected than freshwater systems, which is the main distinction between them (Figure 8.2). The surrounding terrestrial ecosystems, particularly their land use, debris and detrital supply, soils, and vegetation, have a greater influence on freshwater systems. Coasts and estuaries can be impacted by catchment degradation. Additionally, there is mounting proof that modifications to agriculture and urbanization are linked to the depletion of marine ecosystems and the loading of sediment into coastal waters (Airoldi, 2003). When a catchment level approach is used, there have been significant successes that result in the cleaning of headwaters, rivers, and estuaries, which benefits larger coastal waters. On the other hand, significant interventions like the Yangtse dams are probably going to have a significant effect on nearshore and coastal waters in addition to freshwater species. Collaboration between freshwater and marine ecologists as well as conservationists is required for this endeavor.

With the exception of semi-terrestrial saltmarsh, mangroves, and seagrass beds, rooted vegetation is far more significant in freshwater systems. Since algae lack roots, the main ways that their environment is altered during succession are by changes in water flow, substrate cover, and light conditions. Detritus is exported rather than retained locally. Therefore, in freshwater environments and the surrounding saltmarshes, mangroves, and seagrass beds, below-ground/sediment processes are more significant.

Although dispersal capacity can vary even amongst closely related species, marine benthic organisms possess the ability to disperse widely through planktonic larvae or broadcast algal propagules, which facilitates recovery from distant sources. Only freshwater aquatic benthic invertebrates have limited downstream dispersal potential, usually only in one direction. With the exception of ocean skaters, a few collembolids, and some dipteran larvae, insects are essentially nonexistent in the oceans. Insects that live mostly as aquatic larvae can spread farther thanks to their ability to fly and swim, as demonstrated by the adult stages of beetles that have settled in freshwater environments. The flying stages of adult insects are comparable to the dispersive phase of marine benthic invertebrates in terms of connectivity. Certain freshwater bivalve species and fish have developed intricate relationships involving parasitic or commensal larvae. Due to the need to manage interactions between different species, these can make restoration challenging. Regarding restoration recommendations, it is evident that freshwater and marine habitats share the desire for a step-by-step approach that adheres to predetermined targets, stakeholder involvement in decision-making, and an evaluation supported by appropriate monitoring. One difference is that biologically, when pressures are lessened or eliminated, marine systems—which are much more connected—tend to recover more quickly. In contrast, strongly fragmented freshwater habitats, in particular stream ecosystems, often depend on active restoration action (e.g., when it comes to connecting fragmented streams and rivers laterally and longitudinally) and such action often in addition needs to be aided or augmented by restocking of target species to start recolonization and initiate successional processes.

8.4.3. The Way Forward and Future Challenges

The restoration and rehabilitation of aquatic habitats will undoubtedly face great obstacles, but recent developments in this area are encouraging. This is especially true in more developed countries like the USA, Canada, Australia, and most of Europe, where a large amount of applied research-based legislation, planning frameworks, and policy guidance have promoted impact reversal to promote recovery and active habitat and ecosystem restoration. De-industrialization has also contributed to progress because it has created opportunities for urban renewal and working with nature as traditional heavy industry has declined. Environmental protection frequently receives less political attention than issues of public health, food security, and wealth creation in regions experiencing rapid economic development and population growth. In these environments, maintaining the biodiversity that already exists and healthy ecosystems ought to eliminate the need for further restoration work. While Europe, North America, Australasia, and more affluent areas like Singapore and Hong Kong can teach us valuable lessons, these lessons cannot be applied universally to other regions without modification.

Conceptual developments in ecology over the course of research, especially the acknowledgement of the role of ecosystem engineers and constructive interactions, offer a more predictive framework for restoring ecosystems. The significance of biodiversity for ecosystem function in freshwater and ocean environments has been highlighted in a great deal of research. The framework offered by taking ecosystem services into account allows one to weigh the advantages of actively restoring and promoting recovery over the costs of biodiversity loss and ecosystem dysfunction. Furthermore, the importance of habitat connectivity is becoming more widely acknowledged in both freshwater and marine systems, highlighting the necessity of expanding the scope and aspirations of restoration initiatives. Thus, more and more restoration initiatives are rethinking from just structural factors to ecosystem functioning. For example, there is growing recognition of the vital role that flow processes play in stream ecosystems. More and more, connectivity is being acknowledged as a key factor in the fight against the ocean sprawl of expanding man-made structures. Furthermore, there is a growing call for more evidence-based restoration that follows a methodical approach starting with the establishment of specific goals and ending with participative decision-making, evaluation, and the publication of both positive and negative outcomes along with adaptive management.

Emerging general principles include: restoring the hydrodynamic processes and geomorphological template; minimizing impacts and utilizing natural recovery mechanisms; focusing on keystone species, ecosystem engineers, and species or assemblages that form and shape habitats; regulating excessive bottom-up forcing, like eutrophication, and reestablishing top-down control by higher trophic levels. In open systems, minimize impacts and rely as much as possible on natural recovery; you could even try encouraging nature by supporting facilitation of successional processes. Greater scope for intervention and targeted restoration to a defined desired state is found in lakes and saline lagoons, which share many characteristics. More closed systems offer far more opportunities for intervention.

Though we remain optimistic, it is unlikely that universally applicable methods for restoring aquatic habitats that perform similarly well in various habitats and geographical areas can be found. Restoration managers should be cautious when making promises regarding the rehabilitation of highly modified systems, as it is improbable that near-natural biological community structures will be restored. This is especially true given the emergence of novel biological communities as a result of non-native species invasions. It is crucial to acknowledge that anthropogenic stressors cannot be completely eliminated in the majority of aquatic ecosystems.

Perhaps the best outcome of intervention is remediation. This suggests that since restoration is only the second-best option, conservation of the top-quality habitats that remain should be given top priority. ′Ersatz′-ecosystems are frequently the best we can hope for, particularly in urban areas. In general, biological indices should not be the only criteria used to evaluate the effectiveness of restoration initiatives. The rehabilitating of Liverpool's former commercial dock basins, the opening and restoration of a former sewer, the Emscher in northwest Germany, and the restoration of the Isar River inside Munich, Germany's city limits are all excellent examples. These highlight the importance of this kind of restoration in an urban environment, which needs to be evaluated not only on the basis of biological and ecological factors but also on the overall value to the general public. Enhancing biodiversity and restoring scarce but vital ecosystem services in the world's most populated areas are wise investments that frequently yield quick results.

Even today, ecologists, city or landscape architects, or planners are typically involved in the restoration process. Transdisciplinary approaches that simultaneously take into account ecological, technical, and socio-economic factors as well as public perception and engagement are likely to become most successful in making restoration more beneficial and acceptable in the future. However, it must be acknowledged that there is currently no universally applicable guideline on the most effective way to restore. Therefore, rather than being viewed as a failure, attempts that do not produce the desired results should be acknowledged as lessons to be learned and used for future improvement.

However, prevention is always preferable to treatment, so more should be done to preserve the biodiversity that already exists in aquatic ecosystems to guarantee the sustainability of these ecosystems and the provision of essential services. easing restrictions to allow many aquatic ecosystems to recover naturally. Modern conservation tools include actively repairing or correcting deteriorated systems where interventions are known or expected to be effective and enabling, as well as accelerating or enhancing natural recovery processes. To help guide best practices, such an intervention should be supported by evidence and closely observed, allowing for the learning of lessons from both successes and failures (Timothy et al., 2017).

REFERENCES

1. Airoldi, L. (2003). The effects of sedimentation on rocky coast assemblages. *Oceanography and Marine Biology, 41*, 161–236. eBook ISBN9780429217715.
2. Alnafissa, M., Kotb, A., Alamri, Y., Alagsam, F., & Alhashim, J. (2021). The impact of climatic and environmental changes on the sustainable yield of the Saudi's capture fisheries. *Journal of King Saud University–Science, 33*(5), 101458. Available from: https://www.sciencedirect.com/science/article/pii/S1018364721001191 (accessed on 16 October 2024).
3. Block, B. A., et al. (2011). Tracking apex marine predator movements in a dynamic ocean. *Nature, 475*(7354), 86–90. https://www.nature.com/articles/nature10082 (accessed on 16 October 2024).
4. Boerder, K., Bryndum-Buchholz, A., & Worm, B. (2017). Interactions of tuna fisheries with the Galápagos Marine Reserve. *Marine Ecology Progress Series, 585*, 1–15. https://doi.org/10.3354/meps12399.
5. Boersma, P. D., & Parrish, J. K. (1999). Limiting abuse: Marine protected areas, a limited solution. *Ecological Economics, 31*, 287–304.
6. Carr, M. H., et al. (2017). The central importance of ecological spatial connectivity to effective coastal marine protected areas and to meeting the challenges of climate change in the marine environment. *Aquatic Conservation: Marine and Freshwater Ecosystems, 27*(S1), 6–29. https://onlinelibrary.wiley.com/doi/epdf/10.1002/aqc.2800 (accessed on 16 October 2024).
7. Cesar, H. S. J., & van Beukering, P. J. H. (2004). Economic valuation of the coral reefs of Hawai'i. *Pacific Science, 58*(2), 231–242. https://scholarspace.manoa.hawaii.edu/bitstream/10125/2723/1/vol58n2-231-242.pdf (accessed on 16 October 2024).
8. Davidson, K., Hamnett, M., & Minato, C. (2003). *Economic Value of Hawaii's Nearshore Reefs*. Honolulu: Social Science Research Institute, University of Hawaii, Manoa. Available from: www.hawaii.edu/ssri/cron/files/econ_brochure.pdf (accessed on 16 October 2024).
9. Diz, D., Morgera, E., & Wilson, M. (2019). Marine policy special issue: SDG synergies for sustainable fisheries and poverty alleviation. *Marine Policy, 110*, 102860. Available from: https://www.sciencedirect.com/science/article/pii/S0308597X17305493 (accessed on 16 October 2024).
10. Edgar, G. J., et al. (2014). Global conservation outcomes depend on marine protected areas with five key features. *Nature, 506*(7487), 216–220. http://dx.doi.org/10.1038/nature13022.

11. Edgar, G. J., Ward, T. J., & Stuart-Smith, R. D. (2018). Rapid declines across Australian fishery stocks indicate global sustainability targets will not be achieved without an expanded network of 'no-fishing' reserves. *Aquatic Conservation: Marine and Freshwater Ecosystems*. https://doi.org/10.1002/aqc.2934.
12. Falkowski, P. (2012). Ocean science: The power of plankton. *Nature, 483*, S17–S20. http://dx.doi.org/10.1038/483S17a; Broecker, W. S. (1991). The great ocean conveyor. *Oceanography, 4*(2), 79–89. https://doi.org/10.5670/oceanog.1991.07.
13. Game, E. T., et al. (2009). Pelagic protected areas: The missing dimension in ocean conservation. *Trends in Ecology & Evolution, 24*(7), 360–369. https://doi.org/10.1016/j.tree.2009.01.011.
14. Geist, J. (2014). Trends and directions in water quality and habitat management in the context of the European Water Framework Directive. *Fisheries, 39*, 219–220. https://doi.org/10.1080/03632415.2014.903838.
15. Gell, F. R., & Roberts, C. M. (2003). Benefits beyond boundaries: The fishery effects of marine reserves. *Trends in Ecology & Evolution, 18*(9), 448–455. https://doi.org/10.1016/S0169-5347(03)00189-7.
16. Gell, F. R., & Roberts, C. M. (2003). Benefits beyond boundaries.
17. Gleick, P. H., Institute Pacific, Ajami, N., Christian-Smith, J., Cooley, H., Donnelly, K., Fulton, J., Ha, L. N., Heberger, M., Moore, E., et al. (2011). *The World's Water, Vol. 7: The Biennial Report on Freshwater Resources*. Island Press: Washington, DC. ISBN: 15287-7165.
18. Harrison, H. B., et al. (2012). Larval export from marine reserves and the recruitment benefit for fish and fisheries. *Current Biology, 22*(11), 1023–1028. https://doi.org/10.1016/j.cub.2012.04.008.
19. Hawkins, S. J., Allen, J. R., & Bray, S. (1999). Restoration of temperate marine and coastal ecosystems: Nudging nature. *Aquatic Conservation: Marine and Freshwater Ecosystems, 46*, 23–46. https://doi.org/10.1002/(SICI)1099-0755(199901/02)9:1<23::AID-AQC324>3.0.CO;2-C.
20. Hawkins, S., Allen, J., Ross, P., & Genner, M. J. (2002). Marine and coastal ecosystems. In M. R. Perrow & A. J. Davy (Eds.), *Handbook of Ecological Restoration* (pp. 121–148). Cambridge University Press. ISBN: 0521791294.
21. Hering, D., Borja, A., Carstensen, J., Carvalho, L., Elliott, M., Feld, C. K., Heiskanen, A. S., Johnson, R. K., Moe, J., Pont, D., et al. (2010). The European Water Framework Directive at the age of 10: A critical review of the achievements with recommendations for the future. *Science*

of the Total Environment, 408, 4007–4019. https://doi.org/10.1016/j.scitotenv.2010.05.031.

22. Hoegh-Guldberg, O., & Bruno, J. F. (2010). The impact of climate change on the world's marine ecosystems. *Science, 328*(5985), 1523–1528. http://science.sciencemag.org/content/328/5985/1523 (accessed on 16 October 2024).
23. Hoegh-Guldberg, O., et al. (2007). Coral reefs under rapid climate change and ocean acidification. *Science, 318*(5857), 1737–1742. http://science.sciencemag.org/content/318/5857/1737 (accessed on 16 October 2024).
24. Hultberg, H., & Andersson, I. B. (1982). Liming of acidified lakes: Induced long-term changes. *Water, Air, and Soil Pollution, 18*, 311–331. https://doi.org/10.1007/BF02419420.
25. Jones, C. G., Lawton, J. H., & Shachak, M. (1994). Organisms as ecosystem engineers. *Oikos, 69*, 373–386. https://doi.org/10.1890/03-0245.
26. Knauss, J. A., & Garfield, N. (2016). *Introduction to Physical Oceanography* (3rd ed.). Long Grove, IL: Waveland Press Inc.
27. Lester, S. E., & Halpern, B. S. (2008). Biological responses in marine no-take reserves versus partially protected areas. *Marine Ecology Progress Series, 367*, 49–56. https://www.int-res.com/articles/meps2008/367/m367p049.pdf (accessed on 16 October 2024).
28. Lester, S. E., et al. (2009). Biological effects within no-take marine reserves: A global synthesis. *Marine Ecology Progress Series, 384*, 33–46. https://www.int-res.com/articles/meps2009/384/m384p033.pdf (accessed on 16 October 2024).
29. Louzao, M., et al. (2019). Oceanographic habitat of an endangered Mediterranean procellariiform: Implications for marine protected areas. *Ecological Applications, 16*(5), 1683–1695.
30. McDermott, G. R., Meng, K. C., McDonald, G. G., & Costello, C. J. (2019). The blue paradox: Preemptive overfishing in marine reserves. *Proceedings of the National Academy of Sciences, 116*(12), 5319–5325. Available from: http://www.pnas.org/content/116/12/5319 (accessed on 16 October 2024).
31. Mueller, M., Pander, J., & Geist, J. (2011). The effects of weirs on structural stream habitat and biological communities. *Journal of Applied Ecology, 48*, 1450–1461. https://doi.org/10.1111/j.1365-2664.2011.02035.x.
32. Mueller, M., Pander, J., & Geist, J. (2014). The ecological value of stream restoration measures: An evaluation on ecosystem and target species scales. *Ecological Engineering, 62*, 129–139. https://doi.org/10.1016/j.ecoleng.2013.10.030.

33. O'Leary, B. C., et al. (2016). Effective coverage targets for ocean protection. *Conservation Letters, 9*(6), 398–404. https://doi.org/10.1111/conl.12247.
34. Pander, J., & Geist, J. (2013). Ecological indicators for stream restoration success. *Ecological Indicators, 30*, 106–118. https://doi.org/10.1016/j.ecolind.2013.01.039.
35. Roberts, C. M., et al. (2017). Marine reserves can mitigate and promote adaptation to climate change. *Proceedings of the National Academy of Sciences, 114*(24), 6167–6175. http://www.pnas.org/content/114/24/6167 (accessed on 16 October 2024).
36. Russ, G. R., & Alcala, A. C. (1996). Marine reserves: Rates and patterns of recovery and decline of large predatory fish. *Ecological Applications, 6*(3), 947–61. https://www.jstor.org/stable/2269497 (accessed on 16 October 2024).
37. Silliman, B. R., Schrack, E., He, Q., Cope, R., Santoni, A., van der Heide, T., Jacobi, R., Jacobi, M., & van de Koppel, J. (2015). Facilitation shifts paradigms and can amplify coastal restoration efforts. *Proceedings of the National Academy of Sciences of the United States of America, 112*, 14295–14300. https://doi.org/10.1073/pnas.151529711.
38. Sternecker, K., Cowley, D. E., & Geist, J. (2013). Factors influencing the success of salmonid egg development in river substratum. *Ecology of Freshwater Fish, 22*, 322–333. https://doi.org/10.1111/eff.12020.
39. White, T. D., et al. (2017). Assessing the effectiveness of a large marine protected area for reef shark conservation. *Biological Conservation, 207*, 64–71. https://www.sciencedirect.com/science/article/pii/S0006320717300678 (accessed on 16 October 2024)

Chapter **9**

Policy Frameworks and International Agreements

Contents

INTRODUCTION

Governments and international organizations create aquaculture policies, rules, and plans of action to monitor and support the sustainable growth of aquaculture—the farming of aquatic species like fish, shellfish, and seaweed. By ensuring that aquaculture practices do not harm ecosystems or deplete natural resources, this policy framework seeks to strike a balance between environmental protection and economic growth. The emphasis on sustainable practices—which include minimizing the use of chemicals and antibiotics, preventing the escape of farmed species into the wild, maintaining water quality, and sourcing feed responsibly—is at the heart of aquaculture policy. Policies also address the socio-economic aspects of aquaculture, including promoting market access for aquaculture products, guaranteeing fair labor practices, and assisting small-scale fish farmers. Scientific studies and feedback from stakeholders, such as business leaders, environmental organizations, and local communities, are used to inform the creation of aquaculture policies.

This cooperative approach seeks to develop regulations that are workable and efficient, promoting creativity and adaptability to shifting market demands and environmental conditions. Furthermore, since aquaculture operations frequently affect transboundary borders and involve international trade, cooperation between nations is essential. Organizations like the Food and Agriculture Organization (FAO) and regional bodies are important in promoting best practices, facilitating the exchange of knowledge and technology, and harmonizing policies and standards among nations.

Measures to enforce and monitor regulatory compliance are also included in aquaculture policy, guaranteeing that aquaculture operations follow set standards. This could entail systems for licensing and permits, frequent inspections, and sanctions for noncompliance. These regulatory frameworks are increasingly incorporating technology to improve accountability and transparency, such as blockchain for traceability and satellite monitoring.

Aquaculture policy also frequently takes into account adaptation and mitigation strategies for climate change, acknowledging the industry's susceptibility to effects like temperature swings, ocean acidification, and extreme weather. The creation of robust aquaculture systems, like integrated multi-trophic aquaculture, which blends several species to increase productivity and environmental sustainability, may be encouraged by policy.

Aquaculture policy addresses the nutritional and food security benefits of aquaculture, supporting initiatives to increase the availability and affordability of nutrient-dense seafood, in addition to environmental and economic considerations. This entails promoting technological innovation in aquaculture, enhancing the effectiveness of the supply chain, and promoting the use of farmed seafood as a sustainable source of protein.

9.1. FUNDAMENTALS OF AQUACULTURE POLICY

The rapidly growing aquaculture sector is anticipated to be crucial in supplying fish demand. Aquaculture has grown dramatically in recent years, making development planning more and more crucial. Through the provision of incentives and safeguards, attraction of investment, and promotion of development, effective planning will both stimulate and steer the evolution of the sector. Effective planning will also contribute to economic growth and the reduction of poverty by guaranteeing the sector's long-term social, environmental, and economic sustainability.

9.1.1. Aquaculture Governance

The formulation and execution of successful aquaculture development policies depend on good governance. To establish an environment that is conducive to attracting investment in the expansion of aquaculture, good governance built upon a solid and enforceable legal and institutional framework is essential. The aquaculture industry is improving its governance in response to growing concerns, which will inevitably result in higher production outputs. Since it will guarantee sustainable aquaculture development from an ecological, economic, and institutional standpoint, achieving good governance in aquaculture is an equally important goal as aquaculture development.

Since it is now widely acknowledged that challenges and setbacks in the development of aquaculture are no longer exclusively related to technological issues, good governance in the sector will prove advantageous in both the short and long run. The key components of good governance are participation, consensus orientation, strategic vision, responsiveness, effectiveness and efficiency, accountability, transparency, equity, and the rule of law. It consists of legally enforceable guidelines found in international treaties or national laws. At the national level, efforts are becoming apparent to address the sector's development holistically through ecosystem approaches, create national policies and implementation plans centered on the sector's sustainable growth, establish or amend legal frameworks and regulations, acknowledge the significance of stakeholders beyond the aquaculture industry in policy and participatory decision-making, establish producers' organizations and responsive institutions to address aquaculture, and support globally recognized codes of conduct in farming practices. Planning and policy-making properly are the main ways that governance can be strengthened. Effective planning for the growth of aquaculture, including the creation of policies and choices regarding how best to carry them out, supports the sector's sustainable development and good governance.

What is often called "policy-making," or planning processes, produce aquaculture policies and the means to carry them out. Making plans, strategies, and policies is a logical and rational process known as planning. The sector's overall direction, top priorities, and development objectives at all levels—regional, national, and worldwide—are reflected in policy. A strategy serves as a road map for putting a policy into action and includes tools, targets, and specific objectives to address problems that hinder the sector's growth and impact. A plan is a set of instructions for carrying out a strategy in order to accomplish its goals and put strategy instruments into action. It has a deadline, contains targeted programs and activities, and lists the materials needed to complete them. Plans and strategies can be combined into a single document due to their practical nature. On the other hand, policies are occasionally presented as standalone documents that are subsequently enhanced by a plan and strategy.

In the hierarchy of planning, a statement of policy goals is at the top. Typically, these are wide-ranging, qualitative objectives like ending poverty, generating jobs, or supporting sustainable development. The roles that the public, private, and civil society sectors should play in the development of aquaculture should also be specified in policies. The strategies follow the policy. To achieve the policy goals, these offer a set of doable targets. These goals may be both quantitative and qualitative, or they may be both. It is typical for several distinct strategies to support each policy goal.

Third in the hierarchy are agreed plans, which center on potential actions to carry out the policy. Plans comprise the tools and actions that will help each strategy aim be accomplished and ultimately result in the full implementation of the policy by achieving the goals of the policy. Plans should come first, strategies should come second, and policies should be developed based on these factors. Policies pertaining to aquaculture must take into account the main objectives of pertinent national, regional, and global agreements and goals that direct development overall.

The development of aquaculture policies will occur in the midst of other commitments, which decision-makers need to be aware of. As a result, the results of aquaculture planning will be in line with and pertinent to more general policy agendas concerning global trade, gender parity, poverty reduction, environmental preservation, adaptation to climate change, and general economic growth.

9.1.2. National Mariculture Policy

Mariculture contributes significantly to the provision of high-protein foods, raises the incomes of coastal dwellers, and fosters regional economic growth. Government policies and programs are necessary to support the growth of mariculture. Mariculture policies cover a wide range of topics, including mariculture, sea use, the marine environment, and marine resources. They are associated with laws and regulations pertaining to mariculture, as well as national ocean strategies, departmental instructions, and industry standards. These policies should encourage the responsible development, coordination, and management of mariculture production nationwide while guaranteeing environmental sustainability and the socioeconomic advancement of stakeholders, thereby paving the way for the widespread adoption of mariculture technologies to meet the increased demand for seafood.

The goal of the mariculture policy is to improve "widespread adoption of mariculture technologies to meet the additional seafood demand by facilitating responsible development, coordination, and management of mariculture production in the country while ensuring environmental sustainability and socioeconomic upliftment of stakeholders." The vision of the policy is to increase the amount of farmed seafood produced in the nation by creating "a sustainable and responsible mariculture sector that contributes to the food and nutritional security of the country and enhances the quality of life of the stakeholders." A specialized subset of aquaculture known as mariculture involves the economically significant marine plants and animals being raised in the sea or any other naturally occurring water body with tidal influence. It also includes onshore facilities such as hatcheries, brood banks, nursery rearing, and grow-out systems that use seawater.

9.1.3. Status and Opportunities of Mariculture

Over the past 20 years, aquaculture has grown at the fastest rate in the world's food production industry, with an annual growth rate of more than 6%. Inland aquaculture in India is a rapidly expanding industry that has grown steadily over the years to replace the diminishing capture fisheries. Aquaculture's fastest-growing subsector, mariculture, has enormous growth potential. 20 to 30% of the 112 million tons of production from marine waters in 2020 came from mariculture. About 38% of the world's aquaculture production comes from 33.6 million tons of production produced in marine waters. Given the rising demand for marine protein and the restricted potential for increasing wild fishery harvests, mariculture clearly offers a great opportunity to increase seafood production. In the Indian region, the potential for annual mariculture production based on available area is estimated to be between 4 and 8 million tons, while the current estimate is less than 0.01 million tons annually. India's achievements in developing brackish water and inland aquaculture are consistent with the potential for the growth of the mariculture production industry.

Initiated in the 1970s by the Central Salt and Marine Chemical Research Institute (CSIR-CSMCRI) through research and development, mariculture began with the cultivation of *Gracilaria edulis* seaweed on Krusadi Island in Tamil Nadu. In the early 1980s, the Central Marine Fisheries Research Institute (ICAR-CMFRI) followed suit by commercially cultivating bivalves on a small scale. Furthermore, the National Institute of Ocean Technology (NIOT), the Marine Products Export Development Authority (MPEDA), and the Central Institute of Brackishwater Aquaculture (ICAR-CIBA) made substantial contributions to the advancement of mariculture during the 1990s. With funding support from the Department of Animal Husbandry, Dairying and Fisheries (DADF), Ministry of Agriculture, Government of India, ICAR-CMFRI established the first open sea floating cage in Visakhapatnam in 2005, marking the inception of open sea mariculture in India. Self-help organizations, fisher societies, and small-scale entrepreneurs have been instrumental in the rapid expansion of cage mariculture along both coasts in nearshore waters, thanks to further technological advancements and their adoption. Technologies available in India include pearl oyster, seaweed, ornamental shrimp, mussels, oysters, clams, Indian white shrimp, green tiger shrimp, blue swimmer crab, and ornamental fish. Fin fish farms include cobia, pompano, sea bass, groupers, snappers, breams, and ornamental fish farming.

The main Objectives of the national mariculture policy:

- To enhance mariculture production in the country and increase income, employment and entrepreneurship opportunities sustainably and responsibly.
- To promote cooperative partnership in mariculture by encouraging the infrastructural, technical and other inputs.
- To adopt an environmentally sustainable approach for the development of mariculture.
- To provide an enabling environment for the sustainable development of mariculture in India by providing the required policy and legal framework and support to entrepreneurs.

9.1.4. Priorities and Development Goals

- **Mariculture Area Development:** There will be designated areas for various mariculture activities based on suitability. Developing onshore support infrastructure and providing the required logistical support for breeding, nursery rearing, feed storage, fish processing, and domestic/international trade are essential in mariculture zones and technology parks, which have the potential to host a large number of small-scale mariculture units.

- **Leasing and Licensing:** The Union Government's guidelines for small and large-scale mariculture enterprises will be followed by the fisheries departments of the coastal states when leasing out or licensing the waters for mariculture.
- **Mariculture Systems and Species:** The enhancement of current technologies to meet international standards, biosecurity requirements, and code of practices will receive a lot of attention. Once new aquaculture systems have been validated at the farm level, they will be promoted in designated areas. All naturally occurring marine food and nonfood species with potential for mariculture will be supported.
- **Precautionary Approach to Environmental Sustainability:** The application of ecosystem approach principles to mariculture will facilitate improved sectoral management at various scales, considering stakeholder and environmental constraints as well as user interests.
- **Seed and Feed**: To maintain the momentum of the planned expansion of the mariculture sector in the nation, it is imperative to guarantee the availability of seed material for the targeted species. There will be state-run finfish/shellfish hatcheries, nursery units, seed banks, and genetically modified/specific pathogen resistant/specific pathogen-free brood banks established. Ensuring the quality of the final feed and the traceability of the feed ingredients used will help to promote the eco-labeling of appropriate mariculture production systems and guarantee premium prices for the produce.
- **Food Safety and Health Management:** Recording the source of inputs like feed, seed, approved veterinary medications, vaccines and antibiotics, additives, and chemicals will ensure traceability and record-keeping of farming activities and inputs that impact food safety. The World Organization for Animal Health (OIE) and the FAO-CCRF Technical Guidelines on Health Management for Responsible Movement of Live Aquatic Animals are two important resources that mariculture operations should consider when implementing aquatic animal health management programs that are set up in accordance with applicable national legislation and regulations.

- **Capacity Building and Extension:** To promote mariculture and make it more widely known in India, the government will work to improve the abilities of traditional fishermen and other relevant parties. As a result, interested fishermen will be able to switch from fishing to more profitable and productive mariculture endeavors. Frontline participatory demonstrations and technology transfer will be undertaken with support and hand holding to give thrust and impetus to new candidate species/technologies/areas for mariculture.
- **Insurance and Support Services:** Both anthropogenic activities and natural disasters pose a risk to mariculture operations. Appropriate insurance programs must be implemented to close this gap and will motivate private insurance providers to create insurance plans tailored to the industry. Adequate support will be provided to entrepreneurs due to the lengthy gestation period and high investment required for cultured marine pearls.
- **Processing, Value Addition and Market Support:** The promotion of value addition and effective market logistics is intended to reduce post-harvest losses and maintain the nutritional value of fish. Public awareness campaigns will raise the nutritional value of fish.
- **Institutional Mechanisms**: Many organizations dedicated to the nation's mariculture development will collaborate closely with the private sector, fishermen's cooperatives, and self-help groups (SHGs) in a coordinated effort to advance mariculture.
- **Research and Development (R&D):** Mariculture is a new industry, and to fully realize its enormous potential, significant funding will be given to research and development (R&D) initiatives. An implementation plan, guidelines, and timelines for addressing emerging issues will be developed to guarantee the growth that is envisaged for NMP. It is hoped that this policy will also be able to take advantage of the significant opportunities that the nation's Blue Economy initiative is expected to present.

9.1.5. Legal Requirements for Developing Industrial Aquaculture

The operator must have a safe and secure right to water of high quality and to carry out aquaculture operations on the land where the fish farm will be situated, according to the aquaculture legal framework. It must also plan and regulate the conflicting uses of resources like land and water to establish control over the industry through an authorization or licensing system. By assessing the suitability of suggested fish farm locations and the possible environmental effects of their operations, licenses also guarantee the environmental sustainability of aquaculture.

9.1.5.1. Licensing System

The first step in managing aquaculture operations is to mandate that an operator obtain a license or authorization prior to starting any operations. It doesn't matter what the document is titled; the term used will rely on local customs and common usage in that nation, whether it is an authorization, permit, or license. For example, the Ukrainian legislature was hesitant to use the word license in the 2012 Law on Aquaculture because it would conflict with the pre-existing Law on Licenses. But the document needs to accomplish two things in particular. Initially, the permit needs to give the government authorities the legal justification for managing all aquaculture operations and overseeing the effects they have on the environment. As a result, the government can evaluate the proposal's environmental sustainability and set requirements requiring the farm to run sustainably. Second, the permit ought to explicitly grant farmers the authority to run the aquaculture facility, provided that the operator abides by all applicable environmental regulations, permit requirements, and aquaculture practice codes.

The permit may impose conditions for environmental or other reasons, or it may be a requirement in its entirety for the establishment of an aquaculture farm. Systems for granting permits could be general or limited to specific types of farms. The license may come with an additional cost. Any system of public land use control that mandates that the owner of land obtain specific authorization before developing or changing the use of the land for specific purposes is commonly referred to as "development licensing." This suggests that unapproved development on public or private property without the required authorization will result in the application of a sanction. A controlled licensing system is used in the Democratic People's Republic of Korea, Canada, Norway, Australia, Gabon, Madagascar, Zambia, and Cambodia.

9.1.5.2. Access to Land

Each national jurisdiction's general land ownership and holding system affects the acquisition of land for aquaculture operations. Here, there is a difference between nations where private land ownership is permitted and nations where state control over land ownership exists. How much the acquisition of land rights for aquaculture imposes on the land holder in terms of social and environmental requirements is a crucial question. A solid legal title to the property the farm is situated on must be obtained by the farmer. Since land laws usually permit farmers to obtain ownership interests, this is usually not a significant issue in the case of freshwater aquaculture. However, there might be extra considerations if the land is state-owned, in a coastal area, or has a traditional tenure structure. The question of whether rights for using the body of water beneath and next to it should also be granted arises in some nations where land rights are entirely under state control. The laws of these nations, which are frequently found in the Caucasus and Central Asian regions, speak of land leasing rather than licenses for aquaculture facilities.

Competition for resources between fish farming and other farming activities can occasionally lead to issues. Land, water, labor, time, and nutrient inputs are needed for fish farming. It must therefore be incorporated into the current farming system. The efficiency of the farm as a whole can be raised by integrating various farming operations, which will raise the likelihood of adoption. Activities that are complementary to one another and ensure a more effective use of resources can also be included in integration. The best opportunities for success lie in making better and more efficient use of the resources that constrain total farm production. For instance, increasing production may be constrained by the amount of water available, but increasing water use efficiency can raise overall farm output. The goal of integrating aquaculture and agriculture is to improve labor, water use, and farm production diversification in order to increase overall production. In conclusion, integrating agriculture with other uses like aquaculture and inland fisheries can lead to resource-use efficiency. It would become a reality if agricultural planners and policymakers were better informed about the potential and financial advantages of this kind of integration. As resource pressure increases, there will likely be a greater incentive for this kind of integration. Additionally, research that creates novel integrated aquaculture-agriculture systems and other cutting-edge systems that can successfully address resource and environmental constraints needs to be encouraged and supported.

9.1.5.3. Access to Water Resources

While having a sufficient amount of high-quality water is necessary for any aquaculture installation, meeting this requirement may also put other water users in conflict who want to use the same water for drinking water supply, agriculture, fishing, industry, or recreational activities. Due to this competition, access to water for fish farming may become severely problematic, and in developed nations, access to the shore may present significant challenges. Water rights can be awarded with many restrictions, such as a cap on the volume of water that can be taken out with a given license. State governments typically use concessions or environmental licenses to regulate water use.

9.2. AQUACULTURE POLICY CONTEXT

Despite being a relatively recent commercial endeavor globally, aquaculture already holds a prominent place in the seafood production industry. Aquaculture has emerged during a period of increased public scrutiny of industry and government efforts to mitigate potentially negative consequences of industrial activity, even though it offers enormous socio-economic benefits. While people want aquaculture to reach its full potential, they are also worried about things like food safety, aquatic animal health, habitat interactions, water quality, escapes of farmed salmon, and navigational safety. Cabinet endorsed the Federal Aquaculture Development Strategy (FADS) in 1995, realizing the potential benefits of aquaculture for human welfare as well as the necessity of ensuring the sustainable use of aquatic resources. FADS, which offered a roadmap to assist government in facilitating the growth of the aquaculture industry, was the outcome of in-depth discussions with stakeholders in the aquaculture sector. DFO's position as the principal federal agency for aquaculture development was also confirmed by FADS. In response to FADS and more recent opportunities and challenges related to aquaculture development, DFO developed its Aquaculture Policy Framework. Serving as the principal federal agency for the development of aquaculture, DFO is dedicated to boosting the industry's competitiveness on a worldwide scale as well as public trust that aquaculture is being developed sustainably. The purpose of this policy framework is to help DFO accomplish these two crucial goals of public policy.

9.2.1. Global Trends

Developing an effective policy framework for aquaculture requires a thorough understanding of the major international pressures, especially considering the growing significance of the sector's contribution to the global economy and seafood supply. The approach to aquaculture development is shaped by the following global trends.

9.2.1.1. Economic Globalization

The volume of international trade and economic activity has skyrocketed in recent years. Many factors, such as the decline in national trade and investment barriers, the formation of regional free trade agreements, and technological advancements that have dramatically lowered the cost of communications, processing, and transportation, have contributed to the increased globalization of economic activity. Increases in capital market flows, foreign direct investment, and international trade that followed have had a major impact on the socioeconomic landscape and opened up a multitude of business opportunities.

The global economy is becoming more complex, which has increased business risk and made it necessary for nations that want to compete globally to offer stable and encouraging regulatory, policy, and fiscal environments that allow businesses to advance social and economic development. It is anticipated that the global economy will remain dynamic, necessitating creative domestic policies. The aquaculture industry is especially in need of a stable and encouraging environment. Despite having many advantageous developmental characteristics, its share of the world's aquaculture production is still very small—only 0.2 percent. Growing foreign ownership and capital mobility necessitate competition with other jurisdictions that are supportive of and well-suited for aquaculture development. To realize potential in freshwater and marine aquaculture, a stable and encouraging legislative and regulatory framework is essential.

9.2.1.2. Pressures on Natural Resources

The world's natural resources are under strain due to population growth and economic expansion, which is causing environmental and social issues of global proportion like air pollution, climate change, threats to water availability and quality, and declining biodiversity. These issues are especially important for the many coastal communities around the world that are growing more crowded and thus both contributing to and at risk from environmental degradation. Growing environmental deterioration has brought human health and security concerns—including the production and distribution of food worldwide—more attention. Governments and the private sector are expected to take action to ensure the long-term sustainability of the world's natural resources through responsible stewardship, sound environmental management, and scientific innovation, given the potentially dire consequences of environmental degradation. Aquaculture has the potential to significantly impact the development of human wealth, food security, and other facets of society. In order to realize this potential, policies pertaining to safe navigation, clean water, healthy products, and shared resource use must be created and put into action within the framework of sustainability. Both the general public and the aquaculture industry, whose survival depends on healthy resources, stand to benefit from these goals.

9.2.1.3. Knowledge-Based Society

Every aspect of human society has been profoundly impacted by and is still being transformed by scientific and technological advancements, particularly those in the field of information technology. The capacity to create, disseminate, and use knowledge has led to a great deal of significant discoveries that have improved society and raised standards of living. Science and technological advancements have also brought forth a host of new challenges for society on the legal, ethical, cultural, economic, and environmental fronts. With the continued development of both new and established technologies, as well as the increased integration of technological disciplines, it is anticipated that the magnitude and pace of change resulting from scientific and technological innovation will increase (e.g., communications, engineering, materials science, and molecular technologies). Recent scientific and technological advancements in aquaculture have dramatically increased production efficiency, boosted competitiveness, and improved environmental sustainability. Additionally, progress is being made in identifying new uses for aquaculture products outside of the food production sector, in fields like bioengineering and biomedicine. The enormous potential for innovation in the aquaculture industry and the advantages that such innovation will bring to a range of disciplines must be acknowledged in government policy for aquaculture.

9.2.1.4. Modern Governance

Globalization and the convergence of the social, economic, and environmental agendas have made governance more complex. The use of international organizations and agreements to support multilateral negotiations over trade, environmental cooperation, and human development has increased as a result of increased trade and globalization. Governments must respond to demands from for-profit and nonprofit organizations, as well as from citizens in general, for greater government accountability and transparency, in addition to managing and benefiting from the many aspects of globalization. These elements are putting traditional models of governance to the test and forcing governments to alter their methods of operation. A dispersal of power away from governments is often triggered by the same factors that are undermining traditional models of governance. As a result of the power being distributed, there are now expectations in society for non-state actors. This is especially important for private companies that are being asked to prove that they are committed to environmental and social responsibility. This policy framework must not only support the department's regulatory role but also pave the way for innovative governance strategies in high-priority areas like site access (e.g., proactive planning), government initiatives supporting industry development, and scientific research and development (R&D). These strategies need to incorporate procedures for involving stakeholders, Aboriginal groups, and aquaculturists in decision-making processes and raising their level of accountability with regard to important sustainability goals.

9.2.2. Federal Government Priorities

The government is reacting to the challenges as well as the opportunities that these new international trends and domestic pressures bring. Building a stronger, more secure, and better quality of life for all Canadians is its main objective. It will open up and distribute opportunities for Canadians to take part in this joint venture. The federal government is acting in the following priority areas to achieve this goal:

9.2.2.1. Innovation

Understanding that creativity and knowledge are essential for success in today's knowledge-based economy, the federal government has committed to rank among the world's most innovative nations. To accomplish this, it is collaborating with the private sector and provincial governments to take the lead in creating and using new knowledge. By 2010, the federal government intends to have doubled its RandD investment. This investment will prioritize supporting collaborative R&D initiatives and expediting Canada's ability to commercialize new technologies, in addition to bolstering institutional research capacity. Its focus also encompasses high standards of environmental stewardship and support for initiatives aimed at diversification and value-added growth. The government is funding numerous R&D projects because it sees aquaculture as a crucial component of the knowledge-based economy. These include Aquanet, the Network Center of Excellence for Aquaculture, and the Aquaculture Collaborative Research and Development Program (ACRDP) of DFO, which both seek to increase industry innovation and competitiveness. Exciting new pharmacological and medical applications are on the horizon, in addition to new technologies and procedures targeted at eradicating escapes and disease incidence. The federal government is working to fully utilize aquaculture by boosting industry competitiveness, promoting sustainability, and fostering public trust in the process through these and other technological advancements.

9.2.2.2. Strong Communities

To preserve the fundamental ideals of our nation, Canadian communities must grow stronger and more vibrant. The goal of the federal government is to make Canadian communities more affluent and desirable places to work and live. While maintaining traditional ways of life, the federal government is assisting Canadians in taking a more active role in the knowledge-based economy through wise investments in community development and infrastructure. Many coastal communities depend on aquaculture for their future, as it is essential to the expansion of the seafood industry. Aquaculture, as an economic endeavor that yields numerous direct and indirect advantages, can facilitate the retention of rural and coastal Canadians in their communities, thereby fortifying Canada's social cohesion and international recognition. Freshwater and marine commercial aquaculture can also benefit interested Aboriginal communities by decreasing economic inequality and providing a means of engagement with the knowledge-based economy.

9.2.2.3. Healthy Environment

The federal government sees a healthy environment as a vital component of our economy and high standard of living, and it acknowledges its responsibility for safeguarding Canada's natural resource endowment. The preservation of natural areas, clean air, and clean water are top priorities for the federal government. For the aquaculture industry to remain viable, clean water is especially vital when it comes to the culture of fish and shellfish. The federal government of Canada is collaborating with provincial and territorial governments, local communities, Aboriginal groups, industry, and public interest groups to enhance the efficacy of environmental protection and conservation measures through more integrated approaches to aquatic resources management. This is due to the government's recognition of the economic, social, and environmental significance of Canada's freshwater and oceans.

9.2.2.4. A Strong International Voice

In this context, partnerships that foster innovation and collaboration are the key to gaining a competitive edge. These kinds of partnerships foster the emergence of new governance models, the advancement of science and technology, the well-being of communities, and the preservation of the environment. The government is collaborating with other governmental branches to establish itself as a global pioneer in the advancement of aquaculture. It is achieving this in two ways: i) by making sure that Canadian policy instruments are effective, equitable, and suitable and that they represent Canadian values and the principles of sustainable development; and ii) by encouraging innovation in important developmental domains. Apart from collaborating with other national administrations, the Government aims to establish favorable conditions for the prosperous and conscientious growth of aquaculture abroad. It will not only lead the way in their development but also adhere to strict international standards pertaining to aquaculture production.

9.2.3. Aquaculture Opportunity

An essential component of the seafood chain, aquaculture is one of the food production sectors with the quickest rate of growth in the world. In unison, the fishing, aquaculture, and associated service sectors yield substantial financial gains and socio-economic advantages for Canadians across the nation. Over the past 10 years, aquaculture has steadily increased its share of global fisheries production, to the point where it now provides a third of all fish available for human consumption. According to UN Food and Agriculture Organization projections, aquaculture will account for over half of fish consumed by 2030 and control the majority of fish supplies. In fact, the yearly demand for seafood is expected to surpass the yearly production of wild fisheries by approximately 55 million tons by 2025, thereby offering enormous prospects for the aquaculture industry. Given the rising global demand and the fact that natural stocks are currently mostly at or above their maximum capture potential, it is obvious that aquaculture will be crucial to meeting future demand and ensuring the stability of the world food production system. Biomedicine, pharmacology, and bioengineering are three fields in which significant advancements can be made, in addition to the opportunities associated with the production of fish supplies and all associated services supporting the aquaculture industry.

Every province and territory has aquaculture operations, which produce a wide range of species using different culture techniques. The astounding variety of species that are being cultivated is a testament to the industry's creativity and perseverance. Aquaculture accounted for 22.8% of the total value extracted from living aquatic resources in 2000. This percentage is rising yearly, with industry growth rates historically averaging close to 30% annually. Over 14,000 people are employed in the aquaculture industry, which generates close to $1 billion in direct and indirect economic activity yearly. Right now, they can benefit specifically from the following:

- job creation and contribution to economic growth;
- increased social cohesion through diversification and wealth generation in rural and coastal areas;
- opportunities for Aboriginal communities;
- reduced pressure on wild fish stocks, helping to sustain and enhance the wild fishery;
- the assurance of high-quality, safe, competitively priced and nutritious seafood; and
- expansion of domestic and export markets for fish and seafood products.

Based on recent aquaculture sector growth trends and significant aquaculture development potential, the sector's contribution to the economy could reach $3 billion by 2010.

9.2.4. Aquaculture in the Context of Sustainable Development

The World Commission on Environment and Sustainable Development initially defined sustainable development as development that satisfies present needs without jeopardizing the capacity of future generations to satiate their own needs. This definition has been adopted by the government. It is acknowledged that attaining sustainable development requires cooperation from public interest groups, businesses, governments, and private citizens. The following components of aquaculture are included in the context of sustainable development:

- maintaining or enhancing the quality of life and the environment for present and future generations;
- adopting an ecosystem approach and respecting the interests and values of all resource users and considering those interests and values in decision-making;
- identifying, planning, developing, operating, harvesting, processing, and when necessary disposing of aquacultural products in the most efficient, competitive and environmentally responsible manner, using best practices;
- respecting constitutionally protected Aboriginal and treaty rights;
- creating and sharing knowledge to promote innovation, continuous learning and efficiency;
- securing the participation of stakeholders, individuals and communities in decision-making to ensure the best use of aquatic space; and
- making decisions in a fair, transparent and inclusive manner.

9.2.5. Conservation, Restoration and Management

Despite making up only 0.01% of the Earth's surface, rivers, lakes, and wetlands are home to a disproportionately high proportion of the planet's biodiversity. There are an estimated 44,000 scientifically named species of freshwater biota, and freshwater fish alone make up about 25% of all living vertebrate species. Freshwater biodiversity is generally more threatened than terrestrial biodiversity, according to tallies of endangered species. For instance, of the species included in the 2000 edition of the International Union for Conservation of Nature (IUCN) Red List, 20% were amphibians and 30% were fish (mostly freshwater). Numerous threats confront the biodiversity of freshwater ecosystems. These include the indirect effects of land-based activities like logging, agriculture, industry, housing development, and mining, as well as the direct effects of dams, exotic species, overfishing, pollution, stream channelization, water withdrawals, and diversions. The protection of freshwater biodiversity requires the development and application of conservation strategies. The goals of the aquatic conservation strategy are to: conserve and preserve the ecological integrity of watersheds and aquatic ecosystems.

- In order to safeguard the aquatic systems that species, populations, and communities have evolved specifically for, it is important to preserve and protect the distribution, diversity, and complexity of landscape- and watershed-scale features. Preserve the interconnectedness of space and time both within and among watersheds. Wetlands, headwater tributaries, upslope regions, and flood plains are examples of lateral, longitudinal, and drainage network connections. These connections must offer physically and chemically clear paths to vital locations for aquatic and riparian-dependent species to meet their life cycle needs. Preserve and mend the aquatic system's physical characteristics, such as its banks, shorelines, and bottom patterns.
- Keep the water quality at the level required to sustain aquatic, wetland, and riparian ecosystems. Water quality must be kept within a range that supports the migration, growth, survival, and reproduction of the individuals that make up riparian and aquatic communities as well as the biological, physical, and chemical integrity of the system. Preserve the sedimentation regime that facilitated the evolution of aquatic ecosystems. The timing, volume, rate, and type of sediment input, storage, and transport are all components of the sediment regime. Keep stream flows high enough to maintain and create riparian, aquatic, and wetland habitats as well as to preserve patterns of wood, nutrient, and sediment routing (i.e., the flow of woody debris via the water system. Peak, high, and low flow timing, magnitude, duration, and spatial distribution need to be preserved. Preserve the water table elevation in meadows and wetlands, as well as the timing, variability, and duration of floodplain inundation.
- In order to maintain appropriate rates of surface erosion, bank erosion, and channel migration, as well as to supply enough amounts and distributions of coarse woody debris to sustain physical complexity and stability, it is important to maintain and conserve the species composition and structural diversity of plant communities in riparian zones and wetlands. To support evenly distributed populations of native plants, invertebrates, and vertebrate riparian-dependent species, habitat must be preserved.
- The conservation and management of aquatic ecosystems necessitate cooperative research combining natural, social, and multidisciplinary

studies with the goal of understanding. the different elements as a crucial tool for creating long-term conservation plans, such as socioeconomic dependency, biodiversity, water quality monitoring, and other activities. This calls for multidisciplinary professionals with training who can start educational programs targeted at increasing public awareness and comprehension of aquatic ecosystem restoration, goals, and techniques. These professionals can then spread the word about the significance of ecosystems at nearby schools, colleges, and research institutions. In the vicinity of the water bodies, actively involved schools and colleges might appreciate the chance to offer experiential environmental education, which might involve establishing laboratory facilities at the location. Regular water body monitoring (with on-site lab facilities) would supply essential data for management and conservation.

A key component of any conservation program should be watershed restoration, which will promote riparian habitat, habitat recovery, and improved water quality. The restoration of riparian vegetation, the prevention and control of pollution and sediment production, and the complexity of the in-stream habitat are the three most crucial elements of an aquatic restoration program.

9.3. ADMINISTRATION AND LEGISLATION

The majority of nations lack a dedicated aquaculture policy document, and instead include aquaculture primarily in their Fishery Sector Development Policy. Other strategic policy documents also make reference to it; the two main ones that deal with it are those related to industry and environment. Aquaculture development is primarily based on development plans created by the authorities in charge of overseeing the industry, but without official approval, in the absence of specific policies. With three notable exceptions, informal consultations make up the majority of participatory mechanisms involved in policy formulation. First, representatives from the aquaculture industry are part of the official Consultative Committee on Fisheries and Aquaculture that has been established in Cataluña, Spain. The second exception is Greece, where the Ministry of Agriculture oversees the work of the Agricultural Policy Council (APC), a consultative body that includes representatives from the Ministry as well as organizations representing science, farmers, and academic institutions. France is the third country where inter-professional committees exist.

9.3.1. Administrative Framework

Most nations have a centralized administrative structure, with capable Ministries creating policy. For instance, regional districts implement national and regional policy, which is defined by central government and ministries in Israel. Some nations have implemented decentralization, such as Bulgaria, where it was first implemented in December 1998. Bulgaria now has 28 districts instead of just 9. Increasing local (municipal) responsibilities is the goal of this fragmentation. Other nations with decentralized administrations and local autonomy include Spain and Romania. The Prefects oversee resource administration in Romania and speak for the Central Government. In Spain, the Aquaculture sector is solely under the jurisdiction of the Autonomous Communities. A ministry that oversees agriculture typically oversees fisheries and aquaculture. The Ministry of Agriculture and Rural Affairs (MARA) in Turkey is responsible for managing agriculture, fisheries, and husbandry, which includes aquaculture. However, MARA, the Ministry of Transportation (Department of Marine), the Ministry of Health, the Ministry of Forestry, the Department of Water Works, the Ministry of Treasury, and other authorities like the Ministry of Tourism, the Ministry of Internal Affairs, the Ministry of Culture, the Ministry of Commerce and Trade, and the Ministry of Finance share administrative authority over aquaculture. When it comes to the administrative facets of ethical aquaculture, administrative overlap and interference are the most prevalent issues. This issue results from a disorganized administrative environment and imprecise industry regulations. Another major issue is resource user competition, especially when it comes to using coastal land and sea areas (e.g., Cyprus).

9.3.2. Legal Framework

9.3.2.1. Basic Legislative Climate for Aquaculture

- **Access to Public Land is Regulated in All the Countries:** Certain states have enacted laws specifically addressing public land access (i.e., Cyprus: The Republic of Cyprus Immovable Property Regulations; Croatia: The Agricultural Lands Act, 1995). This legislation does not specifically address the use of land for aquaculture in most cases. Certain laws (like those in Croatia) mention state land but permit sales, leases, and concessions of land, including that which has fishponds already in place. Nonetheless, some nations—like Malta—have laws that expressly mention aquaculture. Declarative Decree No. 83–228 (March 22, 1983) governs the issuance of permits for mariculture, whereas Article 23 of Law No. 9413 (January 31, 1994) relates to permits for the establishment of aquaculture, or pêcherie fixes. Turkey's Law No. Aquaculture licenses may be granted in public areas (freshwaters, sea, and land) under Section 1380.

- **In All Countries a Licensing or Similar System is in Place to Regulate the Use of Freshwater:** As an illustration, the Water Works Law No. In Cyprus, Section 341 stipulates that any farm using river or reservoir water must obtain a license. In Greece and Cataluña (Spain), licenses are also required; in Turkey, however, the MARA issues licenses following clearance from the Ministry of Forestry, the Ministry of Energy and Natural Resources, the Department of Water Works, and local authorities. Licensing is another method used to control access to marine waters. In Cyprus, for example, obtaining a license is necessary before engaging in any shoreline activity (Wells Law Cap. 351 and the Cap of the Foreshore Protection Law. Additionally, for the construction of piers (the Piers Law Cap. 78)]; or concessions [as in Croatia, where the departments of maritime affairs of the coastal counties are obliged to grant concessions]. In Malta, concessions are also utilized, subject to the Planning Authority's development permission, and are provided by the Malta Development Corporation. Spain and Italy both use comparable systems. Other nations, like Romania, have laws protecting the environment (Law 137/1995), which makes access to marine waters contingent upon environmental stewardship.

When developing economic and legal instruments, aquaculture specialists are consulted. Nevertheless, the private sector may only have a secondary role in certain situations. Certain laws (such as Law 137/95 of the Romanian Ministry of Water, Forest, and Environment Protection, which stipulates that NGOs, central and local authorities, and industry experts can participate in the decision-making process on various subjects) exist in this regard. Government authorities in the majority of the consulted countries receive advice from advisory committees or councils when developing legislation related to land area management, coastal area management, and water management. There are comparable temporary working groups or ad hoc committees in places where there isn't an advisory committee of that kind. It is standard procedure for experts in fisheries and aquaculture to serve on those committees.

9.3.2.2. Specific Legislation Applicable to Aquaculture

Every nation where reports were received has either a fisheries law that covers aquaculture (i.e., Tunisia), or an exclusive aquaculture law. A draft Fisheries and Water Act for Bulgaria is awaiting parliamentary approval. A Marine Fishery Act No. exists for Croatia. 46/97 that governs mariculture, the Freshwater Fisheries Act No. 16/86, 34/89, and a newly proposed freshwater fishery bill that is not yet enacted. A special aquaculture law has been drafted in Cyprus and is anticipated to be approved shortly. Aquaculture in Malta is subject to several regulations, including those pertaining to marine vegetation licenses, veterinary medicinal products, and environmental protection. Federal and provincial laws are identical in practically every country in the Mediterranean region. The laws governing freshwater and marine aquaculture are essentially the same. However, because different authorities are involved, getting approval for concessions on freshwater and marine aquaculture differs. For instance, permission from the Department of Fisheries, the Town Planning Department, and the Water Development Department is needed for freshwater aquaculture, while approval from the Council of Ministers, the Department of Fisheries, and the Ports Authority are required for marine aquaculture licenses in Cyprus. Similar circumstances exist in Israel, where approval for offshore marine aquaculture is needed from the Department of Fisheries, the Committee of Coastal Waters, the Ministry of Environment, and the military. In Israel, inland aquaculture requires permission from the appropriate regional council in addition to documentation of land and water ownership.

With the exception of France and Greece, the majority of nations acknowledge the lack of clarity surrounding the rights and obligations of aquaculture producers. Morocco argues that the underdeveloped state of the aquaculture industry justifies this lack of clarity. Every nation acknowledges disputes between the various participating authorities and, occasionally, disputes arising from the relevant laws. Conflicts, for instance, exist in Romania between Water Law No. 107/96 as well as Rule No. Government document 390 (24.07). 97) was mentioned. These inconsistencies stem from the fact that water issues are administered by the Ministry of Water, Forestry, and Environmental Protection, while land issues are overseen by the Ministry of Food and Agriculture. The second ministry is in charge of monitoring and maintaining water quality, while the first ministry handles the financial aspects of fishing. With the exception of Israel, Bulgaria, and Romania, the majority of nations have established regulations governing the issuance of permits. It is typical for freshwater and marine aquaculture to have different permitting processes. Regarding the requirement for an Environmental Impact Assessment (EIA) prior to the establishment of an aquaculture farm, there are no shared experiences. In certain situations, the expected production level of the aquaculture farm determines whether an EIA is necessary. For instance, in Croatia, farms with an anticipated production of more than 50 tone are required to complete an EIA. The criteria in Egypt for requiring an EIA are determined by the designated area (Protected Area) in which the farm is to be established. Morocco, on the other hand, lacks specific regulations governing the requirement that new aquaculture ventures submit an EIA. Malta, on the other hand, has very detailed legislation regulating EIAs. Diverse perspectives exist regarding foreign involvement in aquaculture endeavors. In Egypt, foreign investment in aquaculture is actively encouraged, whereas in Cyprus, foreign investment is limited to 49% of total investment.

9.3.2.3. Mechanisms to Control the Environmental Impact of and on Aquaculture

The effects of aquaculture on the environment are not specifically regulated by law. Aquaculture is not given special consideration in any of the reporting nations; instead, it falls under the broad protections that are applied to commercial and industrial activity. With the exception of Croatia and Greece, where the precautionary approach is covered by their regulatory frameworks, the majority of nations do not have legislation incorporating the idea. The notion is incorrectly understood in Turkey to mean preventing adverse environmental effects. Nonetheless, in Cataluña (Spain), the reports created by the relevant environmental authority prior to aquafarm authorization are understood to follow the precautionary approach as intended. In the country of Romania, Law No. The polluter pays principle is applied in 137/95. But there's no mention of rehabilitation—the return to one's prior state. According to reports, the majority of Mediterranean states abide by the international conventions pertaining to the introduction of non-indigenous species. Even in the absence of specific legislation, the introduction of GMOs and non-indigenous species is always subject to control by the appropriate authorities. These include people in charge of animal production, environmental preservation, and veterinary control. For instance, in Croatia, non-native species may only be introduced with the written approval of the State Environmental Protection Administration and by qualified research institutions. The Livestock Husbandry Act in Croatia forbids the introduction of genetically modified organisms. The Bulgarian Academy of Sciences' Zoological Institute is in charge of introductions in Bulgaria. There is specific legislation in Tunisia (Decree 29 August 1994) regarding the introduction of non-indigenous species and genetically modified organisms (GMOs) for aquaculture.

Very few efforts by NGOs and the private sector to support ethical aquaculture by advocating for new or modified laws are mentioned in the Mediterranean region. Communication between government agencies handling aquaculture and the private industry or non-governmental organizations is generally lacking. This rule does not always apply. In Malta, for instance, the private sector intervened to amend the 1934 Fisheries Regulation, and in Croatia, nongovernmental organizations are involved in the legislative process. Certain nations (Croatia, Cyprus, Israel, Malta, Turkey) forbid aquaculture entirely or in part in areas designated as protected (conservation or historically significant). In the cases where it does not, this does not imply that aquaculture is incompatible with environmental preservation; rather, it indicates that the appropriate authorities neglected to expressly forbid this practice. The reports state that there are only three countries with Aquaculture Zones—Greece, Spain, and Tunisia—where aquaculture is protected by law and actively promoted. In Greece, in accordance with Law No. A region (including land and water space) has been set aside for the sole development of aquaculture operations in accordance with Environmental Protection Law No. 1650/86. Certain areas have been designated as being conducive to aquaculture activities by the Tunisian Aquaculture Plan. Ultimately, specific regions for mussel beds have been identified in Cataluña, Spain (Order 25 June 1975).

9.3.2.4. Involvement of Local Communities in the Legislative Process

National laws typically don't contain many provisions that expressly support the growth of aquaculture as a means of assisting local communities. There is a maritime law exception in Cataluña, Spain, where authorizations to conduct aquaculture activities are granted to cooperatives and associations of fisheries first. Greece's legal system is comparable. Furthermore, few agreements exist that shield parties from disagreements over common resources. Nonetheless, there are ad hoc agreements in Cyprus. These include agreements about the location of offshore farms between the Fisheries Department and coastal fishermen and aquafarmers, as well as agreements between land-based trout farm owners and agricultural farmers regarding the water supply to fish farms and the water outflow used for irrigation. In the Mediterranean region, customary laws do not apply to aquaculture activities, with the exception of Greece, where certain laws—such as those allowing amateur or commercial fishing in leased lake or coastal lagoon areas—do apply. Numerous rights that have been acquired over time are not regarded as customary law. The national reports generally pointed out that local communities ought to be more involved in the creation of laws pertaining to fisheries and aquaculture as well as those pertaining to tourism, business, and environmental preservation. Local communities in Cataluña, Spain, are consulted prior to the enactment of laws.

9.3.2.5. Collaboration Between Governments on Transboundary Aquatic Ecosystems

A multitude of agreements have been signed by the majority of nations to safeguard transboundary aquatic ecosystems. Bulgaria is a party to several international agreements, including the Danube Fisheries Agreement, the Black Sea Fisheries Agreement (1959), the Odessa Ministerial Declaration (1992), the Black Sea Environmental Program (1993), and the new draft Convention on the Black Sea Living Resources and Fisheries. Nonetheless, there aren't many agreements with neighboring nations that are especially meant to defend the environment from careless aquaculture operations or to preserve the possibility of aquaculture development from the actions of other resource users. A bilateral agreement on fisheries and aquaculture cooperation exists between Bulgaria and Romania. Additionally, Italy has agreements with Yugoslavia to prevent pollution in the Adriatic Sea and its coastal zones, as well as with Greece to cooperate in protecting the marine environment in the Ionic Sea and its coastal areas (Rome 6/3/79). In the EU member states where the report was received, EU Directive No. With the exception of Greece, national laws do not yet incorporate 990/220/EEC on the purposeful release of genetically modified organisms into the environment. Most nations do not include in their legislation the requirement to set up suitable mechanisms for gathering, sharing, and disseminating data related to aquaculture activities, with the exception of Spain (where data sharing between central administration and autonomous communities is mandated) and Greece (where producers are required to provide statistical data). Generally speaking, the creation of databases and other information sources is optional rather than required.

9.3.2.6. Codes of Practice

The Code of Conduct for Responsible Fisheries was referred to as "the" Code of Practice in almost all of the national reports. Cyprus specifically makes reference to the OIE International Aquatic Animal Health Code and the EIFAC/CECPI Codes of Practice for the introduction and transfers of marine and freshwater organisms. Morocco also makes reference to the EU Directives on shellfish and the section of the Codex Alimentarius that deals with aquaculture products. According to a few reports, national codes of practice for aquaculture and responsible fishing are being developed soon. For instance, a code is being planned in Bulgaria that will cover the Danube, Dnepr, and Dnestr rivers as well as the entire Black Sea basin. The National Action Plan that Turkey submitted to the EU in July 1998, which addresses general hygiene and HACCP regulations for aquaculture establishments, is the source of the Code that needs to be created.

9.3.2.7. Trade in Aquaculture Products

There are significant differences in the national laws that govern the trade of aquaculture products. The majority of the region's non-EU nations lack legislation specifically pertaining to aquaculture products. Morocco is an exception, enforcing certain laws to limit the resale of products from inland aquaculture. The same laws that govern fish products from capture fisheries also apply to aquaculture products in non-EU nations. For instance, the conditions under which fish can be marketed and sold are regulated in Croatia by Article 19 of the Veterinary Act and related laws. In Cyprus, the Food Sale and Control Law regulates hygiene and quality control. In the region's EU member states (e.g., EU directives pertaining to aquaculture products (such as EEC Directive 91/67, which addresses sanitary measures for aquaculture products) have been integrated into national laws in several countries (Greece, Italy, Spam). Some nations have included HACCP practices in their national laws (e.g., HACCP is expressly stated as a necessary system for the protection and control of aquaculture product quality in Bulgaria and Croatia. There is a tendency either towards harmonization with EU legislation or towards the direct incorporation of EU Directives concerning quality control in aquaculture products into national legislation, despite the fact that the majority of the countries involved in the Consultation are not EU members. Regarding the laws governing the import and export of aquaculture products, the same trend is also evident.

A variety of typical issues arise for aquaculture producers when trading their products. First, the issue of an ambiguous regulatory framework exists in EU nations (e.g., Italy), which acknowledges the need for laws to guarantee that aquaculture products bear a clear label identifying their place of origin. The second restriction concerns non-EU nations that must adhere to extremely stringent requirements in order to join the EU market (e.g., tariff barriers of 15%), import taxes, national laws, etc. The reports show a definite desire to harmonize the regulations governing the trading of aquaculture products (e.g., in the Malta report), which noted that harmonization would level the playing field for producers and benefit consumers).

9.3.2.8. Incentives and Penalties

It has been reported by certain Mediterranean countries that aquaculture producers receive incentives. But it's unclear (c/f section 3.4) how much these incentives really promote ethical aquaculture methods. Cyprus offers interest rate subsidies on commercial bank loans for money invested in aquaculture ventures that are authorized. Aquaculture farmers in Morocco can take advantage of several advantages, including tax breaks. In accordance with Article 34 of Law No., Romania also anticipates tax exemptions for individuals who implement ecological conservation and reconstruction measures with the goal of protecting the environment. 137 out of 96. Penalties are typically included in laws pertaining to aquaculture and are generally considered to be reasonable and potentially effective. With the passage of the Environmental Protection Act, Croatia established an Environmental Protection Inspectorate.

But most nations acknowledge that, in reality, enforcing penalties is ineffective (e.g., In Malta and Morocco), the human resources available for enforcement are restricted.

9.3.2.9. Conflict Resolution

Numerous nations that report include, e.g., Morocco, Croatia) declared that there are no particular dispute resolution procedures. Some countries (Egypt, Croatia) claimed that conflicts have not materialized. A few nations have stated that disputes are resolved on a case-by-case basis, including Egypt and Malta. Conflicts are resolved in Cyprus through personal meetings, in-person meetings, the creation of ad hoc committees, letter exchanges between the parties, lobbying, and consultations. Three mechanisms have been set up in Cataluña, Spain, to resolve disputes. The first is a preventive measure (affected parties are notified prior to aquaculture farm authorization being granted). The second is to report any potential issues to the appropriate administration. The involvement of associations related to fisheries and aquaculture in the Consultative Committee on Fisheries and Aquaculture is the third.

9.4. ENVIRONMENTAL MANAGEMENT IN AQUACULTURE

Unchecked aquaculture growth has the potential to destabilize rural communities, pollute the environment, and promote excessive use of natural resources. Environmental management seeks to reduce, eliminate, or lessen the negative effects of human activity on the environment in order to make better use of resources and promote more sustainable practices. If the following criteria are met, it may be possible to assess the effectiveness of environmental management: (i) potential negative environmental effects of human activity must be identified; (ii) variables that can be changed without unacceptable environmental effects must be specified by standards; and (iii) the availability of technology-based management techniques to prevent excessive changes in environmental variables must be ensured. Ensuring environmental quality has traditionally been a shared responsibility between the public and private sectors. The government is in charge of setting environmental management standards and enacting laws and regulations to ensure that these standards are followed. The cost of assessments to identify potential ecological impacts falls on the private sector, but governments are required to set guidelines and approve the findings and recommendations of these assessments. Concern over environmental issues grows as aquaculture operations expand. Accidents or poor management at aquaculture facilities can have serious consequences. Lower water quality, harm to the ecosystem, and a decline in other economic activity are possible outcomes of aquaculture operations. Licensing procedures take environmental concerns into account, ranging from the possible environmental effects of starting an aquaculture enterprise to the effects that result from running an aquaculture farm.

Therefore, if an operation poses a serious risk to the environment, proposed legislation should require an environmental impact assessment. The ecosystem approach to aquaculture (EAA) is a method that is gaining popularity in aquaculture management. The EAA's goals are to manage the aquaculture industry in a way that doesn't endanger the larger ecosystem and to encourage the sustainable development of interconnected ecosystems. The attainment of these objectives heavily depends on environmental considerations. The policy framework of a nation should be the primary source of information about the ecosystem approach. Under the ecosystem approach, legislation can be passed in many ways, most frequently by enacting many provisions. The EAA should be included in the opening sections of an aquaculture act since it can be viewed more broadly as the guiding concept. Strengthening institutions and related management systems is another goal of EAA legislation.

9.4.1. Site Selection and Environmental Impact Assessment (EIA)

Farms situated on subpar or marginal sites are frequently the cause of fish farm failure or negative environmental effects. As a result, in order to ascertain whether the site's features are appropriate for the establishment and long-term operation of a farm, a thorough site evaluation should be carried out. These analyses ought to ascertain freshwater flow, climate, water availability and quality, and other pertinent variables. A comprehensive environmental impact assessment (EIA) should incorporate the information gathered as well. Generally, the granting of a general aquaculture license is contingent upon the submission of an EIA. The proposed aquaculture farm size and the area's sensitivity are just two of the many factors that are covered by the EIA. An aquaculture farm's location at an unfavorable site can have disastrous consequences, so the administration should carefully consider the site's suitability before granting a permit. Governmental organizations can now incorporate plans for managing lakeshores and coastal zones with the location of aquaculture farms thanks to the approval of a permit application.

In certain situations, the proposed aquaculture operation's intended site should be investigated in addition to the proposal's other environmental implications. Determining the appropriate time for an EIA is crucial. Upon receipt of the permit application, a preliminary determination regarding the project's eligibility for an environmental impact assessment must be made. The proposed legislation ought to provide clear standards for figuring out whether an assessment is necessary for a proposed aquaculture project. Only projects that truly pose a risk of environmental damage should be required to undergo an EIA, as they can come with significant financial costs for the applicant. The size of the proposed project, the site's maximum carrying capacity, the possibility of waste discharge into sensitive areas, the protection of exotic products or species from modern technology or genetically modified organisms, and whether the project threatens rare or endangered species should all be taken into consideration when determining whether this risk exists. When an EIA is necessary, the findings of the assessment should be taken into account by the regulator when deciding whether to issue a permit and when drafting the conditions of the permit. When an EIA is not necessary, permits should be granted by the regulator in accordance with clearly defined guidelines. It is imperative that specific decisions regarding the granting or withholding of development be made against a backdrop of explicit policy principles in order to draft good aquaculture legislation.

9.4.2. Effluent Regulation

The potential for aquaculture farms to generate significant amounts of waste products, which, if improperly managed, could have grave negative effects on the environment, is one of the main causes for concern. Waste products present a risk to human health as well as the aquatic environment they enter. Unsatisfactory effluent from fish farms can be controlled using two different regulatory strategies. The first is the implementation of a discharge licensing scheme that establishes maximum levels of contaminants that can be found in fish farm waste water and imposes fines when these levels are exceeded. The second strategy entails outlawing or restricting the use of specific medications and chemicals that are known to negatively impact ecosystems and receiving waters. Regular monitoring at predetermined intervals is required to prove compliance with discharge licensing systems that include water quality and/or volume criteria. Licensing for discharges of one or more effluents is appropriate for large operations. It looks like the best system is to use general discharge permits, which impose minimal basic monitoring requirements and a set of operational Best Management Practices (BMPs) on farmers.

9.4.3. Drug and Chemical Use Control

The aquatic environment is contaminated and polluted by the overuse and misuse of chemicals in the aquaculture sector. This becomes particularly problematic when chemicals created for use in other industries—most notably agriculture—

are adopted by the growing aquaculture sector. Furthermore, improper chemical use can have a negative impact on the quality of aquaculture products, which can then be harmful to human health, and aquaculture farms are particularly vulnerable to low water quality. Three methods exist for regulating chemical contamination through legislation. The first method outright forbids or severely limits the use of particular chemicals that are bad for the environment. It is possible to make it illegal to use, import, distribute, or sell certain dangerous chemicals, while less dangerous chemicals might only be subject to licensing and authorization processes. Typically, this kind of law is not created specifically for aquaculture farming; instead, it can be found in more specialized acts pertaining to the use of chemicals or in general basic environmental laws. Certain nations, including China, Thailand, and Vietnam, have enacted laws that specifically designate certain chemicals for use in aquaculture, along with guidelines for their application. Creating a system for licensing the discharge of wastewater is the second strategy to prevent chemical contamination. The majority of the time, general environmental or water laws will govern licenses. Outright prohibiting the use of certain medications, substances, and hormones that may have a detrimental impact on the physiological processes and general health of fish is a third strategy for controlling the use of chemicals. Sometimes people view drugs, chemicals, and hormones as a way to make up for poorly run aquaculture facilities or a general lack of wellbeing. In response, the Norwegian legislature outlawed the use of chemicals as a stand-in for underperforming aquaculture systems.

9.4.4. Waste and Waste Management

The aquaculture industry can generate waste in a variety of ways, just like other food-related industries. There should be as little of this waste as possible because it could harm the processing environment. Every nation should have procedures in place for storing and disposing of waste, and these procedures should ideally be governed by laws. Numerous nations have enacted national waste disposal legislation. For instance, Chile mandates that aquaculture farms implement safeguards against disposing of residue and solid or liquid waste in a way that could endanger the environment. The Zambia Fisheries Act gives the government the authority to impose requirements on aquaculture licenses in order to reduce fish waste disposal, waste product escape, and water pollution in and around the aquaculture facility. The discharge of waste that could endanger human health or the environment is prohibited by the Namibia Aquaculture Act. It also specifies that waste needs to be disposed of in compliance with the aquaculture license's terms and conditions.

9.4.5. Coastal Aquaculture as Part of Integrated Coastal Management

The notion of merging the aquaculture industry with coastal area management is becoming more and more popular, and some nations have already begun to apply it. For instance, proposals for aquaculture that should be addressed in Coastal Zone Management Plans are expressly included in the Belize Coastal Zone Management Act of 1998. Aquaculture can be integrated into coastal areas using a variety of strategies, including the use of EIAs, the establishment of buffer zones or protected areas, limitations on private ownership, and the acknowledgment of indigenous rights. Zoning, which designates specific land and water areas for aquaculture, is one of the most crucial instruments. Certain coastal zones are designated for aquaculture under the Chilean Fisheries and Aquaculture Law. Aquaculture operations outside of these zones are prohibited. The Tasmanian Marine Farming Planning Act of 1995, which mandates the preparation of Marine Farming Development Plans in regions where marine farming may take place, is another instance of zoning. The public is consulted during the planning process, which considers factors like the physical suitability of possible aquaculture sites, the existing legal framework, and minimizing the impact on other users of the coastal zone.

9.4.6. Escapement

Controlling fish escapement, or fish escaping from an aquaculture facility, has become more and more crucial. The explanation for this is that escaped fish often genetically contaminate their surroundings, leading to the extinction of the species. Regulations pertaining to ecological safety and veterinary control, such as those governing quarantine and the use of alien or introduced aquaculture species, are especially important in this situation. These regulations may lessen the potential impact of alien or introduced species on farmed local species, wild endemic species, and the biodiversity of natural water bodies when they are incorporated into the aquaculture legislative framework. Restrictions that reduce the likelihood of escapement or lessen damages in the event that escapement occurs are examples of potential veterinary control and ecological safety regulations. This covers the guidelines for disclosing and reporting escapement incidents. Examples of nations with comparable laws include Zambia, Canada, and Norway. When fish escape from an aquaculture farm, the owner ought to be able to recover them, which begs the question, "Who is the owner of that fish?" Establishing this is also necessary in order to prevent escapement and take preventative action against the fish's owner. Ownership in an escapement situation is addressed by the Namibia Aquaculture Act. As long as they remain inside the site's boundaries, licensees are the exclusive owners of certain aquaculture products under the Act. As long as the licensee can provide identification, even in the event that the products are released or escape, they continue to be the sole property of the licensee. The Zambia Fisheries Act allows ownership provisions to be included in secondary legislation.

9.5. TRADE AND OTHER ISSUES

The goal of national aquaculture laws should be to apply global trade and quality control standards to the planning, development, and management of aquaculture operations. Adopting a regulatory framework that satisfies EU market standards is imperative for any nation contemplating the export of fish products derived from aquaculture production to EU markets. The Sanitary and Phyto-Sanitary Agreements of the World Trade Organization (WTO) must be complied with, among other WTO regulations, if consideration is to be given to international trade. States that have adopted the Code of Conduct for Responsible Fisheries should work to incorporate it into their national laws, even though it is voluntary and contains some agreed-upon standards of international law. Aquaculture products are a vital source of income for many developing nations. The Code encompasses a range of aspects of global commerce to assist states in fulfilling lawful export obligations and obtaining this revenue. For instance, the Code establishes minimal requirements for quality and safety across the board for the sector. Standards for export to the European Union are established by the European Directive on Hazard Analysis and Critical Control Points (HACCP), which is comparable to the Code. The Codex Code of Practice for Fish and Fishery Products is one example of an internationally accepted standard.

9.5.1. Introduction and Movement of Species

The introduction of exotic species into environments where they may displace or compete with native species is one of the aquaculture industry's potentially dangerous environmental effects. Therefore, precautionary measures about the importation, quarantine, and dispersal of non-indigenous species should be included in laws and policies. These regulations ought to be founded on trustworthy data and specifically take into account the unknowns and unfavorable effects connected to the introduction of new species. Many current codes of conduct and regional guidelines offer models for managing the introduction and transfer of species, with a focus on halting the spread of illness. Numerous effects on the environment, ecology, and disease control can result from species migration. First, native stocks and ecosystems may suffer as a result of the harvesting of aquaculture stocks from the wild for use in aquaculture farms. When non-native species are introduced and have the potential to escape into the wild, the local ecosystems may suffer from invasive habitat competition or negative genetic effects on native stocks. Moving species stocks between farms also puts the aquaculture industry as a whole and individual farms at risk of contracting diseases that could have catastrophic economic effects. Species movement licensing regimes will initially stop movements that are too dangerous and permit individual evaluation before authorization is granted. These policies aim to stop the spread of disease and have a positive ecological impact by outlawing specific types of movement.

The ability of licensing regimes to impose requirements that all major movements be documented and to make it easier to communicate that information to the appropriate authorities is another advantage. It is possible to identify other possibly infected locations by tracking the movements of diseased stock that has recently been moved.

9.5.2. Food Safety and Health

To safeguard the health and safety of consumers, aquaculture products, like other food products, must be governed by public health regulations. Public health is the primary concern of the Code of Conduct for Responsible Fisheries, which also supports states that guarantee the safety of aquaculture products for consumption and encourage initiatives to preserve product quality and value prior to and during harvesting, on-site processing, storage, and transportation. The Code also exhorts states to take the necessary actions to guarantee consumers' rights to pure, healthy, and safe fish and fisheries products. Finally, to safeguard consumer health and stop commercial fraud, the Code encourages states to set up and maintain efficient national safety and quality assurance systems. Similar to this, many nations that import aquaculture products have begun to include the HACCP into their laws governing aquaculture.

Leading the way in these legislative advances are the US and several EU nations. The application and enforcement of safety assurance systems at the level of aquaculture farms is relatively new, even though exporting nations may have advanced these systems in the fish-processing industry. The EU mandated in January 2002 that the nation of origin and species of all fisheries and aquaculture products offered for retail sale to end users must be labeled. Though it can be contentious, so-called eco-labelling can establish a market-based incentive for production that is environmentally friendly. According to the Code of Conduct, states are required to improve the identification of the origin of fish and fishery products treated in order to ensure that both domestic and international trade in fish and fishery products (including aquaculture products) complies with good conservation and management practices.

9.5.3. Mitigating Investment Risks

The regulatory framework must balance the protection of the environment with the promotion of a productive aquaculture sector that serves the interests of all. Environmental regulations are therefore essential to good aquaculture laws; however, aquaculture operations will not be sustainable or commercially successful in the short and long term if their application places an undue burden on the operator. Legislation that guarantees that the aquaculture operation's total costs—including external and social costs—do not surpass the benefits it generates can help prevent this risk. To promote the commercial growth of the aquaculture sector, the legal framework should guarantee that environmental objectives are fulfilled while avoiding needless expenses for applicants. To achieve this, nations should implement a single window system for the multiple permits needed for aquaculture operations as well as for the preliminary screening of applications, limiting the number of applications that require a full environmental impact assessment to those that pose a serious risk to the environment. Nations may also think about establishing a single organization to advance aquaculture and track the development of applications via other government divisions.

9.5.3.1. Aquaculture and Zoning

Growing apprehension about aquaculture's effects on the environment has led governments to reserve public lands for a variety of purposes, including aquaculture. Zoning is a helpful tool for integrating aquaculture in coastal areas, as was previously discussed. Zoning regulates the environment at the farm level, settles disputes about the use, protection, and exploitation of coastal resources, and strikes a balance between conflicting interests. For instance, the Special Administrative Region of Hong Kong (SAR) has established 26 Marine Fish Culture Zones, and all marine fish culture activities must operate within these zones under a license. In Ecuador, shrimp farming is permitted to continue alongside traditional uses thanks to local zoning agreements between local residents and shrimp farmers. Similar conservation-planning zones that dictate the land-use arrangement for fish farms have been established in Zambia. Establishing designated areas for aquaculture leases, such as a Marine Aquaculture Zone (MAZ), is one method to enhance the legal and regulatory framework for offshore aquaculture. In comparison to its land-based counterpart, marine zoning faces many significant challenges, including boundary disputes, difficulties enforcing the law, and a higher frequency of user conflicts. Zoning has proven to be an effective land-based technique for designating regions suitable for industrial development. Establishing a Marine Aquaculture Zone (MAZ) necessitates designating a national agency in charge of overseeing the zone, granting leases to the seabed and water column, and issuing permits that take other pertinent national agencies' concerns into account. Aquaculture zoning ensures that natural resources are accessible because there may only be a limited number of optimal sites, which increases the likelihood that aquaculture operations will be successful.

Additionally, zoning reduces the need for zone perimeter buoys and individual cage lights for site marking, which lessens the risk of theft, vandalism, and liability concerns related to boat-cage collisions. Furthermore, zoning lessens the possibility that anglers will catch reef fish drawn to the cages by chance. Prospective aquaculture farmers in Madagascar have been steered toward the creation of Régime de Zone Franche, or preferential trade zones, utilizing financial and economic incentives. In actuality, though, aquaculture development areas have been identified since the early 1990s, and interested parties are invited to develop aquaculture activities in these areas by obtaining "preferential trade status," which makes it easier to access financial and economic incentives from the government.

9.5.3.2. Simplification of Administrative Procedure: Single Window Approach

The use of official authorizations to exert control over aquaculture establishments is a key component of aquaculture legislation. Permits, licenses, concessions, leases, and other forms of authorization are frequently subject to specific requirements. They provide governments with a solid foundation upon which to regulate the finite natural resources at their disposal and enable them to incorporate aquaculture farm siting into integrated coastal management plans. Since numerous laws and agencies are involved in the approval and management of aquaculture projects, the authorization process can be challenging for aquaculturists. This is especially true when it comes to using public lands and water. Authorities in the fields of health, environmental protection, water quality, land use planning, and fisheries and aquaculture are all involved. An applicant typically needs many documents, including land concessions, water licenses, effluent discharge permits, and other kinds of environmental licenses, before they can open or run an aquaculture farm. The current challenge in aquaculture law is to increase the cost-effectiveness of aquaculture operations while removing existing legal and bureaucratic barriers. By establishing a single, lead aquaculture agency that oversees the application process, there is a trend toward standardizing licensing requirements and expediting approval processes.

Promising initiatives have been put forth to expedite the application process. For instance, Chile's Aquaculture Concessions and Authorizations Regulations stipulate that only one application is needed for aquaculture operations. To establish an aquaculture farm in Mexico, all necessary permits, concessions, and authorizations are handled by a dedicated office dedicated to aquaculture development. A unique administrative division known as the Single Window Office was created in Madagascar and later moved to the Ministry of Economy, Plan and Social Recovery. It served as a one-stop shop for investments within the Offices of the Prime Minister. The Office is essential to the growth and establishment of commercial aquaculture activities because it: (a) works with technical ministries; (b) helps investors comply with investment formalities and procedures; (c) receives applications from businesses seeking to establish facilities in preferential trade zones or to become free trade enterprises; and (d) informs potential investors about technical, administrative, legal, economic, and social environments as well as investment opportunities in the nation. Malawi has taken a stance that promotes unrestricted investment in aquaculture projects. The Malawi Investment Promotion Agency's publications emphasize aquaculture investment. With this modification, aquaculture now shares the same commercial standing as agriculture. Applications for foreign investment are approved through an informal single-window process by a ministerial committee.

9.5.3.3. Financial Incentives

Many laws pertaining to aquaculture are characterized by the recognition of the need to provide security to the aquaculturist, regardless of size, in situations where public lands are involved. In order to achieve this, some nations grant agricultural status to public land used for agricultural purposes, extend or renew leases, and introduce marine tenure systems to encourage private sector participation. Vietnam has established regulations pertaining to the distribution of land among households and individuals, intending to ensure stable and extended use for aquaculture purposes. 56 The land lease is renewable and has a maximum term of 20 years. Certain developing nations in Asia and Latin America are also designing long-term leases or concessions, with terms of up to 10, 20, and 50 years, with the aim of:

- create a more permanent linkage over time between producers and supervisors;
- generalize the concept of sustainability; and
- guarantee future and permanent supplies to the aquaculture industries holding these concessions.

The government's authority to revoke leases or concessions of this kind is restricted and can only be employed in instances of flagrant disregard for applicable laws and/or management plans. Extensive rights to end leases for vague reasons typically vanish. Still, there is still ambivalence about this growing trend. Even though there is increased security, one should consider whether or not: in practice.

- such leases/concessions tend to put aside the monopolistic approach to aquaculture ownership;
- competition is distorted; and
- access to aquaculture by others is being limited.

Aquaculture investments are given priority when it comes to receiving government incentives, as stipulated by the Mozambique Investment Law of 1993. The Mozambican Investment Law does not contain any clauses that limit foreign or domestic private or public investment in aquaculture. Unless the Council of Ministers specifies differently, all economic activities are open to the private sector, according to Articles 11 and 12 of the Act. To help the Ministry of Planning with the coordination, evaluation, and support of both domestic and foreign investment, the Investment Promotion Center was established in 1982. All investors must submit a request to the Investment Promotion Center to be eligible for tax relief and incentives. A 50% reduction in income tax, such as the industrial contribution tax, and exemptions or reductions from import, circulation, and consumption taxes are examples of tax incentives. It has been suggested that the permit and tax procedures in Mozambique are quite bureaucratic due to the numerous bodies involved in obtaining the necessary permits for an aquaculture operation. To further the growth of aquaculture, Malaysia's Minister of International Trade and Industry is able to publish a list of products or activities to be promoted thanks to the Promotion of Investments Act 1986. If a company wants to participate in a promoted activity, the Minister of Finance may grant a "pioneer status" certificate with the company's consent. Businesses granted pioneer status are exempt from income tax on 70% of their corporate income for a period of five years, starting five years after the date production starts. This benefit is known as the investment tax allowance.

9.5.4. Social Impacts

Just as unsettling as the environmental effects of fish farming are the perceived social repercussions of this practice. Social concerns may provide strong justifications for restricting or banning fish farming operations. In many nations, resource access has historically been granted to all users. In these situations, privately owned fish farms are thought to be incompatible with open-access usage customs. Investors and fish farm owners are thought to benefit from fish farming disproportionately. Furthermore, the economic advantages of fish farming are restricted and do not benefit nearby communities because fish are raised for export. Conversion, expropriation, the privatization of mangroves and other lands, soil and water salinization, decreased food security, marginalization of coastal communities, unemployment, urban migration, and social conflicts are some of the significant socioeconomic effects.

These outcomes frequently coincide with the expansion of aquaculture operations across numerous nations. When aquaculture farms cut off a community's traditional routes of access to coastal resources, conflicts frequently result. Large farms imply that communities no longer have easy access to the resources they once took advantage of. For rural communities, securing access to land and water is frequently a crucial component of their survival and livelihood strategies.

When property rights are not clearly defined, social conflicts over property rights frequently arise. Resolving all of the land tenure issues is the main obstacle to establishing order in the aquaculture industry. The argument that such transfers can have major negative social and economic effects on the community has led many rural communities to oppose water transfers to outside parties. Because unskilled workers in aquaculture activities typically originate from the area where the farm is located, employment opportunities may be restricted. There is minimal rivalry for these people's services in the labor market since they have few other options for employment.

9.5.4.1. Land Privatization and Land Security

Mangroves are known to be significant to local communities in addition to their ecological value. Local fishermen lament that there are fewer commercial species being caught in mangrove areas. This decline is frequently linked to the industry practice of capturing larvae and eliminating the supporting fauna. The loss or compromised access to mangroves, lagoons, and estuaries in some places is another social effect of the fish farming industry's changes. Locals are occasionally compelled to relocate.

As a result, some members of the local community lose their source of food, energy (firewood), and eventually money. In certain nations, where shrimp farmers have secured government authorization to utilize publicly owned coastal lands, this is acknowledged by the industry. The locals had previously used this land for agriculture, home sites, hunting, fishing, and other uses, as well as for gathering wood for construction and fuel. Shrimp farms on the property may restrict local residents' access to resources or possibly drive them off. However, from the perspective of the industry, regular employment from shrimp culture development offsets the loss of traditional uses. The people in charge of the production process include bankers, government officials, large rural landowners, and urban entrepreneurs are almost always viewed negatively. They can also be seen as logical economic actors taking advantage of available opportunities and risks to maximize returns on investment. As a result, this could help create excess capital that can be invested in other profitable businesses, which is essential for economic expansion (Dominguez et al., 2004).

REFERENCES

1. Brugère, C., Ridler, N., Haylor, G., Macfadyen, G., & Hishamunda, N. (2010). *Aquaculture Planning: Policy Formulation and Implementation for Sustainable Development* (FAO Fisheries and Aquaculture Technical Paper No. 542). Food and Agriculture Organization of the United Nations.
2. Dominguez, L., Molina, J., & Vergara Martin, J. M. (2004). Aquaculture environmental impact assessment. In V. Popov, H. Itoh, C. A. Brebbia, & S. Kungolos (Eds.), *Waste Management and the Environment II* (pp. 321–333). International Conference on Waste Management and the Environment.
3. Hishamunda, N., & Ridler, N. B. (2008). Sustainable commercial aquaculture: A survey of administrative procedures and legal frameworks. In *Aquaculture Economics & Management* (pp. 167–178). London, UK.
4. Siriwardena, P. G. S. N., & William, R. (2003). Code of good management practices for shrimp aquaculture in Sri Lanka. Report prepared under the World Bank, NACA, WWF, and FAO Consortium Program on Shrimp Farming and the Environment.
5. VanderZwaag, D. L., & Chao, G. (Eds.). (2006). *Aquaculture Law and Policy: Towards Principled Access and Operations*. Routledge Taylor & Francis Group.

Chapter 10

Community-Based Conservation Approaches

Contents

INTRODUCTION

Globally, community-based conservation techniques are becoming more and more acknowledged as successful tactics for protecting biodiversity. These strategies are focused on involving local people in the care and management of their own natural resources. These programs seek to improve livelihoods and promote sustainable development in addition to protecting biodiversity by giving communities the tools they need to actively participate in conservation efforts. Recognizing that local communities frequently possess extensive traditional knowledge about their environments and ecosystems is at the core of community-based conservation.

Understanding ecological patterns, species behaviors, and the dynamics of natural resources all benefit greatly from this knowledge. Projects can take advantage of this abundance of local knowledge by incorporating communities in decision-making processes related to conservation, guaranteeing that interventions are successful and appropriate for the context. The emphasis on inclusive and collaborative methods is one of the main features of community-based conservation.

These initiatives place more emphasis on forming partnerships with local stakeholders, such as indigenous peoples, community organizations, and local leaders, than they do on imposing top-down solutions. This cooperative strategy fosters long-term sustainability and ownership while also strengthening the legitimacy of conservation efforts.

Furthermore, socioeconomic development goals and conservation objectives are frequently integrated into community-based conservation projects. These initiatives provide communities with incentives to actively participate in conservation efforts by showcasing the concrete advantages of conservation to local communities, such as improved access to clean water, sustainable agriculture practices, or enhanced ecotourism opportunities. This integrated approach addresses the interrelated problems of environmental degradation and social inequality by reducing poverty and conserving biodiversity at the same time.

There exist numerous regions and ecosystems where community-based conservation initiatives have proven to be effective. Community-managed marine reserves, for example, enable regional fishermen to safeguard fisheries and coral reefs, promoting ecological resilience and sustainable fishing practices. In a similar vein, community forest management programs enable rural communities to manage forest resources sustainably, preventing deforestation and preserving the livelihoods of those who depend on forest products. Implementing community-based conservation strategies still faces difficulties, such as protecting tenure rights, managing power structures within communities, and making sure benefits are distributed fairly.

Governments, non-governmental organizations, and other stakeholders must remain committed to fostering inclusive governance structures, supporting capacity building, and offering technical assistance in order to overcome these obstacles. A viable route to achieving biodiversity conservation objectives while advancing sustainable development and enhancing livelihoods is provided by community-based conservation techniques. These strategies enable communities to become stewards of their natural heritage, guaranteeing that conservation efforts are both successful and equitable for future generations. They accomplish this by utilizing local knowledge, encouraging collaboration, and incorporating socioeconomic benefits.

10.1. BASIC CONCEPT OF COMMUNITY-BASED CONSERVATION APPROACH

Local communities are actively involved in the management and preservation of natural resources and biodiversity through the Community-Based Conservation (CBC) approach, which is a comprehensive conservation strategy. It acknowledges that communities play a critical role in conservation efforts and seeks to give them more influence through involvement, power over decision-making, and shared gains.

The concept revolves around several key principles and components:

1. **Local Participation and Empowerment:** Local communities should be included in the planning, execution, and oversight of conservation efforts, according to CBC. Through the inclusion of local stakeholders and indigenous peoples in decision-making processes, CBC guarantees that initiatives are relevant, culturally sensitive, and long-lasting.
2. **Utilization of Traditional Knowledge:** Indigenous peoples and nearby communities frequently hold important traditional knowledge about the behavior of species, ecosystems, and sustainable resource management techniques. Through the effective integration of scientific research and this knowledge, CBC informs conservation strategies. In addition to increasing the effectiveness of conservation efforts, this strategy protects and maintains customary cultural practices.
3. **Collaborative Governance and Partnerships:** Communities, governments, non-governmental organizations (NGOs), and other

stakeholders are encouraged to collaborate through CBC. These collaborations encourage shared accountability for conservation objectives, foster trust, and make use of resources and knowledge to tackle difficult conservation problems.

4. **Sustainable Livelihoods and Benefits Sharing:** Ensuring that conservation efforts improve local livelihoods is a fundamental component of CBC. This could entail encouraging the sustainable use of natural resources, creating eco-friendly employment opportunities (like sustainable agriculture or ecotourism), and making sure that the advantages of conservation efforts are distributed fairly.
5. **Adaptive Management and Learning:** The dynamic character of ecosystems and the requirement for adaptive management strategies are acknowledged by CBC. In order to build resilience in communities and ecosystems to adapt to changes in the environment and outside pressures, it promotes learning from both successes and failures.
6. **Policy and Institutional Support:** Institutions and laws that are supportive of CBC are needed on a local, national, and worldwide scale. This entails protecting communities' rights to land tenure, creating legislative frameworks that support community conservation efforts, and offering financial and technical assistance for the development of capacity and sustainable practices.

Community-managed marine protected areas, community-based forest management programs, and cooperative wildlife conservation projects are a few instances of effective Community-Based Conservation (CBC) initiatives. These programs show how local communities can contribute significantly to the preservation of biodiversity and natural resources while also enhancing their own quality of life when given the necessary tools and support.

A paradigm shift toward inclusive and participatory conservation strategies is represented by the CBC approach. In addition to increasing the efficacy of conservation efforts, Community-Based Conservation (CBC) fosters social equity, sustainable development, and resilience in the face of environmental challenges by identifying and utilizing the knowledge, resources, and agency of local communities.

10.1.1. What is Community-Based Conservation?

A method of conserving biodiversity known as community-based conservation (CBC) involves local communities in the maintenance, management, restoration, and/or preservation of the biodiversity of a given area.

The movement got its start in the 1980s as a result of widespread demonstrations that erupted in opposition to global initiatives to conserve biodiversity. This occurred as a result of traditional biodiversity conservation failing to take local and indigenous communities' needs and rights into account.

10.1.2. Decolonizing Conservation: The Yellowstone Tragedy

Western ideologies served as the main inspiration for the early conservation movement. Its subvention in America since the European settlement can be linked to the evolution of policies and mentalities. These ideologies promoted the idea that culture and nature are distinct. Because traditional conservation followed a top-down methodology, local communities were neglected as governments and private organizations acquired ownership of the region.

To "preserve the natural wilderness," European settlers forcibly uprooted an estimated 20 million people from their own land in 1872, during the creation of Yellowstone National Park. This method of implementation was used for the majority of conservation projects until the 1970s when indigenous people began to demand their rights to land ownership and autonomy. The idea of community-based conservation was born out of this.

10.1.3. CBC's Principles and Values: The Biocultural Approach

Community-based conservation differs from traditional conservation in that it acknowledges the relationship between culture and environment. Since biodiversity influences human cultural practices and cultural diversity is driven by biodiversity, the IUCN refers to this biocultural approach as the co-evolution of two ways of diversity.

Community conservation is a rights-based, bottom-up approach that prioritizes community agency, access, and autonomy in decision-making while identifying and addressing injustices (such as land grabs and territorial enclosures) that are committed against them.). In addition, CBC prioritizes supporting fairness in benefit distribution and honoring the contributions and role of women in conservation. The goal is to establish multilevel, cross-cultural relationships that are formed locally to fight global biodiversity loss.

10.1.4. CBC Model

Community conservation has been developed in several countries worldwide since its inception. As a result, different approaches and techniques are used in different areas to incorporate communities into conservation management, depending on a range of variables. Success is largely determined by the type of ecosystem, socioeconomic background, level of government involvement, and cultural and religious beliefs. The conceptual model below illustrates the goals, treatments, and desired results for CBC (Figure 10.1).

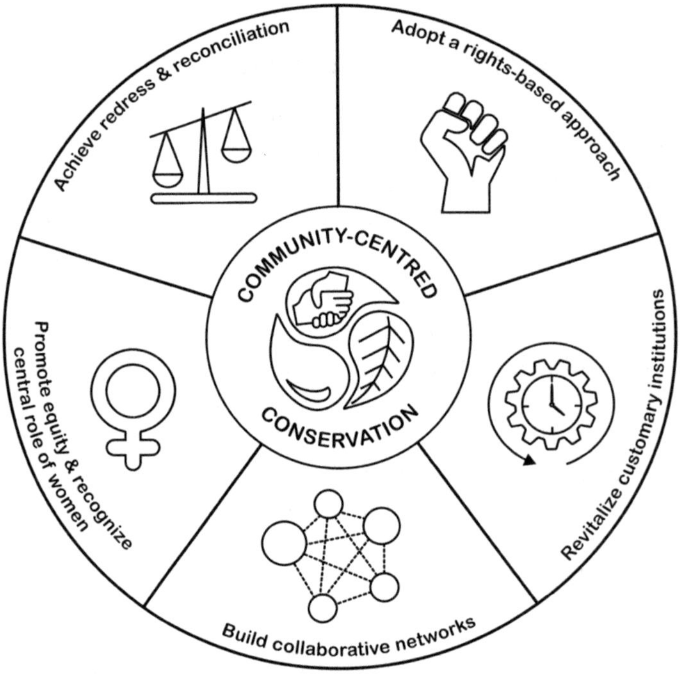

Figure 10.1. *CBC model.*

Source; Singh Kaleka, A., & Preet Kour Bali, G. (2021). Community Conservation. IntechOpen. doi: 10.5772/intechopen.94557.

10.1.5. A Pioneer Project: Community Baboon Sanctuary

The community-based baboon sanctuary (CBS) in Belize is an illustration of a successful community-based conservation effort. One of the first experimental programs in the history of private land conservation was CBS. The largest population of black howler monkeys in Central America was found in the riparian forests along the Belize River, which served as the focal point of the conservation area.

In order to address the local communities' need for alternate livelihoods and land management, Dr. Rob Horwich and community leader Fallet Young established the sanctuary in 1985 with the goal of protecting the habitat of black howler monkeys. The researchers established a formal conservation strategy that included educational awareness campaigns, locally managed eco-tourism, and contacting the local government for support after determining the conservation area and getting in touch with the local populace.

A plan for the development of a sanctuary was also created, encompassing the mapping of properties and vegetation, the censusing of howlers, the creation of land management, and the involvement of landowners. After three years of the project, more than 120 landowners voluntarily committed to managing their properties to protect black howler monkeys. The population of the species rose from 800 in 1985 to 3,500 in 2017 as a result of these conservation efforts. The sanctuary is currently operated by a local Women's Conservation Group and continues to bring in money from small businesses and tourists.

10.1.6. The Role of ICCAs in Conservation

Territories and areas that are conserved by indigenous people and communities are known as Indigenous and Community Conserved Areas (ICCAs).

ICCAs have three main characteristics:

1. A strong relationship between indigenous communities and their territorial area/s. In many cases, the communities' socio-cultural identity, history, health, and well-being are intertwined with their place of residence.
2. Indigenous and local people are major stakeholders in decision-making processes and implementation with regard to environmental protection and management.
3. Community efforts and management have led to an overall positive effect on the conservation of nature.

The IUCN and CBD (Convention on Biological Diversity) have formally recognized ICCAs as key factors in nature conservation. Approximately 20% of ICCAs have already been designated as 'Protected Areas,' making up 40% of global PAs. These numbers are increasing, with countries like Canada developing indigenous-led PAs, some of which are as large as 109,000 sq km.

ICCAs are the forefront of community conservation as they allow the formal involvement of communities as stewards and prevent the large-scale exploitation of culturally and ecologically valuable regions.

10.1.7. Potential Drawbacks

Although CBC is a cost-effective, long term and efficient method of biodiversity conservation, research on long-term projects has discovered factors that could hinder the effectiveness of CBC projects –

10.1.7.1. Poverty

A disparity exists between conservation efforts and economically disadvantaged communities, despite the local population's comprehension of the importance of conservation. The extreme hardships of poverty force individuals to prioritize their basic needs of food and shelter over preserving forests or endangered species. This disconnect has been identified as a challenge in CBC's initiatives in Madagascar, as the majority of the population is poor and their lack of involvement could negatively impact conservation efforts.

The development aid initiatives implemented as part of these programs have proven ineffective in reducing poverty, disease burden, or enhancing public health. Although these issues indirectly contribute to biodiversity loss, community-based conservation projects have struggled to effectively address and mitigate them.

10.1.7.2. Ecological Outcomes vs Social Outcomes

Contradictory outcomes in terms of local communities' socioeconomic benefits have come from community-based conservation projects. Although the ecological effects of CBC interventions in Africa were mainly positive, the social effects were observed to be mixed or negative overall. Reduced social capital and unequal benefit distribution were the key causes of the negative social outcomes.

10.1.7.3. Devolution of Rights

Local community rights devolution is a defining characteristic of CBC. Regretfully, the field has not used it as much as it could. The Malawian national government established the Forest Policy Act in 1997 to give local village forest committees management authority. However, because the national-level minister had to approve it, very few forest committees were able to use their rights. Devolution of community rights is essential to strengthen local governance, encourage leadership, and foster greater community involvement.

10.1.8. Moving Towards a Collaborative Approach

Local communities' assistance with conservation is just one part of the solution. Multi-leveled management and increased cooperation between local community members, NGOs, governments, and international bodies are essential to addressing the current constraints. As a result, stakeholders will be able to handle several goals using a variety of knowledge systems and techniques. Therefore, it is best to think of community-based conservation as a pluralistic strategy that connects with different organizational levels and begins from the ground up. However, the communities that have developed alongside our endangered flora and fauna will determine their fate. To preserve the ecosystems of which these communities are a part, it is our responsibility to uphold their rights.

10.2. COMMUNITY

In ecology, a community—also referred to as a biocoenosis, biotic community, biological community, ecological community, or life assemblage—is a collection or association of populations of two or more distinct species that are coexisting in the same geographic area. There are several ways to use the term "community." In its most basic sense, it describes populations of organisms in a particular location or period of time, such as the Lake Ontario fish community before industrialization.

The study of communities, which are collections of interacting populations of a species that live in a specific region or habitat, is known as community ecology. Biological communities are created when species interact with one another in populations. These communities' abundance of interacting species and the intricacy of their interactions serve as prime examples of biodiversity. Communities develop structures as a result of interactions between species; food chains, food webs, guilds, and other interactive webs are formed. As a result of species' mutual adaptation to one another during the process of coevolution, these relationships alter over the course of evolution.

Community ecology, also known as synecology, is the study of interactions between species in communities across various spatial and temporal scales, including distribution, structure, abundance, demography, and interactions between coexisting populations. It primarily focuses on the interactions between populations based on specific genotypic and phenotypic traits. In addition, abiotic factors, such as soil pH or annual temperature, that influence species interactions or distributions are also considered in community ecology. For example, plant communities in tropical rainforests and deserts differ significantly due to variations in annual precipitation.

Through habitat disturbance, such as the introduction of invasive species, humans can also have an impact on community structure. The significance and utility of the community concept in ecology are debatable at a deeper level. Traditionally, communities have been conceptualized at a fine scale in terms of local processes creating (or destroying) an assemblage of species; this includes the way that the composition of grass communities is expected to change due to climate change. This emphasis on the local community has drawn criticism recently. Drawing on evolutionary taxonomy and biogeography, Robert Ricklefs has argued that it is more useful to consider communities on a regional scale, where some species or clades evolve and others go extinct.

10.2.1. Community Structure

The kinds and quantities of species found in various ecological communities can vary greatly. Certain tropical rainforest communities contain a vast number of species per cubic meter, whereas certain Arctic communities only contain a small number of species. Saying that the communities have different structures is one way to explain this difference.

In essence, a community's composition—including the number of species present and their relative abundances—makes up its structure. Additionally, it can be understood more widely to encompass all of the ways in which these various species interact with one another (Figure 10.2).

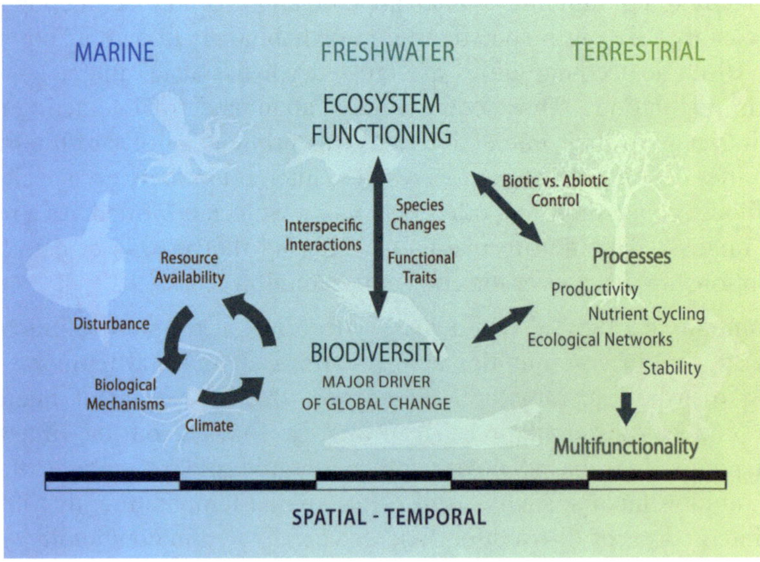

Figure 10.2. *Community structure.*

Source: Correia AM, Lopes LF. (2023). Revisiting biodiversity and ecosystem functioning through the lens of complex adaptive systems. Diversity, 15(8):895. https://doi.org/10.3390/d15080895.

10.2.1.1. How Do We Measure Community Structure?

Two important measures ecologists use to describe the composition of a community are species richness and species diversity.

- **Species Richness:** A community's species richness refers to the variety of species present in it. The species richness of the second community would be significantly higher than that of the first if we discovered 300 species in one community and 30 in the first. Near the equator, where there is plenty of solar radiation (supporting high primary productivity), warm temperatures, copious amounts of rainfall, and minimal seasonal variation, are communities that typically have the highest species richness.

The communities with the lowest species richness are found closest to the poles, where there is less solar radiation and a decreased climate that is unsuitable for life. The mammalian species richness pattern is shown below (species richness calculated only for mammal species, not for all species). The species-richness of a community can also be influenced by a wide range of other variables besides latitude.

10.2.1.2. Species Diversity

One indicator of the complexity of a community is species diversity. It depends on the total number of species (species richness) and the relative abundances of those species (species evenness) in the community. Higher species diversity is a result of both greater species diversity and more uniform species abundances.

For example:

- A forest community with 20 different kinds of trees would have greater species diversity than a forest community with only 555 kinds of trees (assuming that the tree species were even in abundance in both cases).
- A forest community with 20 different kinds of trees in even abundances would have greater species diversity than a forest community with the same number of species in very uneven abundances (for instance, with 90% of the trees belonging to a single species).

Ecologists generally believe that ecological communities with greater diversity are more stable—that is, better equipped to bounce back after a disturbance. In the ecological networks video, you can learn more about the reasons behind this. The relationship between diversity and stability isn't always true, though, and there are situations in which variables other than species diversity have a greater bearing on the stability of communities and ecosystems.

10.2.1.3. What Factors Shape Community Structure?

The structure of a community is the result of many interacting factors, both abiotic (non-living) and biotic (living organism-related). Here are some:

- The climate patterns of the community's location.
- The geography of the community's location.
- The heterogeneity (patchiness) of the environment
- The frequency of disturbances, or disruptive events.
- Interactions between organisms.

Incidents that occur by coincidence throughout a community's history can also influence its structure. Let's say, for example, that one seed blows into the soil in a specific spot. If it succeeds in taking root, the species might establish itself and eventually become dominant (apart from closely related species). If the seed doesn't germinate, it might take another species that is similar to yours to succeed in establishing itself and taking over.

10.2.1.4. Foundation and Keystone Species

Some species have unusually strong impacts on community structure, preserving the balance of the community or even making its existence possible. These "special" species include foundation and keystone species.

10.2.1.4.1. Foundation Species

A foundation species is special and vital to the formation and identity of a community. Foundation species frequently work by altering the surroundings to make it more hospitable for the other creatures that make up the community.

The foundation species of the kelp forests off the coast of California is kelp, or brown algae. The community of organisms that make up the kelp forest is made possible by the environments that kelps create. Another foundation species in a coral reef is the coral itself.

The majority of the reef structure, which shields other species from waves and ocean currents, is made up of the exoskeletons of both dead and living coral. Beavers are another example of a foundation species because they alter their surroundings by creating dams.

10.2.1.4.2. Keystone Species

A keystone species is one that, in relation to its biomass or abundance, has an outsized impact on community structure. Two main characteristics set keystone species apart from foundation species: first, they are more likely to be top predators at higher trophic levels, and second, they exhibit a wider range of behaviors than foundation species, which typically alter their surroundings. Perhaps the most well-known illustration of a keystone species is the intertidal sea star *Pisaster ochraceus*, which is found in the northwest of the United States.

The intertidal zone in which the sea stars resided was purposefully removed from them in a traditional community ecology experiment. As a result, the number of their prey (mussels) increased, changing the community's species composition and drastically lowering species diversity. About 25 different species of barnacles and algae were found in the lower intertidal zone when sea stars were present; however, when they were absent, the population of mussels expanded downward and nearly completely replaced these other species.

The removal of a keystone species frequently results in this kind of abrupt collapse of community structure or sharp reduction in diversity. In this instance, species that would typically be able to survive because sea stars kept the mussels in check were displaced by the mussels, resulting in a loss of diversity.

10.2.2. Patterns of Community Structure

10.2.2.1. Stratification and Gradation

As species adapt to their habitat, community structure can become stratified both horizontally and vertically during the succession process. This fractionation is caused by changes in environmental parameters like light, temperature, or water. The different levels of light that the various strata in a forest receive cause vertical stratification; the taller the plant and the more foliage it produces, the more light it can intercept. A forest typically contains three or more vertical layers of plants: a layer of herbs, a layer of shrubs, a layer of small trees, and a layer of canopy trees. This plant life stratification has an impact on animals. Even though they are quite adept at switching between layers, they frequently stay put in one area for the purpose of feeding, mating, or other activities.

Additionally, gradients in the physical environment can give rise to horizontal patterns among species. The distribution of plant and animal species can be impacted by variations in the region's nutrient or water availability (see biogeographic region). Different plant and animal species can be found on a mountain at different elevations as well as on the east, west, north, and south slopes. While gradual changes can result in a more integrated flow of species, drastic differences in some factors over very short distances can produce sharp boundaries between communities. The regional biodiversity is preserved in part by these gradients.

10.2.2.2. Ecotones

Almost invariably, ecosystems consist of a patchwork of communities existing at various stages of succession. Ecosystems vary in the sizes, frequencies, and intensities of disturbances, which leads to variations in the patch dynamics of communities. Ecotones are regions that are found around the borders of every patch. In addition to some species that have evolved especially to live in these zones, these junction zones frequently contain species from all of the overlapping communities. The edge effect refers to the frequent occurrence of a higher species and population density within the ecotone compared to the neighboring communities.

In ecotones between mature forests and earlier successional patches in North America, brown-headed cowbirds (*Molothrus ater*) are particularly prone to parasitize bird nests. Cowbirds are mostly active in early successional patches and lay their eggs in other birds' nests. Compared to birds living in ecotones, forest birds with nests deep within mature forests are less vulnerable to attacks. Mature forest removal has expanded the area of ecotones, which has led to an increase in cowbird parasitism rates throughout North America.

10.2.2.3. Ecological Niches

An ecological niche includes a species' lifestyle. In essence, it describes how a species interacts with or blends into its surroundings. Natural selection helps a species develop specialized traits that enable it to uniquely utilize the resources in its environment as it adapts to the physical conditions and biota of the community. The niche is shaped by the region's physical characteristics, such as temperature, topography, or nutrient availability, and by biological limitations, such as competition, predation, or resource scarcity, which restrict how a species uses its surroundings. For instance, different plant species have different needs in terms of light, nutrients, and microbes, as well as different capacities to repel rivals and herbivores.

Only a portion of the plants found in a community can be consumed by herbivore species, and only a portion of the numerous possible prey species can be caught by predators. As a result, the species "carves out" a place in the community. One such niche is occupied by the critically endangered Kirtland's warbler (*Dendroica kirtlandii*), which is found in North America. It only builds its nests in uniform stands of young, 2–4 meter (6–13 foot) tall jack pines (*Pinus banksiana*). Periodic fires are allowed to occur on these trees, which is essential for the jack pine seeds to germinate. Additionally, because these fires consistently create large new areas with young trees, the warblers are able to move their nesting locations over time to stay inside the preferred height of jack pine stands.

As biological and physical aspects of the community shift, a species' niche changes as well—as long as the changes happen slowly enough for the species to adjust. The fact that no two species in a community can occupy the same niche is the primary barrier to this evolution. Thus, specialized modes of existence give coexisting species a selective advantage, mitigating direct competition for resources.

10.2.2.4. Biodiversity and the Stability of Communities

Species create niches and guilds as a result of their mutual adaptation and community adaptation. It is possible for more species to coexist when more complex structures are developed. The community becomes more stable as a result of the rise in species richness and complexity acting as a buffer against environmental stresses and natural disasters.

10.2.2.4.1. Community Equilibrium and Species Diversity

Certain environments experience a climax in succession, resulting in a stable community dominated by a few key species. It is believed that this equilibrium, known as the climax community, arises when the network of biotic interactions gets so complex that no additional species can fit in. In other settings, a variety of species coexist in communities created by persistent small-scale disturbances, and any species has the potential to become dominant. This nonequilibrium dynamic demonstrates the impact that unforeseen disruptions can have on the composition and structure of a community. There are hundreds of different tree species in a square kilometer of some species-rich tropical forests. The area left by a dead tree that falls to the ground is open for grabs. In a similar vein, hundreds of fish species can be found in some coral reefs; the species that succeeds in colonizing a new disturbance patch will win. Every minor disruption triggers a fresh attempt at dominance. Communities that are diverse are generally healthier. According to extensive ecological research, communities with a higher species richness can bounce back from disturbances more quickly than communities with a lower species richness. In contrast to species-poor grasslands, species-rich grasslands in the Midwest of the United States continue to have higher primary productivity. The community's ability to withstand drought is progressively impacted by each new species that disappears from these grasslands. Similar to this, more diverse plant communities in Yellowstone National Park exhibit higher species composition stability than less diverse communities during periods of extreme drought. Furthermore, the more diverse communities in Africa's Serengeti grassland exhibit greater biomass stability throughout the seasons and greater capacity for post-grazing recovery.

The necessity to preserve the highest level of richness within biological communities is highlighted by the connection between species diversity and community stability. A rich interactive web of native species that have had time to adapt to one another is very different from a field of weeds that contains species that have only recently been introduced to the community. Communities rich in species that are not disturbed possess the resilience needed to maintain a healthy ecosystem, which is essential for life. Additionally, these communities are better equipped to withstand the effects of introduced foreign species, which, while sometimes unintentionally, can cause significant ecological and economic disruptions in less stable communities. Natural biological communities are made up of intricate networks of interactions that support both community stability and biodiversity.

10.2.2.4.2. Biogeographic Aspects of Diversity

The study of species distribution in a region is known as biogeography (see Biogeographic Region: General Features). Islands have been used to observe the factors that affect species diversity because they offer a controlled study area. The size of an island, its distance from other islands or the mainland, and the variety of species that live there are the three factors that affect how quickly an island is colonized. Based on this data, the theory of island biogeography can be used to estimate how many species will exist on a particular island. The species diversity of "islands" on land, such as mountaintops, lakes, and remnants of forest left over after logging, can also be explained by this theory.

According to the theory of island biogeography, the number of species is proportionate to the size of the island and inversely proportional to its distance from the mainland where immigration and extinction rates are equal. A few bacteria, molds, insects, and birds had colonized the island of Surtsey six months after a volcano erupted there in 1963 off the coast of Iceland. Following the 1883 volcanic eruption on the tropical Pacific Island of Krakatoa, a few grass species, insects, and vertebrates became established within a year or so. It had only taken a few decades for hundreds of species to arrive on both Surtsey and Krakatoa. Although not every species is able to establish itself and remain there for good, the island communities eventually stabilize into a dynamic equilibrium (Ricklefs and Milletr, 2000).

10.3. COMMUNITY CHANGES

In ecology, the makeup of a community varies over time. This transition, which can be impacted by the environment, biotic interactions, and dispersal, is the subject of succession research. You can visit any site over a period of years, be it your backyard, the nearby forest, grassland, pond, abandoned farm (or urban lot), park, or even just a nearby park, and you will probably notice alterations in the creatures that call that area home. It's possible for certain plants to vanish and for new ones to emerge; the animal kingdom may also exhibit these changes. Communities will always change over time in terms of composition, structure, diversity, and productivity—these shifts may be subtle or evident. For thousands of years, human populations have taken advantage of this principle by manipulating the habitat (e.g., utilization of fire) to increase the number of specific plant and animal species for our benefit. Ecologists have studied the nature of these changes in communities for more than a century, recording the reasons and trends of change, pinpointing the mechanisms underlying the change, and creating sets of overarching theories to account for the change. Succession is the term used to describe this gradual change in the makeup of the community. Adjectives for succession such as "orderly," "directional," and "stabilizing" are also included in some definitions. These qualifications, however, have generated a great deal of debate and, for the most part, do not adequately describe the procedure. The majority of the initial research on succession dealt with vegetation, but more recent studies have also examined the interplay of all organisms, including trophic interactions, as well as animal and microbial succession. Ecologists can gain insight into community dynamics through an understanding of succession, and it also benefits other fields such as forensics, conservation, and restoration.

10.3.1. The History of the Study of Succession

Previous European studies describing succession on coastal dunes served as the inspiration for Henry Cowles's dissertation research on sand dunes around Lake Michigan. Even though Cowles did not invent the concept of succession, his work defined the laws that underlie patterns of community change, and as a result, it had an impact on generations of ecologists. Henry Cowles established the still-held belief that plant communities are dynamic in terms of both the species composition and the underlying physical environment. The basic idea was that the ages of the plant communities in sand dunes revealed the ways in which the communities had changed over time. The chronosequence approach has been utilized by Cowles and subsequent scholars to gain a deeper comprehension of the succession process. In order to infer the successional sequence, this method substitutes spatial differences (locations of different ages) for time periods, as it would be challenging to follow a successional sequence that lasted decades or centuries. A sere is a particular stage in a successional sequence. In the field of ecology, particularly plant ecology, the question of how this process operated has always been prominent. Since they dominate the landscape, plants have gotten the majority of attention in the last century. Additionally, ecologists have found that animals exhibit parallel successional patterns and that these patterns can even influence the direction and rate of succession. Nowadays, succession is a widely studied topic that is being looked into in almost every ecosystem (e.g., aquatic, and marine) encompassing almost every taxonomic category (e.g., rodents, fish, insects, and protozoa).

10.3.1.1. Types of Succession

Two types of successional processes have been identified, based on the characteristics of the habitat in which they take place. The colonization of lifeless habitats in a newly created habitat that has never had any vegetation before is known as primary succession. Volcanism, the creation of dunes and barrier islands, and glaciation are the causes of these new land surfaces, and they necessitate the influx of outside species. Because soil development is a laborious process necessitated by both abiotic factors and the activities of organisms, the rate of primary succession is typically slow. For example, lichens frequently colonize rock locations. These lichens work in tandem with physical processes (e.g., wind, and rain), will alter the substrate in a way that makes it possible for additional colonists to survive (e.g., plants that are vascular and bryophytes). When an ecosystem is disturbed and leaves behind seeds, spores, or the underground parts of plants, secondary succession takes place. Any environmental event, usually abiotic, that causes a community to deviate from its equilibrium is referred to as a disturbance. This is an erratic event that may involve mortality or the removal of biomass. Disturbances can alter resource availability, community densities and composition, and occur as a result of fires, severe storms or droughts, landslides, mammal burrowing, flooding, and tree falls. A habitat forms at a faster rate than primary succession because it is easily accessible to many of the existing species as well as those that are colonizing. The level of soil development also varies: in secondary succession, the soil is typically fully developed, whereas in primary succession, it may take thousands of years. The pioneer community refers to the first residents (early serial stage), and the climax community refers to the last stage (Ricklefs and Milletr, 2000).

10.3.1.2. Models of Succession

Numerous models that explain and forecast the succession process have been created. The work of Frederic Clements from the early 1900s was one of the most important. Succession was understood as a serial stage sequence that culminated in the climax community, following a deterministic and orderly order. According to this theory, every stage was made up of a collection of species that were thought of as a single super-organism that collaborated to develop on a location before being replaced by a new stage. The habitat's climax community, which was considered to be the "expression of the climate," was eventually reached by succession. Henry Gleason, on the other hand, offered an alternative interpretation of succession that placed an emphasis on the individualistic characteristics (abiotic tolerances and dispersal) of species and how they interact with their surroundings. Not because they are a component of a cohesive "superorganism," but rather because of their abiotic tolerances, or niche requirements, species coexist in a given area. According to this viewpoint, succession is more dependent on random dispersal events and is less deterministic. The possibility of distinct successional trajectories leading to distinct climax communities is one of the effects.

For many years, Clements' incorrect beliefs dominated ecology, but in the 1950s, research that upheld Gleason's individualistic viewpoint and early criticisms by eminent ecologist Arthur Tansley started to challenge this. This work included the continuum and gradient approaches of John Curtis and Robert Whitaker, which shed light on the idea of individualistic species segregation along environmental gradients that did not match the consistent associations that Clements had predicted. During this period, Clements' influence could also be observed in the advancement of ecological perspectives on succession. Within the context of the Clementsian successional paradigm, Eugene Odum offered an ecosystem perspective on succession. Expectations regarding energy, life history, biogeochemical processing, and selection pressure (i.e., r- and K-selection), quantity, and the composition of communities evolve with time. Connell and Slatyer finally reworked many the succession models and concepts. They offered three models that explained processes and offered theories regarding the course of succession. All three believed that early pioneer species that produced large numbers of seeds, germinated early, and grew quickly were responsible for determining the community's original makeup. The succession process was where the models diverged. One way that succession might happen is through facilitation, in which early species change the environment to make room for later species that take their place. The life-history traits that cause later successional species to grow more slowly and eventually supplant earlier successional species are the driving force behind the second pathway, tolerance. Finally, the inhibition model predicts that long-lived late successional species will eventually supplant early colonists, who will remain dominant until a disturbance or mortality eliminates them. David Tilman offered a different mechanistic explanation of succession using the resource-ratio theory. He provided precise forecasts regarding how adjustments to the resource ratios (e.g., because species became specialized on different ratios of those resources, community composition can change over time due to factors like light or nitrogen availability. The current perspective recognizes the wide range of intricate variables and events that affect succession. The models (e.g., Markov Models) and techniques have advanced through the incorporation of numerous contributing elements, such as (but not restricted to) spatial processes (e.g., dispersal), life cycles and characteristics of species, intricate relationships between species (e.g., gradients in the environment, disturbance characteristics, and direct and indirect).

10.3.1.3. Factors Affecting Change

Disturbances are one of the many reasons why a community's composition can shift over time. The characteristics of disturbances, such as their frequency, extent, and severity, can affect their effects. A disturbance in the succession context can move the community backwards or forwards in time. For instance, fire can change the availability and utilization of nutrients in a community and eliminate the biomass of plants that are above ground. Species that are fire-adapted may have seeds that sprout again from subterranean portions or that germinate in reaction to fire. On the other hand, species that cannot withstand fire might have to relocate from areas that have burned. The rate at which the habitat is altered is another element that may have an impact on succession. For instance, some species' ability to coexist in their habitat may be restricted by the rate at which soil forms.

The rate of succession can also be determined by the rates at which species disperse into a site. In addition, this rate may interact with arrival time and order to create patterns of shifting community composition. Priority effects, for instance, can happen, in which the emergence of a species depends on early arrival — "first come, first served." In fact, there are more similarities than differences between the concepts and procedures covered in the field of community "assembly" and the framework of "succession" when it comes to forecasting how communities will evolve over time in response to biotic interactions, dispersal, and abiotic factors. Though attempts to reconcile these two approaches are relatively new, many of their differences are artificial constructs of the ecology field. Remarkably, both strategies have been used to direct land management and habitat restoration. For instance, both paradigms discuss how species interactions and dispersal affect how communities change over time. Future research ought to keep combining these two theoretical streams.

10.3.2. Species Interactions in Communities

Numerous interactions exist between different species. Certain interactions take place directly, such as when lions kill and devour gazelles or when clown fish hide inside sea anemone tentacles. However, a great deal of interactions are less evident because they are indirect. Lions, for instance, rely directly on the existence of the plant-eating animals they hunt, which in turn depends on the plants they require for sustenance. Thus, despite not eating plants, lions are indirectly dependent on the kind and quantity of plants. Changes in one species, such as those pertaining to population size, health, or geographic range, will almost certainly have an impact on other species as well as other elements of the Earth system. This is a result of both direct and indirect interactions between species (Figure 10.3).

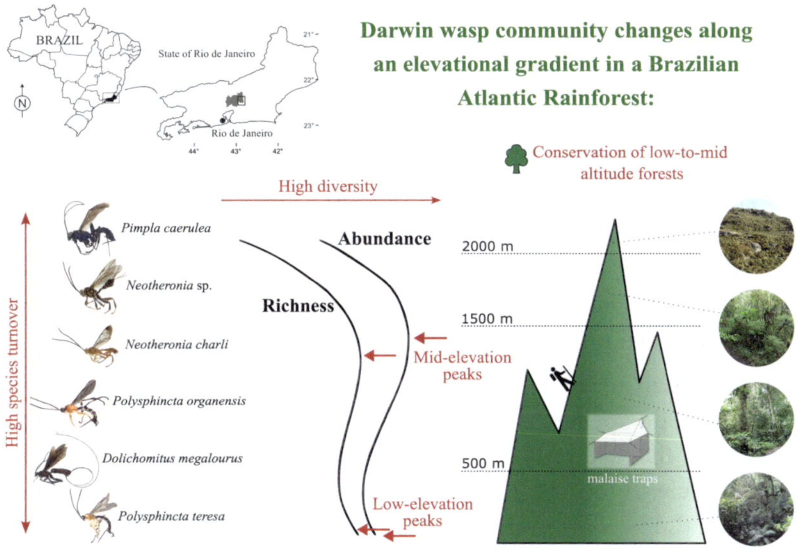

Figure 10.3. *Species interactions in communities.*

Source: Flinte V, Pádua DG, Durand EM, Hodgin C, Khattar G, da Silveira LFL, Fernandes DRR, Sääksjärvi IE, Monteiro RF, Macedo MV, et al. (2023). Variation in a Darwin wasp (Hymenoptera: Ichneumonidae) community along an elevation gradient in a tropical biodiversity hotspot: implications for ecology and conservation. Insects, 14(11):861. https://doi.org/10.3390/insects14110861.

Species influence one another through interaction. Interactions between species can be direct or indirect, with favorable or unfavorable results. The most evident interactions are those that occur directly, like when a giraffe eats a tree's leaves. The giraffe benefits from the interaction because it receives nourishment from the tree, but the tree suffers because it loses leaves that are necessary for photosynthetic energy production. Giraffes do, however, also directly benefit trees by leaving their droppings on the ground, which enriches the soil with nutrients and improves its quality, thereby encouraging the growth of trees and other plants. Giraffes are one of the many herbivores that rely on this plant growth for food.

As a result, giraffes indirectly contribute to the survival of other herbivorous species. More species are added to the web of indirect species interactions as a result of the plant growth, which also offers smaller animals—such as insects and birds—a place to live and nest. The biomass and productivity of an ecosystem are influenced by species interactions, which in turn impact the size of species populations. Evolutionary changes that impact an organism's capacity for long-term survival and reproduction are also influenced by species interactions. Learn more about interactions between species by looking through the examples below.

10.3.2.1. Earth System Models of Species Interactions

Because species interact both directly and indirectly with other species, modeling species interactions within the biosphere can get complicated very quickly. For example, oxygen is produced by photosynthesis, and nearly every organism, including photosynthetic organisms, uses oxygen to break down food into energy through a process called respiration.

Thus, almost all species interact indirectly with photosynthetic species, and thus photosynthetic species are relevant to all Earth system models about the biosphere. Note that we only include interactions that most directly influence the relevant species, ignoring some direct and the numerous indirect interactions that also affect the species in the model.

Earth system focuses on how interactions between species can impact columbine flower evolution. Knowing that columbines, like the majority of flowering plants, have evolved flower traits (colors and shapes) to attract pollinators—insects and birds that move pollen from plant to plant—will help you understand how this works. Pollinators visit columbine flowers in search of food, which the plant provides in the form of nectar. Each pollinator unintentionally brushes up against the stamens as it feeds, which transfers pollen to the pollinator's body. Then, some of this pollen rubs off onto other flowers as the pollinator travels from flower to flower in order to feed. Consequently, the pollinator inadvertently aids in flower pollination as it moves from flower to flower while consuming nectar. To find out more information about each topic, click on the icons or hover over them for a brief explanation.

10.3.2.2. How Human Activities Influence Species Interactions?

Interactions between species are impacted by human activity both directly and indirectly. Large-scale farming and agricultural practices are one of our direct effects; they boost the populations of food species and reduce those of species that occupy the habitats that these practices have altered. The numbers of the exploited species' populations are also decreased by hunting and fishing. Smaller-scale hunting for food can also have an impact on humans, albeit in a negative way. For instance, eating wild-caught animals can spread dangerous animal diseases like HIV and Ebola to human populations. Urbanization and large-scale farming practices alter habitat, which have indirect effects on interactions between species. The impact of climate change through processes like ocean acidification and extreme weather, such as drought and fire, was also considered an indirect effect.

10.3.3. Biological Interaction

A biological interaction in ecology refers to the impact that coexisting organisms have on one another. Interspecific interactions refer to interactions between different species, while intraspecific interactions involve the same species. Both of these effects—which can be long-term or short-term, like pollination and predation—strongly impact the species in question's evolutionary history. A symbiosis is a long-term relationship. Mutualism, which benefits both parties, and competition, which hurts both parties, are two different types of symbioses. Indirect interactions can occur when people interact through intermediaries like common enemies or resources.

This kind of relationship can be demonstrated by the net effect, which is based on the individual effects that the relationship has on both organisms. According to many recent studies, mutualisms and habitat modification among non-trophic species may play a significant role in determining the structures of food webs. It is still unknown, though, if these results hold true for all ecosystems and whether non-trophic interactions have a random effect on food webs or have an impact on particular trophic levels or functional groups.

10.3.3.1. Short-Term Interactions

In ecology and evolution, short-term interactions such as pollination and predation are crucial. In terms of the length of a single interaction, these are brief—a predator kills and consumes its prey, and a pollinator spreads pollen from one flower to another—but they are incredibly persistent in terms of their impact on the evolutionary paths of both partners. The partners consequently coevolve.

10.3.3.1.1. Predation

Predation is the process by which an organism, known as the predator, kills and consumes its prey. Predators have keen senses like vision, hearing, and smell, and they are frequently highly specialized for hunting. Sharp jaws or claws are used by many predatory animals, both vertebrate and invertebrate, to seize, dismember, and kill their prey. Additional adaptations that increase hunting efficiency include aggressive mimicry and stealth.

Predation exerts a strong selective pressure on prey, leading to the development of antipredator adaptations like chemical and/or color changes, warning calls and other signals, camouflage, and defensive spines. Since at least the Cambrian era, predators have played a significant role in the evolution of species. Microbiologists have found several interesting microbes that feed on other people to survive over the past few decades. Members of the genera Daptobacter, Vampirococcus, and Bdellovibrio comprise some of the best examples. Bdellovibrios are highly mobile, active hunters that swam around in search of vulnerable Gram-negative bacterial prey.

A belovibrio cell detects such a cell and swims faster until it runs into the prey cell. It then penetrates its prey's outer membrane to reach the periplasmic space. It develops into a long filament as it grows, which eventually splits into septae and releases baby bacteria. New Bdellovibrio cells are released during the prey cell's lysis. It has never been observed that Gram-negative prey bacteria develop resistance to bdellovibrios, and the bacteria will not assault mammalian cells.

This has sparked interest in treating infected wounds with these bacteria as a probiotic. Though it hasn't been tested yet, it makes sense that, given the rise in pathogens resistant to antibiotics, these kinds of treatments might be seen as workable substitutes.

10.3.3.1.2. Pollination

In order to facilitate fertilization, pollinators—such as insects (entomophily), certain birds (ornithophily), and certain bats—transfer pollen from a male flower part to a female flower part in exchange for pollen or nectar. Over geological time, the partners have coevolved; for insects and flowering plants, this coevolution has lasted for more than 100 million years. Flowers that are pollinated by insects have developed shaped structures, vivid colors, patterns, scents, nectar, and sticky pollen to draw in insects, assist them in gathering and depositing pollen, and compensate them for their labors. Pollinator insects, such as bees, are specialized to recognize color, pattern, and scent in flowers; to gather and carry pollen (e.g., by using bristles on their hind legs to form pollen baskets); and to gather and process nectar (e.g., honey, which bees make and store). Natural selection has shaped the adaptations on both sides of the interaction based on how well they pollinate. These adaptations match the adaptations on the other side.

10.3.3.1.3. Seed Dispersal

The act of moving, spreading, or transporting seeds away from their parent plant is known as seed dispersal. Due to their restricted mobility, plants rely on a range of dispersal vectors, including both living (biotic) vectors like birds and abiotic vectors like the wind, to spread their propagules. Seeds can be distributed both spatially and temporally, as well as individually or collectively from the parent plant. The dispersal mechanism plays a major role in determining seed dispersal patterns, which has significant effects on the genetic makeup and demography of plant populations, as well as on migration patterns and interactions between species. Seeds can be dispersed by five main methods: wind, gravity, ballistics, water, and animals. Certain plants are serotinous, meaning they only scatter their seeds in reaction to external cues. A diaspore must release itself or separate from its primary parent plant in order to disperse.

10.3.3.2. Symbiosis: Long-term Interactions

The six possible types of symbiosis are mutualism, commensalism, parasitism, neutralism, amensalism, and competition. These are distinguished by the degree of benefit or harm they cause to each partner.

10.3.3.2.1. Mutualism

When two or more species interact, mutualism occurs when each species gains something from the other, like a higher carrying capacity. Cooperation describes similar interactions that occur within a species. Symbiosis, which is sometimes mistaken for mutualism, is the closest association that can be used to categorize mutualism. One or both of the interacting species may be obligate, meaning that they would not be able to endure either temporarily or permanently without the other species. Even though mutualism has historically gotten less attention than other interactions in ecology, like predation, it is still a significant topic. Gut flora, Müllerian mimicry, cleaning symbiosis, and nitrogen fixation by bacteria in legume root nodules are a few examples.

10.3.3.2.2. Commensalism

Commensalism is advantageous to one organism while having no negative effects on the other. It happens when an organism interacts with another and gains something from it without affecting the host organism. A remora coexisting with a manatee is a prime example. Remogras consume the excrement of manatees. This interaction has no effect on the manatee because the remora does not deplete its resources.

10.3.3.2.3. Parasitism

A relationship between species known as parasitism occurs when one organism, called a parasite, lives on or inside another organism, called a host, harming it occasionally and developing a structural adaptation to this way of life. When it comes to intestinal parasites, the parasite either feeds on the host or partly consumes its food.

10.3.3.2.4. Neutralism

The relationship between two species that interact but do not affect one another is referred to as neutralism, a term that was first used by Eugene Odum. It is nearly impossible to find examples of true neutralism; in reality, the term is used to characterize circumstances in which interactions are minimal or inconsequential.

10.3.3.2.5. Amensalism

An interaction known as amensalism occurs when one organism harms another without incurring any expenses or reaping any rewards for itself. The negative impact one organism has on another is referred to as amensalism. This is a one-way process that occurs when an organism releases a particular compound that harms another. The microbial production of antibiotics, which can inhibit or kill other, susceptible microorganisms, is a classic example of amensalism. When sheep or cattle trample the grass, it is an obvious example of amensalism.

Although the presence of the grass has very little negative impact on the hoof of the animal, the grass becomes crushed. Amensalism is frequently used to characterize highly asymmetrical competitive interactions, like those that have been noted between the Spanish ibex and Timarcha weevils, which both consume the same kind of shrub. Weevils hardly affect the availability of food, but their presence greatly reduces the number of weevils because ibex eat a lot of plant matter and inadvertently ingest the weevils that are present in it. Amensalisms are often very intricate. An actinomycete and a parasitic fungus in the genus Escovopsis interact in a way that Attine ants (a tribe of ants from the New World) can exploit. The ant and members of the fungal genus Leucocoprinus can maintain a mutualistic relationship because of their amensalistic relationship. It is astounding to learn that these ants grow a garden of Leucocoprinus fungi for their own food. The ants also encourage the growth of an actinomycete of the genus Pseudonocardia, which generates an antimicrobial compound that inhibits the growth of the parasitic fungus Escovopsis, to keep it from completely devasting their fungal garden.

10.3.3.3. Non-trophic Interactions

Competition for space, mutualism, and habitat modification are a few instances of non-trophic interactions. Recent research suggests that by influencing the species in the network and the strength of trophic links, non-trophic interactions may have an indirect impact on the topology of the food web and trophic dynamics. Ecological network analyses must incorporate both trophic and non-trophic interactions, according to many recent theoretical studies. The few empirical studies that address this claim that interactions between species outside of the trophic network can have a significant impact on food web structures, or network topologies. It is unclear how far these results can be applied to other coastal systems because these studies only cover a small number of them. It is still unclear if non-trophic interactions normally influence particular trophic levels, functional groups, or species within the food web, or if they indiscriminately mediate species and their trophic interactions across the network. Non-trophic facilitation appears to benefit sessile species with generally lower trophic levels more than other species, according to some studies, while higher trophic and more mobile species also benefit from it.

In 2018, Borst conducted a study to test the general hypothesis that foundation species, or spatially dominant habitat-structuring organisms, modify food webs by increasing their size, as indicated by the number of species, and their complexity, as indicated by the number of links. The study also examined whether this modification occurs randomly throughout the entire food web or specifically for specific species that fall into particular trophic levels or functional groups.

A longer average chain length and a higher mean trophic level were the results of the finding that carnivores are more strongly facilitated in the food webs of foundation species than would be expected based on random facilitation, and that species at the base of the food web are less strongly facilitated. This suggests that foundation species, by non-trophic facilitation of species throughout the entire trophic network, significantly increase the complexity of the food web. Despite being a component of the food web just like any other species (e.g., as predator or prey), many studies have demonstrated that they significantly support the related community by generating new habitats and reducing physical strain.

It has been discovered that foundation species engage in this type of non-trophic facilitation in a variety of ecosystems and environmental settings. Corals, kelps, mussels, oysters, seagrasses, mangroves, and salt marsh plants all help organisms in harsh coastal zones by lowering waves and currents, supplying aboveground structure for attachment and shelter, concentrating nutrients, and/or lessening the stress of desiccation during low tide exposure. It has also been discovered that foundation species, such as grasses, shrubs, and trees in savannahs, as well as macrophytes in freshwater systems, significantly contribute to the habitat-structuring process in less hazardous systems. In the end, all foundation species improve the availability and complexity of their habitat, which divides and expands the niche space that is available to other species.

10.3.4. Competition

More broadly, competition can be defined as the direct or indirect interaction of organisms that results in a change in fitness when the organisms share a resource. Typically, competition is thought of as the interaction of individuals that compete for a common resource that is in limited supply. The weaker competitors are typically negatively impacted by the outcome. Three main types of competition exist. Real competition is defined as two of them: exploitation competition and interference competition. Apparent competition, the third form, is not. While apparent competition and exploitation competition happen indirectly between individuals, interference competition happens directly between them (Figure 10.4).

Figure 10.4. *Competitive interactions.*

Source: By lubye13 – IMG_1300, CC BY-SA 2.0, https://commons.wikimedia.org/w/index.php?curid=17637811.

Interference competition occurs when one person intentionally modifies another person's resource-attaining behavior. For instance, a dominant male gorilla directly modifies the mating behavior of other males when he uses physical aggression or other aggressive displays to prevent other males from accessing a mate. Another illustration of an intra-specific interaction is this. The phenomenon of exploitation competition arises when people engage in indirect interactions while vying for shared resources such as territory, prey, or food.

In other words, the resource will become less available for use by others when it is used by one person. A superior competitor can eventually drive an inferior one out of the market through manipulation or outright exploitation, leading to competitive exclusion. Equations can be used to predict the results of competition between two species; the Lotka-Volterra model is one of the most well-known examples. This model incorporates both the overall impact of two species on each other as well as the relationship between population density and carrying capacity.

The four outcomes of this model are:

- Species A competitively excludes species B;
- Species B competitively excludes species A;
- Either species wins based on population densities; and
- Coexistence occurs. Species can survive together if intra-specific is stronger than inter-specific competition.

This means that each species will inhibit their own population growth before they inhibit that of the competitor, leading to coexistence. Adopting alternative life history and dispersal strategies—which are typically reinforced by natural selection—is another way to avoid competitive exclusion. Through this mechanism, chances for new colonization and nutrient acquisition are increased while competitive interactions are decreased.

Events (like tidal, flood, or fire disturbances) that produce chances for dispersal and nutrient acquisition are frequently necessary for this to succeed. Let us consider the following: Plant B is a better disperser, but Plant A is more effective at absorbing nutrients. In this case, nutrients are the resource that is under competition, but obtaining nutrients is dependent on their availability. Plant B is anticipated to move in first and stay in the community until Plant A shows up and starts to compete with Plant B if a disturbance creates new space for colonization. Plant A will eventually surpass Plant B in terms of growth rate, possibly due to Plant A's superior ability to absorb nutrients. The Plant B population will decrease as the Plant A population grows, and with enough time, it will be able to be eliminated from that area. If a local disturbance (like prairie fires) continuously creates new opportunities (space) for colonization, Plant B's exclusion can be prevented. Since this frequently occurs in nature, disturbance can prevent competitive exclusion and balance competitive interactions by forming patches that are more easily colonized by species with more effective dispersal mechanisms. However, the frequency and spatial proximity (or closeness) of disturbance events in relation to the dispersal rates of individuals of the competing species determines the outcome of the dispersal versus nutrient acquisition trade-off. When disturbances happen frequently enough or far enough apart to keep the weaker but frequently more effective dispersing competitor within a habitat, coexistence can be attained. The inferior competitor (better disperser) prevails if the disturbance occurs too frequently; however, if the disturbance occurs infrequently, the superior competitor gradually outcompetes the inferior competitor, leading to competitive exclusion. The intermediate disturbance hypothesis is the name given to this (Ricklefs and Milletr, 2000).

When two people who do not directly compete with one another for resources nevertheless have an indirect impact on one another through sharing a predator, this is known as apparent competition. Think about a hawk that eats mice and squirrels. In this relationship, an increase in the squirrel population may have a positive effect on the mouse population because more squirrels will be available for the hawks to prey on. Nevertheless, as the squirrel population decreases, a greater number of hawks may eventually need more prey, which would negatively impact the mice through increased predation pressure. The predator's food supply may decline, which would have the opposite effect. Since mice will be the hawks' more plentiful food source, a decline in the squirrel population may also indirectly result in a decline in the mouse population. In nature, apparent competition can be hard to spot, frequently due to the intricacy of indirect interactions involving several species and shifting environmental factors (Ricklefs and Milletr, 2000).

10.4. ECOLOGICAL SUCCESSION

The process through which an ecological community's species structure changes over time is known as ecological succession. The duration may vary from a few decades (post-wildfire, for instance) to several decades. The community starts out as a relatively small group of pioneering plants and animals and grows more complex over time until it reaches a climax community, which is stable and self-sustaining. The engine of succession, the cause of ecosystem change, is the impact of established organisms upon their own environments.

Living has the effect of changing one's surroundings, sometimes subtly and other times overtly. It is the process through which an ecological community changes, after a disturbance or after the first colonization of a new habitat, in a relatively orderly and predictable manner. A community may be disturbed by a fire, a strong windstorm, logging, or the creation of new, unoccupied habitat as a result of a landslide or lava flow.

These events can also start a succession. Primary succession refers to a succession that emerges in new habitats independently of pre-existing communities, while secondary succession is a succession that occurs after a pre-existing community is disrupted. One of the earliest ecological theories to be proposed was succession. The Indiana Dunes in Northwest Indiana are where ecological succession was first observed, and it is still a crucial ecological subject of research.

10.4.1. History

The concept of ecological succession has roots that date back to the early 1800s. The French naturalist Buffon observed in 1742 that poplars naturally come before oaks and beeches in the development of a forest. Later, the University of Paris's theological committee forced Buffon to renounce many of his theories because they ran counter to the biblical account of Creation.

It was the French naturalist Adolphe Dureau de la Malle and the Swiss geologist Jean-André Deluc who first used the term "succession" to describe the development of vegetation following forest clear-cutting. Henry David Thoreau discussed succession in an oak-pine forest in an 1859 speech titled The Succession of Forest Trees. Although observers have long known that squirrels burrow nuts in the ground, no one has used this information to explain why forests regularly change.

An investigation into the plant succession in the Danube River basin was published in 1863 by the Austrian botanist Anton Kerner. According to Ragnar Hult's 1885 study on the stages of forest development in Blekinge, heath emerges from grassland before forests do. In the early phases of forest development, birch predominated, followed by pine (on dry soil) and spruce (on wet soil). It eventually becomes beechwood if oak replaces the birch. Swamp vegetation progresses from moss to sedges to moor vegetation, then birch to spruce.

10.4.1.1. H. C. Cowles

Henry Chandler Cowles, at the University of Chicago, created a more formal understanding of succession between 1899 and 1910. Cowles investigated the growth of vegetation on sand dunes along the shores of Lake Michigan, specifically the Indiana Dunes, after being inspired by Eugen Warming's studies of Danish dunes. He realized that the vegetation on different-aged dunes could be seen as distinct phases of a larger trend in the development of vegetation on dunes (a method of studying vegetation change that was later dubbed space-for-time substitution, or chronosequence studies). The ecological relations of the vegetation of the sand dunes of Lake Michigan was the title of the paper he first published on this subject in the Botanical Gazette in 1899. He developed the concepts of primary succession and a sere—a recurring series of community changes unique to certain environmental conditions—in this seminal work and other papers that followed.

10.4.1.2. Gleason and Clements

However, the theories of Frederic Clements, a contemporary of Cowles, who maintained that seres were highly predictable and deterministic and converged on a climatically determined stable climax community regardless of starting conditions, dominated knowledge of succession from roughly 1900 to 1960. The pseudo-organismic theory of community ecology is the term used to describe Clements' model, which clearly compares the ontogenetic development of individual organisms with the successional development of ecological communities. Clements and his adherents created a sophisticated classification system for communities and lineages of succession. A different framework was provided by Henry Gleason as early as the 1920s. Compared to the Clementsian, the Gleasonian model was considerably less deterministic and more complex. The two main ways it deviates from the Clementsian perspective are by arguing that chance factors play a much larger role and by rejecting the idea that there are coherent, sharply bounded community types. Gleason maintained that communities were best understood as artifacts of the juxtaposition of species distributions and that species distributions responded individualistically to environmental factors. Until the late 1950s, Gleason's theories—which were first published in 1926—were mainly disregarded.

10.4.1.3. Modern Era

The work of Robert Whittaker and John Curtis in the 1950s and 1960s generally marked the beginning of a more rigorous, data-driven testing of community theory and successional models. Since then, succession theory has evolved to become more nuanced and less rigid. J. Connell and R. Slatyer made an attempt to use mechanisms to codify successional processes. The idea of a stable climax vegetation has largely been abandoned by ecologists in North America and Britain, and successional processes are now viewed as considerably less deterministic, with significant roles for historical contingency and alternative pathways in the actual development of communities. There is still debate regarding the relative significance of equilibrium vs. random processes and the general predictability of successional dynamics. irregular procedures. Fakhri A., a former Harvard professor. Bazzaz brought up the topic of scale, arguing that while processes are random and patchy at the local or small area scale, there are undeniable trends when larger regional areas are taken into account.

10.4.2. Factors

Site characteristics, the nature of the events that start succession (perturbations), interactions between the species present, and more random elements like the availability of colonists or seeds or the meteorological conditions at the time of disturbance can all have an impact on the successional change trajectory.

While some of these factors add more probabilistic elements, others help to make succession dynamics more predictable. Climate change and human activity are two major perturbation factors in the modern world. Early succession communities are typically dominated by quickly expanding, widely distributed species (opportunist, fugitive, or r-selected life histories). These species will typically give way to more competitive (k-selected) species as succession moves forward.

Although some trends in ecosystem and community properties have been proposed, most of them don't seem to be widespread. For instance, species diversity almost always rises in the early succession as new species appear, but it may fall in the later succession when opportunistic species are eliminated by competition and locally superior competitors take the lead. Depending on the specific system and site, net primary productivity, biomass, and trophic properties all exhibit variable patterns over succession.

The climax, also known as the "potential vegetation" of a site, was once thought to be the stable end-stage of ecological succession that was largely influenced by the local climate. Modern ecologists have mostly given up on this theory in favor of nonequilibrium theories of ecosystem dynamics. The majority of natural ecosystems undergo disturbances at a pace that prevents the emergence of climax communities. Climate change frequently happens frequently enough to keep things from reaching a peak state. Communities can also be continuously reshaped by the introduction and expansion of new species into existing pools.

Community characteristics impact the development of some ecosystem attributes, like nutrient cycles and soil qualities, which in turn impact subsequent successional development. This process of feedback may only happen over millennia or centuries. In addition to the random character of disruption occurrences, and other extended (e.g., climatic) shifts, it is questionable if the "climax" idea ever applies or is especially helpful when taking into account real vegetation.

10.4.3. Types

10.4.3.1. Primary Succession

Successional dynamics that start when an ecological community settles in a previously uninhabited area, like recently exposed sand or rock surfaces, lava flows, recently exposed glacial tills, etc., are known as the principal succession.

Pioneer microorganisms, plants (lichens and mosses), the grassy stage, smaller shrubs, and trees are among the stages of primary succession. When there is food available, animals start to come back. It has reached the pinnacle community stage when it is an entirely functional ecosystem.

10.4.3.2. Secondary Succession

Secondary succession refers to successional dynamics that occur after a pre-existing community is severely disrupted or eliminated. Pre-disturbance conditions, such as soil development, seed banks, remaining organic matter, and residual living organisms, have a significant impact on secondary succession dynamics. Early on in secondary succession, community change can occur relatively quickly due to pre-existing organisms and residual fertility.

Compared to primary succession, secondary succession is observed and studied far more frequently. Responses to both human-caused and natural disturbances, such as logging and agriculture, as well as natural disturbances like fire, flood, and strong winds, are particularly prevalent forms of secondary succession. To allow the new material to rebuild itself, the soils and organisms must be left intact during secondary succession. For example, woody plants colonized large, nearby patches more quickly (per unit area) in a fragmented old field habitat created in eastern Kansas.

A landscape can shift dramatically through secondary succession. A wildfire in Acadia National Park in the 1900s obliterated much of the surrounding area. At first, the terrain was covered in evergreen trees. It took the area at least a year to sprout shrubs after the fire. Gradually, deciduous trees began to take the place of the evergreens.

10.4.3.3. Seasonal and Cyclic Dynamics

These kinds of vegetation changes, in contrast to secondary succession, are periodic and result from shifting species interactions or recurrent events rather than being dependent on disturbance. These models shift the idea of a climax to a dynamic state.

10.4.3.4. Causes of Plant Succession

Changes in the soil brought about by the organisms living there can result in autogenic succession. These alterations can be caused by plants growing in the soil, shifting soil nutrients, or building up organic matter in the humic layer or litter. The community can also be changed by the plants' own structural makeup. Larger species, like trees, for instance, cast shade on the growing forest floor as they mature, which tends to keep out species that need light. Shade-loving plants will spread throughout the region. The cause of allogenic succession is not the vegetation itself, but rather factors from the surrounding environment. For instance, alterations to the soil brought about by leaching, erosion, or the deposition of silt and clays can change the ecosystems' water and nutrient relationships. Animals are herbivores, seed dispersers, and pollinators, which means they contribute significantly to allogenic changes. Additionally, they have the power to alter the soil's composition (as termites, ants, and moles do) and add more nutrients to specific regions of the ecosystem. Regeneration sites that favor particular species may result from this. Perhaps more so than any other factor, but on a far longer timescale, are climate-related factors. Variations in rainfall and temperature will encourage changes within communities. After each ice age, significant successional changes occurred as the climate warmed. Mixed deciduous forest succeeded bare glacial till deposits and tundra vegetation. In the upcoming century, significant Allogenic changes are probably going to occur due to the greenhouse effect's increased temperature. Allogenic changes are also brought about by meteors, earthquakes, avalanches, floods, fires, and strong winds, among other geological and climatic disasters.

10.4.3.5. Mechanisms

A descriptive theory of succession was published by Frederic Clements in 1916, and it was then promoted as a broad ecological idea. His succession theory had a significant impact on ecological philosophy. The idea of Clements is commonly referred to as classical ecological theory.

According to Clements, succession is a process involving several phases:

- **Nudation**: Succession begins with the development of a bare site, called Nudation (disturbance).
- **Migration**: refers to the arrival of propagules.
- **Ecesis**: involves the establishment and initial growth of vegetation.
- **Competition**: as the vegetation becomes well established, grows, and spreads, various species begin to compete for space, light and nutrients.
- **Reaction**: during this phase autogenic changes such as the buildup of humus affect the habitat, and one plant community replaces another.
- **Stabilization**: a supposedly stable climax community forms.

10.4.4. Climax Communities

A community of plants, animals, and fungi that have reached a steady state through the process of ecological succession in the development of vegetation in an area over time is referred to as a climax community, also known as a climatic climax community in scientific ecology. It was previously believed that this equilibrium would arise because the species in the climax community are the most suited to the local average conditions. The phrase is occasionally used in reference to soil development. However, it has been discovered that a steady state is more observable than actual, especially over extended periods of time. Nevertheless, it's still a good idea.

10.4.4.1. Frederic Clements' Use of "Climax"

According to Clements, the ontogenetic development of individual organisms and the successional development of ecological communities are similar. Clements only proposed analogies with extremely basic organisms. The notion that an ecological community is a superorganism was developed by later ecologists, who also attempted to identify a single climax-type for each region and occasionally asserted that communities could be homologous to complex organisms. This concept was developed by the English botanist Arthur Tansley, who discovered the polyclimax—multiple steady-state end-points in a given climatic zone that are determined by edaphic factors. These endpoints were referred to by different names by Clements, who believed that they were unstable because climax vegetation is inherently the most suited to a particular region's climate. Research in the 1950s and 1960s significantly supported Henry Gleason's early criticisms of Clements' organism simile and other methods of characterizing vegetation, which were largely ignored for many years (see below). In the meantime, vegetation management and theoretical ecology both heavily incorporated climax theory. Pre-climax, post-climax, plagioclimax, and disclimax—all terms coined by Clements—are still used to characterize the multitude of communities that continue to exist in states that deviate from the ideal climax for their specific region.

Although he is sometimes credited with expressing certain opinions, Clements never maintained that there must always be climax communities, that species within an ecological community are physiologically tightly integrated, or that plant communities have distinct boundaries in terms of time or space. Rather, as a conceptual beginning point for characterizing the vegetation in a given area, he used the idea of a climax community—of the type of vegetation best adapted to some idealized set of environmental conditions. There are solid grounds for thinking that the species most suited to a given set of circumstances may show up there when those circumstances materialize. However, a large portion of Clements' research was devoted to describing the outcomes that arise when those ideal conditions are not met. Under those conditions, vegetation that isn't the ideal climax will frequently grow in its place. However, those various vegetation types can still be considered as departures from the ideal climax. As a result, Clements created an extensive lexicon of theoretical terms that describe the different non-climax states that vegetation adopts as well as the different potential causes of vegetation. He approached ecological complexity by characterizing the climax community as the ideal form of vegetation and characterizing other forms of vegetation as variations from it (Ricklefs and Milletr, 2000).

10.4.4.2. Continuing Usage of "Climax"

Even though climax theory was generally abandoned, some theoretical ecologists started using climax concepts again in the 1990s. A lot of writers and environmentalists still use a watered-down version of the term "climax" to describe what are otherwise known as mature or old-growth communities. Another use of the term "climax" is to refer to a stage of late succession in marine macroinvertebrate communities.

Furthermore, some modern ecologists continue to refer to an ecosystem dominated by invasive species that compete to prevent the reintroduction of once-native species as a disclimax. This idea is derived from Clements' original definition of climax, which is related to an ecosystem that is resistant to alien species colonization. Although disclimax is an anthropogenic phenomenon that hinders the facilitation and succession to a true climax community, Clements used the term in context, and it is one of the few examples of climax that can be observed in nature.

Climax has the following- important characteristics:

- **Unitary:** Climax is a unit. All the species are taken as an organized unit to show the climate. In other words, climax is a unit, which acts as the index of the climate of an area.
- **Stability:** The form of the climax community is stable with the climate. The climax community cannot be replaced through competition.
- **Origin and Organism Concept:** Climax community is interpreted as a super organism. This community takes birth, grows and develops like an organism. Climax communities undergo changes with the changes of time and climate.

There are three theories about the climax community:

- Mono-climax theory
- Poly-climax theory
- Climax-pattern hypothesis

i. **Mono-Climax Theory:** This theory was proposed by Fedeick Climents in 1916. According to this theory, there develops only one type of climax community in a climatic region of a geographical area. The other communities existing besides the climax community in that area, are called as subordinate communities.

The subordinate communities may be:

- Pro-climax
- Post-climax,
- Pre-climax.

It would be incorrect to consider climate as a determining factor for climax formation since different types of climax communities can develop in an area of uniform climate. Furthermore, the climax state is not the final modified form of the succession; rather, it is a form of minimum changes. Thus, the mono-climax theory is not without its detractors.

ii. **Poly Climax Theory:** In 1935, Braun-Blanquet and Tansley put forth this theory in response to criticism of the monoclimax theory. This theory states that in addition to the local climate, a variety of other factors also influence the climax communities. So the climax stage may be named depending upon the nature of the factor in stabilization:

- **Topographic Climax:** The climax community stabilized by topographical factors of an area like hills, mountains, slopes, etc., is known as a topographical climax.
- **Biotic Climax:** The climax of biotic community stabilized by living factors including man, herbivores, and animals are known as the anthropogenic climax, grazing climax and zootic climax respectively.
- **Edaphic Climax**: The climax community stabilized due to soil factor is known as edaphic climax.
- **Fire-Climax:** The climax resulted from repeated fire is called fire climax.

iii. **Climax-Pattern Hypothesis:** This hypothesis was proposed by R.I-1. Whittaker in 1953. According to this hypothesis, many climax communities may develop in an area under a particular set of climatic conditions. That is, communities change their living pattern related to the different environmental conditions of the habitat by adjusting themselves (Ricklefs and Milletr, 2000).

REFERENCES

1. Audia, P. G., Sorenson, O., & Hage, J. (2001). Tradeoffs in the organization of production: Multiunit firms, geographic dispersion, and organizational learning. *Advances in Strategic Management*, *18*, 75–105.
2. Begon, M., et al. (2014). *Ecology: From Individuals to Ecosystems*. Wiley-Blackwell.
3. Cadotte, M. W., Cavender-Bares, J., Tilman, D., & Oakley, T. H. (2009). Using phylogenetic, functional, and trait diversity to understand patterns of plant community productivity. *PLoS One*, *4*(5), e5695.
4. Cam, E., Link, W. A., Cooch, E. G., Monnat, J. Y., & Danchin, E. (2002). Individual covariation in life-history traits: Seeing the trees despite the forest. *The American Naturalist*, *159*, 96–105.
5. Carroll, G. R., & Swaminathan, A. (2000). Why the microbrewery movement? Organizational dynamics of resource partitioning in the U.S. brewing industry. *American Journal of Sociology*, *106*, 715–762.
6. Chase, J. M., & Leibold, M. A. (2003). *Ecological Niches: Linking Classical and Contemporary Approaches*. University of Chicago Press.
7. Chave, J., & Leigh, Jr, E. G. (2002). A spatially explicit model of beta-diversity in tropical forests. *Theoretical Population Biology*, *62*, 153–168.
8. Chave, J., Muller-Landau, H. C., & Levin, S. (2002). Comparing classical community models: Theoretical consequences for patterns of diversity. *The American Naturalist*, *159*, 1–23.
9. Chesson, P. L. (2000). Mechanisms of maintenance of species diversity. *Annual Review of Ecology, Evolution, and Systematics*, *31*, 343–366.
10. Cross, W. F., Baxter, C. V., Donner, K. C., Rosi-Marshall, E. J., Kennedy, T. A., Hall Jr, R. O., Wellard Kelly, H. A., & Rogers, R. S. (2011). Ecosystem ecology meets adaptive management: Food web response to a controlled flood on the Colorado River, Glen Canyon. *Ecological Applications*, *21*, 2016–2033.
11. Csillag, F., Boots, B., Fortin, M. J., Lowell, K., & Potvin, F. (2001). Multiscale characterization of boundaries and landscape ecological patterns. *Geomatica*, *55*, 509–522.
12. Currie, D. J. (2007). Disentangling the roles of environment and space in ecology. *Journal of Biogeography*, *34*, 2009–2011.
13. Dale, M. R. T., & Fortin, M. J. (2010). From graphs to spatial graphs. *Annual Review of Ecology, Evolution, and Systematics*, *41*, 21–38.
14. Diniz-Filho, J. A. F., & Bini, L. M. (2005). Modelling geographical patterns in species richness using eigenvector-based spatial filters. *Global Ecology and Biogeography*, *14*, 177–185.

15. Dormann, C. F., et al. (2007). Methods to account for spatial autocorrelation in the analysis of species distributional data: A review. *Ecography*, *30*, 609–628.
16. Dray, S., Legendre, P., & Peres-Neto, P. R. (2006). Spatial modeling: A comprehensive framework for principal coordinate analysis of neighbor matrices (PCNM). *Ecological Modelling*, *196*, 483–493.
17. Richardson, J. E., Pennington, R. T., Pennington, T. D., & Hollingsworth, P. M. (2001). Rapid diversification of a species-rich genus of neotropical rain forest trees. *Science*, *293*, 2242–2245.
18. Ricklefs, R. E. (2003). A comment on Hubbell's zero-sum ecological drift model. *Oikos*, *100*, 185–192.

INDEX

C

H

I

J

K

L

Y

Z

Printed in the United States
by Baker & Taylor Publisher Services